T0294914

Assessing Risk to National Critical Functions as a Result of Climate Change

2023 Risk Assessment Update

MICHELLE E. MIRO, SUSAN A. RESETAR, ANDREW LAULAND, DAVID METZ, VANESSA WOLF, RAHIM ALI, JAY BALAGNA, JASON THOMAS BARNOSKY, R. J. BRIGGS, EDWARD W. CHAN, SHIRA H. FISCHER, QUENTIN E. HODGSON, GEOFFREY KIRKWOOD, CHELSEA KOLB, JENNA W. KRAMER, JOHN LEE, KRISTIN J. LEUSCHNER, SHANNON PRIER, MARK STALCZYNSKI, PATRICIA A. STAPLETON, SCOTT R. STEPHENSON, TOBIAS SYTSMA, KRISTIN VAN ABEL, MICHAEL J. D. VERMEER, BRIAN WONG

Prepared for the Cybersecurity and Infrastructure Security Agency

HS AC
HOMELAND SECURITY
OPERATIONAL ANALYSIS CENTER

This research was published in 2024.

Approved for public release; distribution is unlimited.

Climate Risk Assessment Analyst Team

National Critical Function	Analyst
Health Care, Food, Housing, and Education	
Maintain Access to Medical Records	Shira Fischer
Provide Medical Care	Edward Chan
Support Community Health	Edward Chan
Produce and Provide Agricultural Products and Services	Patricia Stapleton, Karen Sudkamp,* Timothy Gulden*
Produce and Provide Human and Animal Food Products and Services	Patricia Stapleton, Karen Sudkamp,* Timothy Gulden*
Provide Housing	Mark Stalczynski, Michael Wilson*
Educate and Train	Jenna Kramer, Timothy Gulden*
Research and Development	Jenna Kramer
Governance, Public Safety, and Security	
Enforce Law	Jason Barnosky, Geoffrey Kirkwood, Richard Donohue*
Prepare for and Manage Emergencies	Jason Barnosky, Jay Balagna, Edward Chan*
Provide Public Safety	Jason Barnosky, Jay Balagna, Edward Chan,* Richard Donohue*
Provide Materiel and Operational Support to Defense	Quentin Hodgson, Geoffrey Kirkwood
Perform Cyber Incident Management Capabilities	Quentin Hodgson, John Lee
Protect Sensitive Information	Quentin Hodgson, John Lee
Conduct Elections	Quentin Hodgson
Operate Government	Jason Barnosky, Geoffrey Kirkwood
Preserve Constitutional Rights	Geoffrey Kirkwood
Energy and Infrastructure	
Exploration and Extraction of Fuels	Kristin Van Abel, Liam Regan*
Fuel Refining and Processing Fuels	Kristin Van Abel
Store Fuel and Maintain Reserves	Kristin Van Abel
Generate Electricity	Kristin Van Abel, Liam Regan*
Transmit Electricity	Rahim Ali, Liam Regan*
Distribute Electricity	Kristin Van Abel, Liam Regan*
Manage Hazardous Materials	Chelsea Kolb, Michael Wilson*
Manage Wastewater	Chelsea Kolb, Michelle Miro*
Supply Water	Chelsea Kolb, Michelle Miro*
Develop and Maintain Public Works and Services	Rahim Ali

National Critical Function	Analyst
Energy and Infrastructure, continued	
Provide and Maintain Infrastructure	Rahim Ali
Transportation and Material Goods	
Transport Cargo and Passengers by Air	David Metz, Liisa Ecola*
Transport Cargo and Passengers by Road	Scott Stephenson, Liisa Ecola*
Transport Materials by Pipeline	Scott Stephenson
Transport Passengers by Mass Transit	Scott Stephenson, Liisa Ecola*
Transport Cargo and Passengers by Rail	Tobias Sytsma, Liisa Ecola*
Transport Cargo and Passengers by Vessel	David Metz, Liisa Ecola*
Maintain Supply Chains	Tobias Sytsma
Manufacture Equipment	Tobias Sytsma
Produce Chemicals	Tobias Sytsma
Provide Metals and Materials	Tobias Sytsma
Finance, Information Technology, and Telecommunications	
Provide Capital Markets and Investment Activities	R. J. Briggs
Provide Consumer and Commercial Banking Services	R. J. Briggs
Provide Funding and Liquidity Services	R. J. Briggs
Provide Payment, Clearing, and Settlement Services	R. J. Briggs
Provide Wholesale Funding	R. J. Briggs
Provide Insurance Services	R. J. Briggs, Timothy Gulden*
Provide Internet Based Content, Information, and Communication Services	Michael Vermeer
Provide Identity Management and Associated Trust Support Services	Michael Vermeer
Provide Information Technology Products and Services	Michael Vermeer
Provide Internet Routing, Access, and Connection Services	Michael Vermeer
Provide Positioning, Navigation, and Timing Services	Michael Vermeer
Provide Radio Broadcast Access Network Services	Michael Vermeer
Provide Satellite Access Network Services	Michael Vermeer
Provide Wireline Access Network Services	Michael Vermeer
Provide Wireless Access Network Services	Michael Vermeer
Provide Cable Access Network Services	Michael Vermeer
Operate Core Network	Michael Vermeer

* = Author on the prior year's risk assessment (Miro et al., 2022).

About This Report

On January 27, 2021, President Joseph R. Biden, Jr., signed Executive Order (EO) 14008, "Tackling the Climate Crisis at Home and Abroad." EO 14008 describes the threat that the climate crisis poses to the United States and to the world and the roles of various federal agencies in addressing it. It directs the Secretary of Homeland Security to "consider the implications of climate change in the Arctic, along our Nation's borders, and to National Critical Functions" (Section 103[e]). The National Critical Functions (NCFs) represent "the functions of government and the private sector so vital to the United States that their disruption, corruption, or dysfunction would have a debilitating effect on security, national economic security, national public health or safety, or any combination thereof" (Cybersecurity and Infrastructure Security Agency [CISA], undated-b, p. 1).

To fulfill the objectives of the EO, CISA asked the Homeland Security Operational Analysis Center (HSOAC) to (1) develop a risk management framework for integrating climate-driven changes to the strategic operating environment, building on the NCF risk architecture to identify, understand, and manage climate-driven effects on the NCFs;[1] (2) identify the NCFs facing the greatest vulnerability to climate change and, as a result, facing disruption or degradation in the future; and (3) complete a full climate change risk assessment for the NCFs identified as being most vulnerable.

This report describes these three activities, presenting the risk management framework as an ongoing capacity for assessing climate-related risk moving forward and highlighting the results of the climate risk assessment that identified NCFs at risk due to climate change.[2] The findings should be of interest to CISA and other federal agencies and partners managing risk to U.S. critical infrastructure.

This research was sponsored by CISA and conducted in the Infrastructure, Immigration, and Security Operations Program of the RAND Homeland Security Research Division, which operates HSOAC.

About the Homeland Security Operational Analysis Center

The Homeland Security Act of 2002 (Public Law 107-296, § 305, as codified at 6 U.S.C. § 185) authorizes the Secretary of Homeland Security, acting through the Under Secretary for Science and Technology, to establish one or more federally funded research and development centers (FFRDCs) to provide independent analysis of homeland security issues. The RAND Corporation operates

---

[1] The NCF risk architecture "break[s] down each NCF into its sub-functions and lower-level activities to enable CISA to quickly evaluate, identify, and assess both operational and strategic risks to the Nation's infrastructure" (CISA, undated-b, p. 2).

[2] For two NCFs—*Provide Information Technology Products and Services* and *Support Community Health*—risk assessments differ between CISA's annual update response to EO 14008 and what is contained in this report.

HSOAC as an FFRDC for the U.S. Department of Homeland Security (DHS) under contract 70RSAT22D00000001.

The HSOAC FFRDC provides the government with independent and objective analyses and advice in core areas important to the department in support of policy development, decisionmaking, alternative approaches, and new ideas on issues of significance. HSOAC also works with and supports other federal, state, local, tribal, and public- and private-sector organizations that make up the homeland security enterprise. HSOAC's research is undertaken by mutual consent with DHS and organized as a set of discrete tasks. This report presents the results of research and analysis conducted under task order 70RCSA21FR0000052, Assessing Risk to the National Critical Functions as a Result of Climate Change. The results presented in this report do not necessarily reflect official DHS opinion or policy.

For more information on the RAND Homeland Security Research Division, see www.rand.org/hsrd. For more information on this publication, see www.rand.org/t/RRA1645-8.

Acknowledgments

We would like to thank the team of subject-matter experts that carried out peer reviews of each NCF's risk assessment, as well as our peer reviewers—Debra Knopman and Christy Foran—for their reviews of the overall study. Our subject-matter expert peer reviewers are listed in the table below.

NCF	Reviewer
Health Care, Food, Housing, and Education	
Maintain Access to Medical Records	Carrie M. Farmer
Provide Medical Care	Carrie M. Farmer
Support Community Health	Carrie M. Farmer
Produce and Provide Agricultural Products and Services	Flannery Dolan
Produce and Provide Human and Animal Food Products and Services	Flannery Dolan
Provide Housing	Michael Wilson
Educate and Train	Laura Bellows
Research and Development	Beth E. Lachman
Governance, Public Safety, and Security	
Enforce Law	Daniel Egel
Prepare for and Manage Emergencies	Patrick S. Roberts
Provide Public Safety	Patrick S. Roberts
Provide Materiel and Operational Support to Defense	Jeremy M. Eckhause
Perform Cyber Incident Management Capabilities	Sasha Romanosky
Protect Sensitive Information	Daniel Egel
Conduct Elections	Bryan Boling

NCF	Reviewer
Operate Government	Daniel Egel
Preserve Constitutional Rights	Daniel Egel
Energy and Infrastructure	
Exploration and Extraction of Fuels	Ellen M. Pint
Fuel Refining and Processing Fuels	Ellen M. Pint
Store Fuel and Maintain Reserves	Ellen M. Pint
Generate Electricity	Kelly Klima
Transmit Electricity	Kelly Klima
Distribute Electricity	Kelly Klima
Manage Hazardous Materials	Ellen M. Pint
Manage Wastewater	Beth E. Lachman
Supply Water	Beth E. Lachman
Develop and Maintain Public Works and Services	Ellen M. Pint
Provide and Maintain Infrastructure	Ellen M. Pint
Transportation and Material Goods	
Transport Cargo and Passengers by Air	Liisa Ecola
Transport Cargo and Passengers by Road	Liisa Ecola
Transport Materials by Pipeline	Kelly Klima
Transport Passengers by Mass Transit	Liisa Ecola
Transport Cargo and Passengers by Rail	Liisa Ecola
Transport Cargo and Passengers by Vessel	Liisa Ecola
Maintain Supply Chains	Adam C. Resnick
Manufacture Equipment	Adam C. Resnick
Produce Chemicals	Ellen M. Pint
Provide Metals and Materials	Adam C. Resnick
Finance, Information Technology, and Telecommunications	
Provide Capital Markets and Investment Activities	James V. Marrone
Provide Consumer and Commercial Banking Services	James V. Marrone
Provide Funding and Liquidity Services	Natalie Cox
Provide Payment, Clearing, and Settlement Services	Natalie Cox
Provide Wholesale Funding	Natalie Cox
Provide Insurance Services	Michael Dworsky
Provide Internet Based Content, Information, and Communication Services	Edward Balkovich
Provide Identity Management and Associated Trust Support Services	Edward Balkovich

NCF	Reviewer
Provide Information Technology Products and Services	Edward Balkovich
Provide Internet Routing, Access, and Connection Services	Edward Balkovich
Provide Positioning, Navigation, and Timing Services	Edward Balkovich
Provide Radio Broadcast Access Network Services	Edward Balkovich
Provide Satellite Access Network Services	Edward Balkovich
Provide Wireline Access Network Services	Edward Balkovich
Provide Wireless Access Network Services	Edward Balkovich
Provide Cable Access Network Services	Edward Balkovich
Operate Core Network	Edward Balkovich

Summary

Issue

Critical infrastructure systems form the backbone of the United States, meeting communities' need for safe drinking water, reliable electricity, dependable internet access, and countless other functions, and enabling such services as health care, education, agriculture, and supply chains. Climate change represents a significant threat to critical infrastructure, driving unprecedented changes in the frequency and severity of many natural hazards, posing potentially new or growing risks of disruption, and challenging the assumptions used to design and protect these systems.[3] Understanding the future risk posed by natural hazards affected by climate change can help critical infrastructure owners and operators, as well as stakeholders at local to federal levels, allocate resources, make investment decisions, and prepare these systems for future, rather than historical, hazards.

To this end, on January 27, 2021, President Joseph R. Biden, Jr., signed Executive Order 14008, "Tackling the Climate Crisis at Home and Abroad." The order directs the Secretary of Homeland Security to provide an annual response to "consider the implications of climate change . . . to National Critical Functions" (Section 103[e]). The National Critical Functions (NCFs) represent "the functions of government and the private sector so vital to the United States that their disruption, corruption, or dysfunction would have a debilitating effect on security, national economic security, national public health or safety, or any combination thereof" (Cybersecurity and Infrastructure Security Agency [CISA], undated-b). To respond to this requirement, CISA, a component of the U.S. Department of Homeland Security, asked the Homeland Security Operational Analysis Center, a federally funded research and development center operated by the RAND Corporation, to conduct a climate change risk assessment for the 55 NCFs. The current analysis builds on the team's prior risk assessment, which is documented in Miro et al. (2022).

Approach

This assessment examines the current and future risk that a variety of climate hazards, referred to as *climate drivers*, pose to the 55 NCFs on a national scale. This analysis adapts the risk management framework described in Miro et al. (2022),[4] focusing on three primary steps: (1) identifying climate drivers that span the variety of climate-related hazards that could degrade or disrupt NCFs across the United States and characterizing how they might change by 2100; (2) determining the impact mechanisms by which the climate drivers could affect NCFs, such as through physical damage or

---

[3] For example, in many locations, infrastructure design and regulations are focused on withstanding the 100-year storm event. Yet the 100-year storm is typically defined based on a historical record that no longer represents present-day climatology and does not reflect future change. This disconnect creates vulnerabilities for these infrastructure systems.

[4] The methodology was originally described in Miro et al. (2022). This methodology has been refined for the current analysis.

workforce vulnerability, and the consequences (such as operational failure) that these effects could have on an NCF; and (3) assigning a rating to the risk that a climate driver poses to an NCF, based on the characterization of impact mechanisms and consequences. We assessed risk with a scale that CISA uses that ranges from a rating of 1 (no disruption or normal operations) to 5 (critical disruption). We used this scale and projected changes in eight climate drivers (drought, extreme cold, extreme heat, flooding, sea-level rise, severe storm systems, tropical cyclones and hurricanes, and wildfire) to assess risk to NCFs from climate change at a present-day baseline and in two future time periods (2050 and 2100) and two future scenarios of greenhouse gas (GHG) emissions. The two scenarios include a current emissions scenario that reflects global mean temperature change using the current global GHG emissions levels and international commitments to emissions reductions, referred to as representative concentration pathway 4.5 and a high emissions scenario (representative concentration pathway 8.5). In addition to assessing the risk rating for each NCF, we conducted a separate analysis of interdependencies among the NCFs to characterize potential for cascading disruptions due to climate change.

Limitations

This assessment also has several limitations, including ratings that are based on direct effects on individual NCFs without consideration of dependencies on other NCFs; a rating scale that assesses risk at the national level, although regional effects might be significant; and ratings that are dependent on how the underlying subfunctions of each NCF are defined. In addition, the analysis assumes that NCFs will follow current trends into the future, without incorporation of other factors that might affect NCFs, such as major population shifts, technological advancements, or significant climate adaptations, and uses only two emissions scenarios to capture the effects of climate change over time. Finally, the evidence base used to assess risk varies widely across NCFs.

Findings

Risk Ratings

Table S.1 depicts risk to NCFs due to one or more climate drivers under the current emissions scenario for 2050 and 2100. These NCFs are grouped by risk levels. The risk assessment found the following:

- We assessed 45 of the 55 NCFs to be at risk of at least minimal disruption on a national scale from climate change by 2050 under a scenario that follows current GHG emissions trends. *Minimal disruption* indicates that these NCFs are expected to meet routine operational needs across the United States even though operations are experiencing some effects from climate change.
- Twenty-five of the 55 NCFs are at risk of a moderate disruption or greater by 2050 in the current emissions scenario. *Moderate disruption* indicates that these NCFs are expected to be

able to meet routine operational needs in only most of the country. Historical examples of moderate disruption include major disasters, such as Hurricane Katrina.

- Three NCFs—*Provide Information Technology Products and Services, Maintain Supply Chains,* and *Supply Water*—are at risk of *major* or *critical disruption* due to climate change by 2100 in the current emissions scenario. For *Provide Information Technology Products and Services,* the geographic concentration of manufacturing capacity for key hardware products (e.g., microchips) leads to single points of failure that are vulnerable to regional natural disasters. *Supply Water* has already experienced disruptions due to drought, which are expected to continue and worsen. *Maintain Supply Chains* relies on key hubs that make it vulnerable to climate change. If these risks are realized, these NCFs could be disrupted to the point that they cannot meet routine operational needs in most or all of the country.

- Seven NCFs are not at risk of direct effect from climate change in either 2050 or 2100. These are primarily NCFs related to financial services and network (i.e., internet-based) operations and services. These NCFs either rely on decentralized service provision or virtual systems or have infrastructure with sufficient redundancy or that is sited in areas that are not particularly prone to the climate drivers.

- Four of the eight climate drivers—flooding, sea-level rise, tropical cyclones and hurricanes, and wildfire—will have an outsized effect on the NCFs. Although there are important regional distinctions in how and where climate drivers will change, these four drivers pose the greatest risk of disruption to the NCFs on the national level.

Table S.1. National Critical Function Risk Ratings, Current Emissions Scenario

Category	NCF	2050	2100
MODERATE TO CRITICAL	Provide Information Technology Products and Services	4	5
MODERATE TO CRITICAL	Supply Water	3	4
MODERATE TO CRITICAL	Maintain Supply Chains	3	4
MODERATE TO CRITICAL	Fuel Refining and Processing Fuels	3	3
MODERATE TO CRITICAL	Exploration and Extraction of Fuels	3	3
MODERATE TO CRITICAL	Transport Cargo and Passengers by Vessel	3	3
MODERATE TO CRITICAL	Transport Cargo and Passengers by Air	3	3
MODERATE	Provide Housing	3	3
MODERATE	Provide Materiel and Operational Support to Defense	3	3
MODERATE	Produce and Provide Human and Animal Food Products and Services	3	3
MODERATE	Provide Public Safety	3	3
MODERATE	Provide Medical Care	3	3
MODERATE	Support Community Health	3	3
MODERATE	Prepare for and Manage Emergencies	3	3
MODERATE	Manage Wastewater	3	3
MODERATE	Manage Hazardous Materials	3	3
MODERATE	Enforce Law	3	3
MODERATE	Develop and Maintain Public Works and Services	3	3
MODERATE	Transport Materials by Pipeline	3	3
MODERATE	Transport Cargo and Passengers by Road	3	3
MODERATE	Transmit Electricity	3	3
MODERATE	Distribute Electricity	3	3
MODERATE	Provide Wireline Access Network Services	3	3
MODERATE	Provide Wireless Access Network Services	3	3
MODERATE	Provide Cable Access Network Services	3	3
MODERATE	Provide Metals and Materials	2	3
MODERATE	Produce Chemicals	2	3
MODERATE	Produce and Provide Agricultural Products and Services	2	3
MODERATE	Manufacture Equipment	2	3
MODERATE	Generate Electricity	2	3
MODERATE	Provide Insurance Services	2	3
MODERATE	Provide and Maintain Infrastructure	2	3
MODERATE	Transport Cargo and Passengers by Rail	2	3
MODERATE	Educate and Train	2	3
MODERATE	Transport Passengers by Mass Transit	2	3
MINIMAL	Store Fuel and Maintain Reserves	2	2
MINIMAL	Provide Radio Broadcast Access Network Services	2	2
MINIMAL	Protect Sensitive Information	2	2
MINIMAL	Preserve Constitutional Rights	2	2
MINIMAL	Perform Cyber Incident Management Capabilities	2	2
MINIMAL	Operate Government	2	2
MINIMAL	Conduct Elections	2	2
MINIMAL	Provide Payment, Clearing, and Settlement Services	2	1
MINIMAL	Provide Funding and Liquidity Services	2	1
MINIMAL	Provide Capital Markets and Investment Activities	2	1
LOWEST RISK	Research and Development	1	2
LOWEST RISK	Provide Internet Routing, Access, and Connection Services	1	2
LOWEST RISK	Operate Core Network	1	2
LOWEST RISK	Provide Wholesale Funding	1	1
LOWEST RISK	Provide Identity Management and Associated Trust Support Services	1	1
LOWEST RISK	Provide Consumer and Commercial Banking Services	1	1
LOWEST RISK	Maintain Access to Medical Records	1	1
LOWEST RISK	Provide Satellite Access Network Services	1	1
LOWEST RISK	Provide Positioning, Navigation, and Timing Services	1	1
LOWEST RISK	Provide Internet Based Content, Information, and Communication Services	1	1

NOTE: 1 = no disruption or normal operations. 2 = minimal disruption. 3 = moderate disruption. 4 = major disruption. 5 = critical disruption. See Chapter 2 for further explanation of ratings.

Overall, the risk ratings are higher than in last year's assessment, including many more NCFs at risk of moderate disruption at certain time periods under certain scenarios. Climate change is the primary mechanism driving the risk assessed in this study. Because this analysis relied on the same projections of future climate change, changes in ratings from last year's to this year's assessment reflect a combination of developing a more nuanced interpretation of the risk rating scale in the context of a national risk assessment, enhancing interrater reliability in applying it, the use of new scientific information to guide assessments of NCF exposure and vulnerabilities to climate drivers, and enhancements in the analysis over time.

National Critical Functions at Risk Because of Interdependencies

NCFs do not operate in isolation and depend on a variety of other NCFs that provide critical inputs or key services. To examine the effect of these dependencies, we conducted an interdependence analysis based on dependencies or mutually reliant relationships that could affect operations. We used information provided by CISA on the functional dependencies between NCFs. We aimed to identify two main groups of NCFs: (1) those that were assessed as being at lower levels of risk because of climate change but depend on a high number of NCFs assessed as being at higher levels of risk and (2) those that were assessed as being at higher levels of risk and have a high number of NCFs that depend on them. These two groupings allowed us to understand which NCFs might be indirectly affected by climate change via other NCFs and which NCFs have the potential to affect a large number of other NCFs, respectively.

Our analyses found the following:

- NCFs with the same risk ratings from our risk assessment (which considered only the direct effects of climate change) could have a greater exposure to climate change through the indirect effects from dependencies. The ten NCFs assessed as being at no risk of disruption due to climate change in 2050 are dependent on at least six other NCFs assessed as being at risk of moderate disruption or greater.
- NCFs with high numbers of upstream dependencies (i.e., that rely on critical inputs from other NCFs) could be at additional risk from climate drivers, although the number of dependencies does not mean that effects will necessarily cascade. This suggests that resilience planning and analysis that incorporate the effects to other NCFs will be essential to lessen the potential for disruption to normal operations for these NCFs.
- Four NCFs are at risk of moderate or greater disruption on which all other NCFs depend: *Provide Information Technology Products and Services*, *Maintain Supply Chains*, *Prepare for and Manage Emergencies*, and *Provide Medical Care*. An additional five NCFs also have a high number of downstream NCFs. Because these NCFs have the potential to affect many other NCFs and face risk of disruption into the future, mitigation measures that reduce risk to these NCFs could be integral to mitigating overall national risk from climate change.

The interdependence analysis has several limitations. It allowed us to identify only whether a relationship existed between NCFs; it did not provide details of that relationship or clues as to how or where effects might cascade. We were also not able to determine which types of interdependence might be most relevant to risk from climate change. Characterizing the strength of the interdependencies between NCFs could lead to a more complete understanding of the ways in which climate effects could propagate through the network of NCFs and might illuminate points of intervention for reducing these effects.

Interpreting the Results

Because of the complex nature of climate change and the broad scope and scale of the NCFs, there are several important considerations to bear in mind when interpreting the results of this assessment:

- **The NCFs are diverse in ways that can affect the potential for national disruption.** Differences include whether an NCF's footprint is more local or more regional and the amount of time an NCF might be disrupted without causing serious consequences.
- **Our assessment was at the national scale, but climate drivers can have varying effects at local and regional levels.** Because we conducted our analysis at the national level, region-specific risk assessments for each NCF and climate driver could vary widely from the national-focused results.
- **Steady risk ratings between 2050 and 2100 or across the two emissions scenarios could obscure changes in the underlying risks from climate change.** There could be important changes to the frequency, duration, geographic location, or other aspects of how a climate driver interacts with an NCF from one time period to another or under a different emissions scenario that are not reflected in a change in the national-level risk rating.
- **Climate risks can compound over time.** The assessments thus do not fully reflect risk from the cumulative effects of repeated exposure to a climate driver, nor do they include the effects of multiple climate drivers occurring concurrently or in quick succession.
- **Interdependencies can have varying effects that are not captured in the risk ratings.** Although this analysis included some consideration of the functional relationships between NCFs, the necessary information was not available to assess the strength of the relationships among all 55 NCFs.

Implications and Considerations

The findings of our analysis led to the following suggestions for CISA's consideration.

Prioritize Specific National Critical Functions for Further Assessment

CISA should consider prioritizing specific NCFs for further assessment, communication, and risk mitigation. CISA might prioritize those NCFs identified as being at greatest risk to undergo more-granular risk assessments at the local and regional levels, including further assessment of the mechanisms by which climate drivers could disrupt these NCFs. These NCFs should also be prioritized for communication and outreach to stakeholders and the general public and for development of mitigation strategies that address risks posed by the highest-risk drivers identified in this report. The four drivers we identified as posing the greatest risk of disruption to the NCFs—flooding, sea-level rise, tropical cyclones and hurricanes, and wildfire—should also be prioritized for further assessment and risk mitigation activities.

Factor the Consequences of National Critical Function Disruption into Future Assessments

Although an NCF might not be disrupted at the national level, regional disruption of an NCF can create national-level consequences, including significant threats to health and safety, economic loss, and risks to national security. A more complete analysis of the consequences of the level of disruption should be conducted to inform the prioritization of future risk mitigation activities.

Communicate Risk from Climate Change and Climate Drivers

This report illustrates the difficulty in conveying risk information about the NCFs, which, by their nature, do not often reach the level of disruption needed to receive the highest ratings (major or critical disruption) on a national scale but that, given their criticality, can—and do—create severe life, safety, and economic effects when disrupted at the local or regional level. Effectively communicating the risk at the actual spatial scale that climate change poses to NCFs is likely to be an ongoing challenge.

Update Assessments of Climate Change's Risk to National Critical Functions

The science underlying climate projections for 2050 and 2100 will be refined over time as understanding increases about the global climate system and about the dynamics of how climate change affects built and natural systems. At the same time, the provision of these functions could change with technological innovation, investments in adaptation or resilience measures, demographic trends, or other determinants. The basis for this analysis, the Fourth National Climate Assessment, draws from the best available information as of 2017. The Fifth National Climate Assessment (NCA5) was under revision at the time of our analysis and contains updated climate information. For these reasons, we recommend that these assessments be revisited frequently.

Contents

Figures and Tables

Figures

Tables

Chapter 1

Introduction

Critical infrastructure systems form the backbone of the United States. They provide and enable the essential functions and services on which communities across the country rely. This includes the provision of safe drinking water, reliable electricity, and dependable internet access and services. Critical infrastructure also enables many other essential services, such as health care, education, agriculture, and supply chains. When critical infrastructure systems are disrupted in some way, the inability to provide these essential services can have significant and negative effects, including injury and death, property damage, and economic losses. The daily cost of closures at the Ports of Los Angeles and Long Beach, for example, have been estimated at $0.5 billion per day (Inforum, 2022). Disruptions to education from the coronavirus disease 2019 (COVID-19) pandemic shut down in-person schooling for months and, in some places, years, resulting in long-term effects on academic achievement, mental health, and overall well-being that are still being realized (Kuhfeld et al., 2022).

Evidenced in these examples and beyond, critical infrastructure systems face a wide variety of threats and hazards, including both natural hazards (such as hurricanes and wildfire) and targeted threats (such as terrorism). To protect against disruption, decisionmakers in investment, management, and operations rely on assumptions about how often and how severe these threats and hazards might be. However, climate change is driving unprecedented changes in the frequency and severity of many of the natural hazards facing U.S. critical infrastructure systems, posing potentially new or growing risks of disruption, and challenging the assumptions used to protect these systems. The past five years, for example, have seen nearly 18 billion-dollar disasters each year at a cost of nearly $160 billion in damage per year (National Centers for Environmental Information, undated-a). By these numbers, this is more than three times what the country experienced in the 1990s and more than six times what it was in the 1980s (National Centers for Environmental Information, undated-a). In addition, much of the critical infrastructure in the United States that is in use today was designed and built according to a historical climatology that no longer reflects current conditions or future climate change. Understanding the future risk posed by natural hazards affected by climate change can help critical infrastructure owners and operators, as well as stakeholders at local to federal levels, allocate resources, make investment decisions, and prepare these systems for future, rather than historical, hazards.

To this end, on January 27, 2021, President Joseph R. Biden, Jr., signed Executive Order (EO) 14008, "Tackling the Climate Crisis at Home and Abroad." EO 14008 discusses the threat to the United States and to the world posed by the climate crisis and the roles of various federal agencies in addressing it. It also directs the Secretary of Homeland Security to "consider the implications of climate change to National Critical Functions." The National Critical Functions (NCFs) represent "the functions of government and the private sector so vital to the United States that their disruption, corruption, or dysfunction would have a debilitating effect on security, national economic security, national public health or safety, or any combination thereof" (Cybersecurity and Infrastructure Security Agency [CISA], undated-b).

CISA, a component of the U.S. Department of Homeland Security (DHS), identified the set of 55 NCFs in 2019 to enable more-robust situational awareness of U.S. critical infrastructure. The NCFs constitute a functional lens to examine and characterize risks to critical infrastructure systems that go beyond traditional sectors. NCFs cover broad infrastructure services, such as education and training or research and development and focus on the system that supports these services. The full list of NCFs is shown in Figure 1.1.

Figure 1.1. The 55 National Critical Functions

SOURCE: Reproduced from CISA, 2021, p. 4.

Study Objectives

This report highlights the results of a climate risk assessment for the 55 NCFs and describes the climate changes considered in the risk assessment. The study includes an assessment of

- the future risk posed to NCFs by a set of climate drivers
- some of the critical interdependencies among NCFs that could lead to cascading disruptions of NCFs
- the uncertainty in future climate change risk that is captured through two future scenarios of climate change.

The analysis reported here built on a risk assessment completed in 2021 and submitted in the 2022 annual updated response by CISA (Miro et al., 2022). The risk assessment from 2022 served as the basis for this year's assessment. This analysis then built on the team's previous work in two ways:

- First, as recommended in the prior assessment (Miro et al., 2022), the current assessment addressed all 55 NCFs rather than just the 27 higher-vulnerability NCFs, which were the focus of the earlier work. By focusing on all 55 NCFs, we provide a fuller picture of the risk profile that each climate driver poses to the country's variety of critical infrastructure functions. In this analysis, we established a baseline risk for all 55 NCFs, which will allow CISA to track and identify emerging risk related to climate change for the NCFs.
- Second, as also recommended in our prior assessment, we carried out an additional analysis on the interdependencies among all 55 NCFs, which can lead to cascading effects across NCFs.

We note that, because this report provides an update of our previous report (Miro et al., 2022), the descriptions of some components of the research, particularly the methods discussed in Chapter 2, draw heavily from the prior report, with omissions and adjustments to reflect changes in the current analysis.

Organization of This Report

The remainder of this report is organized as follows:

- Chapter 2 describes the methods used in and limitations of this analysis.
- Chapter 3 describes high-level findings of the climate risk assessment.
- Chapter 4 provides an assessment of interdependencies across the NCFs.
- Chapter 5 details the implications of our findings and identifies areas for future work.
- Appendix A presents the analysis of future changes in climate drivers.
- Appendix B contains the findings from our individual assessments of the 55 NCFs.

Chapter 2

Methods

This chapter details the methods we employed to carry out the climate risk assessment detailed in this report, as well as the assumptions and considerations that are key to understanding this assessment.[5]

For this analysis, we adapted the risk management framework described in Miro et al. (2022), focusing on the three primary steps:

1. Identify climate drivers that span the variety of climate-related hazards that could degrade or disrupt NCFs across the United States and understand how they might change over the course of the 21st century.
2. Determine the impact mechanisms by which the climate drivers pose a threat to NCFs, such as through physical exposure or workforce vulnerability, and the consequences of these effects on an NCF, such as operational failure.
3. Rate the risk that a climate driver poses to an NCF, based on the characterization of impact mechanisms and consequences.

In the rest of this chapter, we discuss the methods used to conduct the three steps in this analysis. A full discussion of the framework is provided in Miro et al. (2022).[6]

Characterizing Projected Climate Changes

It is widely recognized that rising global temperatures are producing changes in the climate, including in the intensity and frequency of extreme weather events that can directly disrupt NCFs (Masson-Delmotte et al., 2021). Although many definitions exist, the federal government has defined *climate change* as follows:

> Changes in average weather conditions that persist over multiple decades or longer. Climate change encompasses both increases and decreases in temperature, as well as shifts in precipitation, changing risk of certain types of severe weather events, and changes to other features of the climate system. (USGCRP, undated)

To characterize these future changes, we focused on those changes in climate phenomena and climate-related hazards that would have direct effects on NCFs. Relying on categories used in other

---

[5] This discussion draws from the team's previous report (Miro et al., 2022, pp. 3–12) with omissions and minor adjustments.

[6] Because we completed an assessment of all 55 NCFs for fiscal year 2023, for the current analysis, we did not conduct the first step in the framework, which is screening to identify a subset of high-vulnerability NCFs. We also did not complete the fifth step, identifying potential risk mitigation strategies, for this assessment, although we will address that step for a future assessment report. A discussion of mitigation strategies identified for high-priority NCFs is provided in Regan et al., 2023.

federal climate change efforts that reflect authoritative sources of future climate changes—specifically, the Fourth National Climate Assessment (NCA4)[7]—we identified a set of eight climate drivers, shown in Table 2.1, which represent six groupings of extreme weather events plus sea-level rise and wildfire. For the purposes of this report, we refer to these climate-related hazards as *climate drivers*.

Table 2.1. Climate Drivers

Climate Driver	Description
Drought	An extended period of moisture deficiency with effects on dependent systems; can occur on seasonal to multidecadal time scales
Extreme cold	Extreme cold temperatures
Extreme heat	Extreme high temperatures and heat waves
Flooding	Riverine, coastal, and flash flooding from extreme rainfall events
Sea-level rise	Coastal inundation from sea-level rise and tidal events
Severe storm systems (nontropical)	Convective storm systems, extratropical cyclones, nor'easters, and associated hazards, including hail, extreme rainfall, snow, and ice
Tropical cyclones and hurricanes	Tropical cyclones and hurricanes, as well as their associated weather effects (e.g., storm surge and waves, high winds, extreme rainfall)
Wildfire	Wildfire in wildlands or at the wildland–urban interface

SOURCE: Adapted from National Centers for Environmental Information, undated-b.

Although, in this analysis, we examined extreme weather events, such variables as changes in average daily temperature and precipitation are underlying mechanisms that influence extreme weather events and are thus considered indirectly.[8] For example, warmer temperatures and reduced precipitation can increase the incidence of drought. Instead, the climate drivers, as framed, focus on those changes likely to be directly responsible for future degradation or disruption of NCFs. Indeed, although drought can occur over long time periods, it is generally considered a distinct extreme event. The one exception is sea-level rise, which is conceptualized as a slower-moving climate driver but one that has direct and lasting effects on NCFs. Finally, part of the reason we selected these drivers is that sufficient evidence from a consistent evidence base exists about each to articulate future changes across different climate change scenarios and future time periods.

The analysis considered changes in climate drivers from a baseline time period (a 30-year climatology reflecting climate at the time of research) to two future time periods that are commonly used in regional and national climate assessments:

- the medium term (2050), representing projected changes during approximately the next 30 years

---

[7] As of this writing, the Fifth National Climate Assessment (NCA5) is being carried out and so was unavailable at the time of analysis.

[8] Additionally, the set of drivers developed for this risk assessment includes only direct effects of climate drivers, not the secondary effects of climate hazards, such as landslides or debris flows.

- the long term (2100), representing projected changes during approximately the next 80 years.

This assessment relied on two future scenarios established by the international modeling community to characterize some of the uncertainty in future climate changes by 2050 and 2100. Projected changes in future climate drivers were defined by two scenarios of future greenhouse gas (GHG) emissions:

- a current emissions scenario that reflects global mean temperature change given the current global GHG emissions levels and recent international commitments to emissions reductions
- a high emissions scenario that reflects the global mean temperature change in the event of a significant increase in future GHG emissions or a higher climate sensitivity that results in higher magnitudes of climate change per unit of emissions.[9]

To characterize the projected amount of change for each climate driver, time period, and climate scenario at the national scale, we relied on information on future change contained in NCA4. NCA4 represents a nationally consistent, consensus-based climate assessment intended to support research and adaptation for the United States, drawing from the best available information (as of 2017).[10] It also includes historical baselines, future time periods, emissions scenarios, and datasets necessary for ensuring a consistent basis for risk ratings at the national level.

We first examined national-level changes for each climate driver both quantitatively, when possible, and qualitatively based on information in NCA4 Volume I and the supplementary NCA4 viewer that contains the underlying downscaled global climate model data analysis (Wuebbles et al., 2017). As needed, we referred to NCA4 Volume II or other scientific publications to refine estimates or descriptions of projected changes. Initially, we summarized the national-level changes for each driver, time period, and future climate scenario, noting regional nuances. If sufficient information was not available in NCA4 to describe future climate changes across all time periods, we consulted other widely cited and nationally relevant scientific publications, consulting only those publications published within the past ten years. A summary description for each driver, time period, and climate scenario intended for the risk assessment is contained in Appendix A.

Assessing Climate Risk to the National Critical Functions

We assessed the risk to each NCF based on projected changes in the climate drivers and their expected effect on the NCF. Risk was assessed as being at the national level for each NCF across the two time periods and both emissions scenarios. The team assessed risk to each NCF independently of other NCFs, with ratings based on direct effects to an individual NCF. These assessments relied on available evidence of a given climate driver's expected effect on an NCF, the degree of change in a given climate driver, and our judgment of the amount of risk this produced.

---

[9] These scenarios are representative concentration pathway scenarios 4.5 and 8.5, respectively, that have been adopted by the Intergovernmental Panel on Climate Change (Intergovernmental Panel on Climate Change, 2014).

[10] The national assessments are intended to be carried out on a four-year cycle. NCA5 is on a six-year cycle. For a description of the process and timeline used for NCA4, see Avery et al. (2018).

Homeland Security Operational Analysis Center (HSOAC) subject-matter experts working in small teams that were focused on comparable NCFs assessed the available evidence to determine the NCF risk ratings. The ratings and supporting rationale were also peer-reviewed for correctness by other subject-matter experts from across RAND. Additionally, the HSOAC team held multiple rating calibration sessions in which we discussed differences in ratings among similar NCFs and ensured analyst concordance on how information was applied to determine the risk ratings. This process drew from prior HSOAC work on risk assessment with the NCFs (Lauland et al., 2022).

In the rest of this section, we discuss the key components of the risk assessment—the NCF subfunctions, impact mechanisms, and risk rating scale—in more detail.

National Critical Function Subfunctions

Each NCF consists of between two and ten subfunctions that define the key systems within it. CISA identified these subfunctions as part of the NCF risk architecture and provided that information to us.

For example, *Educate and Train* is made up of two subfunctions:

- *Provide Formal Education*
- *Provide Workforce Training.*

Maintain Supply Chains is made up of six subfunctions:

- *Maintain Supply Chain Operations*
- *Manage Product Development and Manufacturing*
- *Manage Product Marketing*
- *Manage Product Distribution*
- *Manage Retailers*
- *Manage Purchasing.*

At CISA's request, these subfunctions served as the basis for our NCF risk ratings. Thus, for *Educate and Train*, we rated risk for each of the two subfunction–driver combinations across the two time periods and two future scenarios. For *Maintain Supply Chains*, we rated risk for each of the six subfunction–driver combinations.

After we rated the risk that each driver posed to each subfunction, we summarized risk ratings at the NCF level. To do this, we calculated the maximum risk rating across all subfunctions for each driver, time period, and future scenario. For example, if, for the current emissions scenario for 2050 for sea-level rise, one subfunction was assigned a rating of 1 and the second subfunction was assigned a 3, the NCF-level risk for sea-level rise for 2050 under the current emissions scenario would be a 3. We used this type of aggregation to capture the highest level of underlying risk, rather than an average of ratings of risk, to an NCF. For NCFs with nine subfunctions, for example, seven of them could be relatively insensitive to climate change (e.g., subfunctions related to human resources or communication), but two could be at high risk, with disruption or failure of either of these two affecting the NCF at large.

Although this process provided a high level of granularity for risk ratings within an NCF, it also meant that risk ratings depended on how NCFs were characterized by their subfunctions. For example, the subfunctions for *Support Community Health* focus on the systems and functions that provide community health (operations, emergency response, and communications) but not on the health of the community itself. As a result, risk to this NCF focuses on the effect that climate change will have on these types of functions.

Impact Mechanisms

Assessing risk to NCFs requires determining how climate drivers could disrupt or degrade NCF operations. A climate driver might have a direct effect through one or more impact mechanisms, such as physical damage during a flood, or through creating an increase in demand for electricity during an extreme heat event. We identified four prominent impact mechanisms for this analysis: physical damage or disruption, lack of resources, workforce shortages, and demand changes. Table 2.2 describes these four mechanisms.

Table 2.2. Impact Mechanisms

Impact Mechanism	Description
Physical damage or disruption	The effects of climate change, or a specific climate driver, cause physical damage to facilities or equipment necessary for the functioning of the NCF or to NCF operations. This mechanism also includes cases in which a climate driver might disrupt operations but not necessarily damage infrastructure (e.g., high temperatures that ground airplanes).
Lack of resources	The effects of climate change, or a specific climate driver, cause an interruption in the supply of inputs to the NCF, preventing the NCF from functioning. These could be raw materials, or they could be goods or services.
Workforce shortage	The effects of climate change, or a specific climate driver, create a shortage of workers needed to operate the NCF. This includes conditions in which the climate driver makes such work unpleasant or dangerous. It also includes situations in which workforce productivity may be reduced and therefore NCF operations would be degraded. Note that we did not consider interdependencies among NCFs, so workforce shortage includes only those conditions that prevent enough workers from being able to work. If roads are inundated and workers unable to travel, we considered that an interdependence with *Transport Cargo and Passengers by Road*.
Demand change	The climate driver causes changes in the demand for the NCF, resulting in the NCF not being able to appropriately meet the demand. This could be an increase in demand that goes beyond what the NCF can supply, a decrease in demand that affects the viability or efficiency of the NCF, or volatile fluctuations encompassing both increases and decreases.

To assess risk, HSOAC analysts considered the way in which a climate driver affects a particular NCF subfunction, as well as the consequences a driver might have on that NCF as a whole. For our

8

purposes, the particular combination of the climate driver, subfunction, and impact mechanism is termed an *impact pathway*. An example impact pathway that depicts one effect that flooding would have on *Supply Water* is shown in Figure 2.1. This illustrates that flooding can directly affect a drinking water system when floodwaters cause physical damage to water supply infrastructure, bringing contaminants into drinking water sources. For a drinking water provider, this would affect operations by requiring additional treatment or requiring the operator to switch to an emergency or backup water source. For drinking water consumers, contamination of water sources could also mean a degradation of drinking water quality or the need to boil water or obtain bottled water.

Figure 2.1. An Example Impact Pathway for *Supply Water*

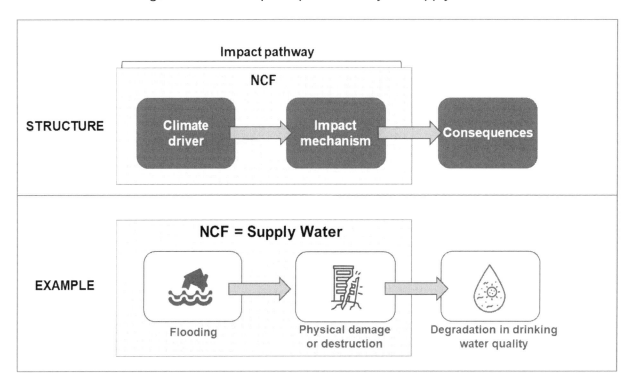

We characterized the impact mechanisms for each climate driver and the consequences to the operations of each NCF, drawing from published scientific literature on NCF operations in the face of extreme events and climate change. Characterizing these types of impact mechanisms for climate drivers and NCFs provided an understanding of how severe the effects of climate change could be on an NCF and was a key input to our risk assessments.

Risk Ratings

We assigned risk ratings on a scale of 1 to 5, as shown in Table 2.3, based on the amount of anticipated disruption at the national scale, with 1 being no disruption or normal operations and 5 being critical disruption. Analysts were also able to assign an unknown rating if sufficient evidence was not available to assess risk for a given subfunction–driver combination. Risk rating definitions are based on CISA's operational-level framework[11] and, although they were tailored to the context of this risk assessment, were developed to ensure alignment with other CISA NCF risk assessments. Using this scale, we determined a risk rating for each subfunction–driver combination across all three time periods and two emissions scenarios. For each NCF, risk ratings and the supporting evidence are provided in Appendix B.

---

[11] Unpublished framework provided to the HSOAC research team by CISA staff.

Table 2.3. Risk Rating Scale

Definition	Application	High-Level Example
• A climate driver's effect on this NCF or subfunction is unknown, or there was insufficient evidence to make a risk assessment.	• There is no direct evidence in the research literature, historical or analogous experiences, or plausible connections based on subject-matter expertise.	• Not applicable
• The NCF or subfunction is anticipated to meet all routine operational needs.	• The NCF is not vulnerable to the climate driver.	• Hurricane Ian and *Provide Satellite Access Network Services*
• Climate change is expected to affect the NCF or subfunction, but the function is expected to meet most routine operational needs.	• Routine operational needs are met, although disruptions could occur from time to time and at limited scales.	• 2017 hurricane season for *Prepare for and Manage Emergencies* • Hurricane Maria for *Research and Development*
• The NCF or subfunction is anticipated to meet all routine operational needs in most but not all of the country.	• Routine operational needs are met in most of the country, but the NCF is disrupted significantly in some part the country. • A complete failure of an NCF is anticipated in an area where the NCF is concentrated, and the NCF is moderately disrupted across the country.	• Hurricane Katrina for *Provide Public Safety* and *Prepare for and Manage Emergencies* • Hurricane Sandy for *Manage Wastewater, Supply Water, Transport Materials by Pipeline, Transmit Electricity*, and *Distribute Electricity* • Extreme cold snap of 2021 in Texas for *Distribute Electricity* and *Supply Water*
• The NCF or subfunction is anticipated to be unable to meet routine operational needs in most of the country.	• Routine operational needs are not met, and the NCF is significantly disrupted in most parts of the country. • A complete failure of an NCF is anticipated in an area where the NCF is concentrated, and the NCF experiences major disruption across the country.	• COVID-19 for *Maintain Supply Chains* and *Educate and Train*

11

Risk Rating	Definition	Application	High-
5: Critical disruption	• The NCF or subfunction is anticipated to be unable to meet any of its routine operational needs across the country.	• Routine operational needs are not met, and the NCF is significantly disrupted throughout the country. • A complete failure of an NCF is anticipated in an area where the NCF is concentrated, and the NCF experiences major disruption across the country.	• 9/11 for *Tra Passenger Consumer Services* • COVID-19 *Passenger*

There is a high degree of subjectivity inherent in applying these ratings at a national scale, across a wide variety of NCFs, and with a large team of analysts. For example, the amount of time that an NCF is disrupted before there is an effect on routine operations varies by NCF. Similarly, NCFs are provided at different geographic scales, so the effects of a disruption will also vary. To reduce subjectivity and ground the analysts, as described above, risk ratings were calibrated across the analyst team through interrater reliability training sessions,[12] team discussions, structured guidance on each step of the risk assessment, and examples of each risk rating. During multiple rounds of review, risk ratings were compared across like NCFs and subfunctions to ensure consistency. Finally, in Appendix B, analysts for each NCF describe each level of risk for an NCF to provide transparency into risk ratings.

Table 2.4 provides an example of the interpretation of these ratings for *Provide Information Technology Products and Services.*

Table 2.4. Example Interpretations of Risk Ratings for *Provide Information Technology Products and Services*

Risk Rating	Description of *Provide Information Technology Products and Services*
1: No disruption or normal operations	Supply of hardware and software products can meet all routine needs.
2: Minimal disruption	Some areas might experience price shocks on certain hardware components or other challenges, but they are able to acquire products as needed.
3: Moderate disruption	Some regions are unable to acquire certain hardware products needed to meet their routine operational needs, and extended backlogs routinely form.
4: Major disruption	Extended backlogs on key hardware products routinely exist, nationwide, and many other domestic functions and industries could become nonviable as a result.
5: Critical disruption	The country cannot acquire adequate supply in multiple key hardware product categories, with cascading devastating effects on nearly all domestic industries and functions that are highly dependent on information technology (IT).

Assumptions and Considerations

Because of the complex nature of climate change, continual scientific advancements in the understanding of how the climate is changing, and the broad scope and scale of the NCFs, there are several important assumptions and considerations to bear in mind when interpreting the results of this assessment. To carry out this analysis under the study's time frame and with data available to the research team, we had to make some assumptions, including the following:

---

[12] We held a session with all the analysts to walk through an assessment of a case example. We provided information on an NCF, its subfunctions, its geographic distribution, and its vulnerabilities. Additionally, information on the climate driver for each time period and emissions scenario was provided. Analysts then worked independently to assess the risk of disruption to the NCF and entered their assessments into a master spreadsheet. The group reconvened, and the results were displayed for structured discussion and clarification of the approach. After the initial session, additional discussion sessions were held with the small teams.

- Climate change is uncertain for a variety of reasons, including the limits of scientific understanding of earth systems and the models used to simulate them, incomplete observational data, natural variability, and human response. These uncertainties grow over time as projections look further into the future. The two climate scenarios that this assessment uses to capture some of this uncertainty represent only two of many plausible futures and the state of scientific knowledge from NCA4.

- In this analysis, we assumed that NCFs follow current trends into the future. We did not account for any major population shifts, technological advancements, or significant climate adaptations. We assumed that the NCFs of today would be affected by the climate of the future.[13]

- In the study, we assessed the risk to each NCF independently of other NCFs. Ratings are based on direct effects to an individual NCF without considering dependencies on other NCFs. Rating risk independently allowed us to characterize which NCFs are most directly at risk due to climate change and which NCFs might pose the most risk to other NCFs in the system. Interdependencies are discussed in a separate section.[14]

- Risk ratings for an NCF are based on ratings for each of the NCF's subfunctions. NCF risk ratings are therefore dependent on the definitions of these subfunctions and how they collectively characterize an NCF. Definitions of each NCF are publicly available (CISA, 2020).

- National-scale risk ratings do not fully capture regional differences in risk, which can occur because of the geographic orientation of NCFs or the geospatial nature of climate change.

- Finally, the strength of the evidence base used to assess the effects of climate change varies widely across NCFs.

Underlaying these assumptions are complexities about the nature of climate change and the nature of the NCFs. These factors are also important to interpreting the results of this analysis:

- The NCFs are diverse in ways that can affect the potential for national disruption. Some NCFs, such as *Exploration and Extraction of Fuels*, have critical regional hubs that, if disrupted, could have effects at the national level. Other NCFs, such as *Supply Water* and *Manage Wastewater*, have local footprints, making disruption across the entire country at the same time unlikely. For example, if a hurricane damages a wastewater treatment plant in Biloxi, Mississippi, the same hurricane is unlikely to affect wastewater treatment in Chicago, Illinois, or lead to disruption of the NCF on a national scale. The NCFs also vary in the amount of time they can be disrupted without causing serious consequences. For example, seconds or minutes of downtime for *Operate Core Network* at key locations in the United States could have significant national implications. However, hours or days of disruption to *Produce*

---

[13] NCF vulnerabilities to climate drivers are based on the status of assets, operations, and services for a given NCF at the time of assessment. An NCF is assumed to follow current trends into the future.

[14] For example, *Provide Medical Care* (like many other NCFs) relies on *Distribute Electricity* and would be vulnerable to disruption if *Distribute Electricity* failed for an extended time period. However, for the purposes of this assessment, the risk rating for *Provide Medical Care* reflects only the direct risk that *Provide Medical Care* faces as a result of climate change and does not incorporate risk from the failure of *Distribute Electricity* as a result of that NCF being disrupted by climate change.

Chemicals might cause only limited disruption on a national scale thanks to the availability of strategic reserves.

- Our assessment was at the national scale, but climate drivers can have varying effects at local and regional levels. For example, hurricanes and sea-level rise are concentrated in specific U.S. regions. However, we assessed risk to the NCFs at the national level. This means that region-specific risk assessments for each NCF and climate driver could vary widely from the national-focused results.

- The fact that a risk rating stays the same over the analysis time periods or scenarios could obscure changes in the underlying climate change risk because the ratings reflect the maximum level of disruption to the NCF at the national scale from a single event for a given time period and scenario. However, there could be important changes to the frequency, duration, temporal or geographic distribution, or other aspects of how a climate driver interacts with an NCF from one time period to another or under a different emissions scenario. Because we focused on the maximum level of disruption within a time period, these changes might not result in a change to the risk rating. For example, a risk of moderate disruption in both 2050 and 2100 indicates the maximum level of disruption during either time period but does not indicate whether this level of disruption will occur more frequently by 2100.

- Climate risks could compound over time. Although climate change can result in multiple, complex natural hazards occurring simultaneously within multiple parts of the United States, we assessed risk for each climate driver separately. For example, successive storm systems that result in heavy precipitation could contribute to greater flooding. Or multiple drought events could exacerbate the severity of wildfires. The assessments thus do not fully reflect risk from the cumulative effects of repeated exposure to a climate driver, nor do they include the effects of multiple climate drivers occurring concurrently or in quick succession.

- Interdependencies can have varying effects that are not captured in the risk ratings. Although this analysis included some consideration of the functional relationships between NCFs, the necessary information was not available to assess the strength of the relationships among all 55 NCFs. For example, comprehensive information on the length of time between an interruption and an effect or resilience in an NCF that could prevent risk from cascading between NCFs is in development but not readily available. These factors provide critical context for understanding how risk could be transferred between NCFs that are functionally interconnected. Further study would be required to properly quantify the effects that NCF interdependencies could have on climate risks.

Chapter 3

Synthesis of Results

This chapter presents the high-level findings from our assessment of the risk posed to NCFs by climate change at the national scale. We provide an overview of our key findings across all 55 NCFs, then describe groupings of NCFs, organized by risk rating. We conclude with a summary of the climate drivers that pose risk to the NCFs. For each NCF, risk ratings and the supporting evidence are provided in Appendix B.

Overview of Climate Risk to National Critical Functions

Table 3.1 shows the number of NCFs at each risk rating for the three time periods and two future emissions scenarios. Table 3.2 depicts risk to NCFs due to one or more climate drivers under the current emissions scenario for 2050 and 2100. These NCFs are grouped by risk level, and we provide more detail on each category of risk later in this chapter.

Table 3.1. Number of National Critical Functions, by Risk Rating

Risk Rating	Baseline	Current Emissions 2050	Current Emissions 2100	High Emissions 2050	High Emissions 2100
1: No disruption or normal operations	18	10	10	10	10
2: Minimal disruption	30	20	10	12	10
3: Moderate disruption	7	24	32	32	31
4: Major disruption	0	1	2	1	3
5: Critical disruption	0	0	1	0	1

NOTE: See Chapter 2 for further explanation of ratings.

16

Table 3.2. National Critical Function Risk Ratings: Current Emissions Scenario

	NCF	2050	2100
MODERATE TO CRITICAL	Provide Information Technology Products and Services	4	5
	Supply Water	3	4
	Maintain Supply Chains	3	4
	Fuel Refining and Processing Fuels	3	3
	Exploration and Extraction of Fuels	3	3
	Transport Cargo and Passengers by Vessel	3	3
	Transport Cargo and Passengers by Air	3	3
MODERATE	Provide Housing	3	3
	Provide Materiel and Operational Support to Defense	3	3
	Produce and Provide Human and Animal Food Products and Services	3	3
	Provide Public Safety	3	3
	Provide Medical Care	3	3
	Support Community Health	3	3
	Prepare for and Manage Emergencies	3	3
	Manage Wastewater	3	3
	Manage Hazardous Materials	3	3
	Enforce Law	3	3
	Develop and Maintain Public Works and Services	3	3
	Transport Materials by Pipeline	3	3
	Transport Cargo and Passengers by Road	3	3
	Transmit Electricity	3	3
	Distribute Electricity	3	3
	Provide Wireline Access Network Services	3	3
	Provide Wireless Access Network Services	3	3
	Provide Cable Access Network Services	3	3
	Provide Metals and Materials	2	3
	Produce Chemicals	2	3
	Produce and Provide Agricultural Products and Services	2	3

	NCF	2050	2100
MODERATE	Manufacture Equipment	2	3
	Generate Electricity	2	3
	Provide Insurance Services	2	3
	Provide and Maintain Infrastructure	2	3
	Transport Cargo and Passengers by Rail	2	3
	Educate and Train	2	3
	Transport Passengers by Mass Transit	2	3
MINIMAL	Store Fuel and Maintain Reserves	2	2
	Provide Radio Broadcast Access Network Services	2	2
	Protect Sensitive Information	2	2
	Preserve Constitutional Rights	2	2
	Perform Cyber Incident Management Capabilities	2	2
	Operate Government	2	2
	Conduct Elections	2	2
	Provide Payment, Clearing, and Settlement Services	2	1
	Provide Funding and Liquidity Services	2	1
	Provide Capital Markets and Investment Activities	2	1
LOWEST RISK	Research and Development	1	2
	Provide Internet Routing, Access, and Connection Services	1	2
	Operate Core Network	1	2
	Provide Wholesale Funding	1	1
	Provide Identity Management and Associated Trust Support Services	1	1
	Provide Consumer and Commercial Banking Services	1	1
	Maintain Access to Medical Records	1	1
	Provide Satellite Access Network Services	1	1
	Provide Positioning, Navigation, and Timing Services	1	1
	Provide Internet Based Content, Information, and Communication Services	1	1

NOTE: 1 = no disruption or normal operations. 2 = minimal disruption. 3 = moderate disruption. 4 = major disruption. 5 = critical disruption. See Chapter 2 for further explanation of ratings.

Key takeaways from these tables and the risk assessment include the following:

- Of the 55 NCFs, 45 are at risk of at least a minimal disruption from climate change by 2050. This means that, because of the direct effects of climate change, 45 NCFs are at risk of being affected by climate change in some way, but many will be able to meet routine operational needs. Within this group of 45, 13 NCFs are expected to face an increased risk due to climate change by 2100 under the current emissions scenario.

- Of the 55 NCFs, 25 are at risk of a moderate disruption or greater by 2050 in the current emissions scenario. This means that half of the NCFs are at risk of being unable to meet routine operational needs in some parts of the country. This number grows to 35 by 2100 in the current emissions scenario. Later in this chapter, we discuss the potential consequences—which can be very significant—of a moderate disruption to any of the NCFs.
- Three NCFs—*Provide Information Technology Products and Services*, *Maintain Supply Chains*, and *Supply Water*—are at risk of major or critical disruption due to climate change by 2100. These NCFs are at risk of being disrupted to the point that they are unable to meet routine operational needs in most or all of the country.
- Seven NCFs—primarily related to financial services and network (i.e., internet-based) operations and services—are not at risk of direct effects from climate change in either of the two time periods.
- In a future with higher emissions or higher climate sensitivity to emissions, risk ratings to NCFs are generally higher overall.
- Overall, the risk ratings are higher than in the previous year's assessment, including many more NCFs at risk of moderate disruption at certain time periods under certain scenarios. Climate change is the primary mechanism driving the risk assessed in this study. Because, for this analysis, we relied on the same projections of future climate change, changes in ratings from last year's to this year's assessment reflect a combination of a refined application of the risk rating scale, the use of new scientific information to guide assessments of NCF exposure and vulnerabilities to climate drivers, and enhancements in the analysis over time.[15]

National Critical Functions, by Risk Rating

This report groups NCFs into four categories based on risk assessment results—those at risk of moderate to critical disruption,[16] moderate disruption, minimal disruption, and no disruption or normal operations. This section describes the risk ratings for the NCFs in each category. We discuss notable climate drivers, impact mechanisms, and geographic characteristics of the NCFs within these categories.

National Critical Functions at Risk of Moderate to Critical Disruption

Seven NCFs are at the greatest risk of disruption due to climate change, as shown in Table 3.3. All seven of these are at risk of moderate disruption at baseline, which suggests that they could experience

---

[15] This risk assessment is part of an iterative process and reflects our best understanding of the tools and information available at the time of analysis.

[16] We include in the moderate to critical grouping those NCFs that were rated as at risk of major or critical disruption in any future time period or scenario, as well as those NCFs rated at risk of moderate disruption in the baseline time period. Those NCFs in the moderate grouping in Table 3.2 were all assessed as being at risk of minimal disruption in the baseline time period, with risk growing to moderate in at least one future time period.

significant disruptions in the future. Some of these NCFs have already experienced such disruptions. In the case of *Supply Water*, severe drought is already disrupting the provision of water supply across many parts of the West, resulting in elevated risk in the baseline time period (Kammeyer et al., 2021; Williams, Cook, and Smerdon, 2022). We assessed that the effects of climate change will stress water supplies, increase demand, and damage or disrupt water infrastructure systems and that all climate drivers except for extreme cold will pose risk to the provision of water supply into the future.

Table 3.3. Ratings for National Critical Functions at Risk of Moderate to Critical Disruption

NCF	Baseline	Current Emissions 2050	Current Emissions 2100	High Emissions 2050	High Emissions 2100
Provide Information Technology Products and Services	3	4	5	4	5
Supply Water	3	3	4	3	4
Maintain Supply Chains	3	3	4	3	4
Fuel Refining and Processing Fuels	3	3	3	3	3
Exploration and Extraction of Fuels	3	3	3	3	3
Transport Cargo and Passengers by Vessel	3	3	3	3	3
Transport Cargo and Passengers by Air	3	3	3	3	3

NOTE: 3 = moderate disruption. 4 = major disruption. 5 = critical disruption. See Chapter 2 for further explanation of ratings.

One NCF—*Provide Information Technology Products and Services*—is at risk of critical disruption by 2100, meaning that this NCF could be unable to meet routine operational needs across the country.[17] This NCF is associated with the development and provision of a wide variety of hardware and software products. Hardware manufacturing is often geographically concentrated and vulnerable to natural disasters (Romanosky et al., 2022; Varas et al., 2021). The geographic concentration of manufacturing capacity for key hardware products (e.g., microchips) leads to single points of failure that are vulnerable to regional natural disasters. This would remain true anywhere that manufacturing capacity for critical hardware components is geographically concentrated, regardless of whether that capacity were concentrated in foreign or domestic locations. Many distinct single points of failure can and do exist in regions that are vulnerable to different climate drivers, so nearly all climate drivers could be major sources of risk.

The networked nature of *Transport Cargo and Passengers by Air*, *Transport Cargo and Passengers by Vessel*, and *Maintain Supply Chains* and their reliance on key hubs within their networks leave them vulnerable to climate change. For example, a majority of enplanements in the United States take place at 61 airports, and delays and cancellations from the effects of, for example, a hurricane can cascade through the national aviation system even if only a few airports (e.g., New York area) are directly affected (B. Miller et al., 2020). *Maintain Supply Chains* and *Transport Cargo and Passengers by Vessel*

---

[17] Our ratings for *Provide Information Technology Products and Services* and *Support Community Health* differ from those that CISA submitted in its annual update response to EO 14008.

are extremely vulnerable to flooding, sea-level rise, and tropical cyclones and hurricanes because of the location of and their dependence on coastal ports. Two trends in cargo shipping—increased specialization in ports (e.g., container, bulk cargo, petroleum) and ports with shipping lanes deep enough to accommodate increasing vessel size—mean that vessels cannot easily load or offload at other ports if operations are disrupted (Verschuur, Koks, and Hall, 2020). For this reason, disruptions at ports moving the highest volume of goods—the ports of Los Angeles and Long Beach and of New York and New Jersey (for containerized goods), the port of South Louisiana (for bulk cargo), and the port of Houston (for petroleum)—could have a disproportionate economic effect (U.S. Department of Transportation [DOT], 2022). These disruptions are likely to be more widespread for *Maintain Supply Chains*.

Fuel Refining and Processing Fuels and *Exploration and Extraction of Fuels* are both at risk of moderate disruption from present into the future. *Fuel Refining and Processing Fuels* is a critical node in stabilizing energy markets (Jing et al., 2020). Because refinery and processing infrastructure are regionally concentrated, climate drivers that have local direct effects could have wider-reaching consequences (U.S. Energy Information Administration [EIA], 2021a; EIA, 2022a; EIA, 2022b). For *Exploration and Extraction of Fuels*, vulnerabilities to climate drivers stem from the geographic concentration of exploration and extraction activities and the sector's lack of investment in hardening infrastructure against climate change (Carlson, Goldman, and Dahl, 2015; Cruz and Krausmann, 2013; Office of Energy Policy and Systems Analysis, 2015).

National Critical Functions at Risk of Moderate Disruption

We assessed that 28 NCFs are at risk of moderate disruption from at least one of the eight climate drivers across the time periods and emissions scenarios. A risk of moderate disruption means that the NCF will meet routine operational needs in most of the country. The consequences of an NCF not meeting routine operational needs in parts of the country can be significant. For example, Hurricane Katrina disrupted the provision of multiple NCFs, including *Transmit Electricity*, *Educate and Train*, and *Enforce Law*, in a large region along the Gulf Coast. Although these NCFs continued to meet routine operational needs for most of the country, the health, safety, economic and other consequences in the affected areas were significant. Under the current emissions scenario, just under half of the NCFs are at risk of moderate disruption by 2050 (see Table 3.4), which could include events with similar levels of consequences. Under the high emissions scenario, all but two—*Transport Passengers by Mass Transit* and *Educate and Train*—are at risk of moderate disruption by 2050. Under both emissions scenarios, all NCFs are at moderate risk by 2100.

Table 3.4. Ratings for National Critical Functions at Risk of Moderate Disruption

NCF	Baseline	Current Emissions 2050	Current Emissions 2100	High Emissions 2050	High Emissions 2100
Provide Housing	2	3	3	3	3
Provide Material and Operational Support to Defense	2	3	3	3	3
Produce and Provide Human and Animal Food Products and Services	2	3	3	3	3
Provide Public Safety	2	3	3	3	3
Provide Medical Care	2	3	3	3	3
Support Community Health	2	3	3	3	3
Prepare for and Manage Emergencies	2	3	3	3	3
Manage Wastewater	2	3	3	3	3
Manage Hazardous Materials	2	3	3	3	3
Enforce Law	2	3	3	3	3
Develop and Maintain Public Works and Services	2	3	3	3	3
Transport Materials by Pipeline	2	3	3	3	3
Transport Cargo and Passengers by Road	2	3	3	3	3
Transmit Electricity	2	3	3	3	3
Distribute Electricity	2	3	3	3	3
Provide Wireline Access Network Services	2	3	3	3	3
Provide Wireless Access Network Services	2	3	3	3	3
Provide Cable Access Network Services	2	3	3	3	3
Provide Metals and Materials	2	2	3	3	3
Produce Chemicals	2	2	3	3	3
Produce and Provide Agricultural Products and Services	2	2	3	3	3
Manufacture Equipment	2	2	3	3	3
Generate Electricity	2	2	3	3	3
Provide Insurance Services	2	2	3	3	3
Provide and Maintain Infrastructure	2	2	3	3	3
Transport Cargo and Passengers by Rail	2	2	3	3	3
Educate and Train	2	2	3	2	3
Transport Passengers by Mass Transit	2	2	3	2	3

NOTE: 2 = minimal disruption. 3 = moderate disruption. See Chapter 2 for further explanation of ratings.

Many of the NCFs in this group rely heavily on physical infrastructure to function, much of which is aging, and therefore not only are exposed to climate drivers but have documented physical vulnerabilities that make them susceptible to those drivers. Other NCFs in this group, such as *Enforce Law*, *Prepare for and Manage Emergencies*, and *Provide Public Safety*, are not tied as heavily to physical infrastructure assets. These NCFs face a risk of moderate disruption due to an anticipated increase in demand for the NCF or potential workforce shortages because of climate drivers.

NCFs in this group account for a variety of geographic distributions, including those that are primarily provided and concentrated locally, others that are clustered within regional hot spots, and a few that are provided nationally. This suggests that geographic concentration alone is not a proxy for the potential risk of climate-driven NCF disruption. Other factors that influence potential risk from climate drivers include system redundancy, a given sector's pace of preparation for and investments in hardening against disruption, and potential exposure to more than one climate driver, which is a function of geographic location (for example, coastal versus inland).

Transmit Electricity and *Distribute Electricity* are vulnerable to seven of the eight climate drivers because of the sheer breadth of infrastructure across the country. Flooding, sea-level rise, and tropical cyclones and hurricanes threaten key transmission and distribution substations, transformers, and power lines and poles (Office of Energy Policy and Systems Analysis, 2015; Shield et al., 2021). Extreme heat reduces the operating capacity of transmission and distribution equipment and causes increased demand for cooling, thus stressing the system (Fant et al., 2020; Rusco, 2014). Wildfire is increasingly a concern for power grid infrastructure, particularly in the western United States (Dale et al., 2018). *Generate Electricity* is similarly at risk due to flooding, sea-level rise, and tropical cyclones and hurricanes. A facility with more than 25 gigawatts of operating or proposed capacity is at risk of flooding and sea-level rise because of the equipment's physical location within floodplains and low-lying coastal areas (Bierkandt, Auffhammer, and Levermann, 2015). Although this number is a small percentage of existing generation capacity across the United States (about 2 percent), without these power plants online, some states might struggle to meet their average power load needs (EIA, 2022c). For example, the generation capacity at risk from sea-level rise in Delaware is 80 percent of the state's average power load; in New York, it is 63 percent, and, in Florida, it is 43 percent (Bierkandt, Auffhammer, and Levermann, 2015). Decreasing freshwater availability caused by sustained drought is of growing concern because of the sector's reliance on water to generate hydroelectricity or used in cooling processes for thermal plants (Zohrabian and Sanders, 2018).

We assessed three IT- and telecommunication-related NCFs—*Provide Cable Access Network Services*, *Provide Wireless Access Network Services*, and *Provide Wireline Access Network Services*—to be at moderate risk by 2050 under both emissions scenarios, primarily because of the potential for climate drivers to physically damage equipment that is critical for providing these functions (e.g., local stations, cables and wires, signal transmission and receiver equipment, and cell towers).[18] Flooding, severe storm systems, tropical cyclones and hurricanes, and wildfire already cause local disruptions to these NCFs (Currie, 2022; Kuligowski et al., 2014; Kwasinski et al., 2009; Mena, 2020). Sea-level rise and extreme heat could also exacerbate the effects that other climate drivers have on these NCFs.

Provide Materiel and Operational Support to Defense is at risk of moderate disruption by 2050 under both scenarios and from all the climate drivers except extreme cold. Although the NCF is managed and provided nationally, some redundancies and adaptive capacities are built into the system because of the distributed nature of U.S. Department of Defense (DoD) installations. Physical damage and disruption, workforce shortages, and demand changes could nevertheless affect this NCF. For example, Hurricane Michael in 2018 caused damage to more than 400 buildings at Tyndall Air Force Base and led to the displacement of F-22 aircraft in Florida (Reeves, 2019).

Finally, we assessed that *Provide Housing* is also at risk of moderate disruption by 2100. Flooding, sea-level rise, tropical cyclones and hurricanes, and severe storm systems have the capacity to cause

---

[18] Shortages of semiconductors are considered in risk ratings for only *Provide Information Technology Products and Services.*

widespread damage and destruction to homes along the coasts and waterways (Buchanan et al., 2020; Dahl et al., 2018; Public and Affordable Housing Research Corporation and National Low Income Housing Coalition, 2021). This will increase demand for the provision of repair and maintenance activities, as well as new construction, potentially beyond existing capacities in affected areas.

National Critical Functions at Risk of Minimal Disruption

Seven NCFs are at risk of minimal disruption from any of the eight climate change drivers across all time periods and scenarios (Table 3.5). Although these NCFs might not always be able to meet routine operational needs, any disruption should be relatively limited, and the NCFs are not at risk of being disrupted simultaneously across large parts of the country.

Table 3.5. Ratings for National Critical Functions at Risk of Minimal Disruption

NCF	Baseline	Current Emissions 2050	Current Emissions 2100	High Emissions 2050	High Emissions 2100
Store Fuel and Maintain Reserves	2	2	2	2	2
Provide Radio Broadcast Access Network Services	2	2	2	2	2
Protect Sensitive Information	1	2	2	2	2
Preserve Constitutional Rights	1	2	2	2	2
Perform Cyber Incident Management Capabilities	1	2	2	2	2
Operate Government	1	2	2	2	2
Conduct Elections	1	2	2	2	2

NOTE: 1 = no disruption or normal operations. 2 = minimal disruption. See Chapter 2 for further explanation of ratings.

There were several reasons for NCFs to be at minimal risk of disruption at the national level. First, many of the NCFs in this group are provided locally. Although some NCFs have critical regional or national nodes, such as *Operate Government* in the Washington, D.C., area, many are provided on a sufficiently distributed basis that we assessed that local disruptions would not rise to the minimum. Second, many of the NCFs in this group have redundancies across the NCF, such as alternative broadcast facilities for the *Provide Radio Broadcast Access Network Services* NCF or the flexibility to accommodate short-term disruptions for *Conduct Elections*, for example, by delaying elections or using alternative systems, such as voting by mail. Finally, although several NCFs in this group have physical infrastructure that is exposed to climate drivers, they rely on electronic systems that are not tied to a single physical location and, as a result, are not likely to be disrupted at the national level.

National Critical Functions at Lowest Risk of Disruption

The 13 NCFs listed in Table 3.6 have no or minimal risk of disruption from any of the climate drivers across all time periods and emissions scenarios, which means that the NCFs can continue to meet all operational needs even if they are exposed to climate drivers. Five of these NCFs are centrally managed at the national scale and two at the local level; the remaining are hybrid.

Table 3.6. Ratings for National Critical Functions at Lowest Risk of Disruption

NCF	Baseline	Current Emissions 2050	Current Emissions 2100	High Emissions 2050	High Emissions 2100
Provide Payment, Clearing, and Settlement Services	1	2	1	2	1
Provide Funding and Liquidity Services	1	2	1	2	1
Provide Capital Markets and Investment Activities	1	2	1	2	1
Research and Development	1	1	2	1	2
Provide Internet Routing, Access, and Connection Services	1	1	2	1	2
Operate Core Network	1	1	2	1	2
Provide Wholesale Funding	1	1	1	1	1
Provide Identity Management and Associated Trust Support Services	1	1	1	1	1
Provide Consumer and Commercial Banking Services	1	1	1	1	1
Maintain Access to Medical Records	1	1	1	1	1
Provide Satellite Access Network Services	1	1	1	1	1
Provide Positioning, Navigation, and Timing Services	1	1	1	1	1
Provide Internet Based Content, Information, and Communication Services	1	1	1	1	1

NOTE: 1 = no disruption or normal operations. 2 = minimal disruption. See Chapter 2 for further explanation of ratings.

We assessed that seven of these NCFs will be unaffected by the climate drivers for both time periods and emissions scenarios. Two NCFs—*Provide Positioning, Navigation, and Timing* [PNT] *Services* and *Provide Satellite Access Network Services*—which are nationally provided—require ground station infrastructure that a climate driver could theoretically damage. However, this infrastructure is intentionally sited in areas that are not particularly prone to the climate drivers and involve sufficient redundancy to protect against disruption from climate drivers. In contrast, *Maintain Access to Medical Records* is locally provided at sites dispersed throughout the country. Although networked electronic record–sharing systems are in use, medical facilities do not rely on these systems, nor do they typically share records electronically across a large network of sites. Instead, facilities generally maintain on-site copies or rely on patient interviews to provide medical histories. Should climate drivers affect electronic systems, the effects are likely to cascade from other NCFs, such as *Distribute Electricity*, which would not be direct. In addition, climate drivers might increase the demand for medical records, but these effects are likely to be local with no appreciable degradation in function. Other NCFs that are unlikely to experience disruption from climate drivers are *Provide Internet Based Content, Information, and Communication Services* and *Provide Identity Management and Associated Trust Support Services*. These could be provided at multiple levels where, for example, some of the activities or assets are found locally but have some integration with assets or systems at the state, regional, or national level.

We assessed that three NCFs will likely experience minimal disruption from climate drivers by the end of the century as the frequency and severity of flooding, severe storm systems, tropical cyclones and hurricanes, and wildfire increase over time. These climate drivers could damage physical infrastructure with local effects. For example, *Research and Development* is locally provided at sites dispersed throughout the country. *Research and Development* activities and physical assets are often found at universities, federal laboratories, and private facilities. These activities are unlikely to experience climate-driven disruption until late in the century, and it would not involve a large geographic area.

Three financially oriented NCFs could experience some disruption as climate drivers become severer by the middle of the 21st century. However, the risks to these NCFs will likely return to baseline by the end of the century. In particular, flooding and tropical cyclones and hurricanes have the potential to affect *Provide Capital Markets and Investment Activities*, *Provide Funding and Liquidity Services*, and *Provide Payment, Clearing, and Settlement Services* by the middle of the 21st century, but the combined effects of federal entities, such as the Federal Reserve System and the Federal Deposit Insurance Corporation (FDIC), working with geographically decentralized service providers supported by technologically robust systems are likely to enable continuing function with only minimal disruptions (American Banker, 2012; Climate-Related Market Risk Subcommittee, 2020; C. Cooper, Labonte, and Perkins, 2019; FDIC, 2017; McAndrews and Potter, 2002). By the end of the century, existing trends toward decentralized service provision and distributed, virtual systems will have progressed enough that they will return the potential for disruption from climate drivers back to baseline, at which the NCF can meet all operational needs.

Key Risks from Climate Drivers

Climate change is increasing the frequency, magnitude, and duration of many of the climate drivers that pose a risk to NCFs (Wuebbles et al., 2017). Although there are important regional distinctions in how and where climate drivers will change, we assessed that, at the national scale, four of the eight climate drivers will have an outsized effect on NCFs. This can be seen in Table 3.7, which shows the number of NCFs assessed as being at risk from each climate driver. The rationale for risk ratings for each NCF to each climate driver is described in Appendix B.

Table 3.7. Number of National Critical Functions at Risk, by Climate Driver

		Baseline	Current Emissions 2050	Current Emissions 2100	High Emissions 2050	High Emissions 2100
Drought	1 - No Disruption	43	33	30	33	30
	2 - Minimal Disruption	10	17	12	13	10
	3 - Moderate Disruption	2	5	12	9	14
	4 - Major Disruption	0	0	1	0	1
	5 - Critical Disruption	0	0	0	0	0
Extreme Cold	1 - No Disruption	43	43	42	43	45
	2 - Minimal Disruption	12	11	10	9	7
	3 - Moderate Disruption	0	1	3	3	3
	4 - Major Disruption	0	0	0	0	0
	5 - Critical Disruption	0	0	0	0	0
Extreme Heat	1 - No Disruption	39	21	16	19	16
	2 - Minimal Disruption	16	27	24	26	20
	3 - Moderate Disruption	0	7	15	10	19
	4 - Major Disruption	0	0	0	0	0
	5 - Critical Disruption	0	0	0	0	0
Flooding	1 - No Disruption	27	14	14	14	12
	2 - Minimal Disruption	24	27	17	17	15
	3 - Moderate Disruption	4	14	22	23	26
	4 - Major Disruption	0	0	2	1	1
	5 - Critical Disruption	0	0	0	0	1
Sea-level Rise	1 - No Disruption	45	22	19	22	18
	2 - Minimal Disruption	10	23	12	17	12
	3 - Moderate Disruption	0	10	23	16	24
	4 - Major Disruption	0	0	1	0	1
	5 - Critical Disruption	0	0	0	0	0
Severe Storm Systems (non-tropical)	1 - No Disruption	28	17	15	17	15
	2 - Minimal Disruption	25	27	26	25	24
	3 - Moderate Disruption	2	11	14	13	16
	4 - Major Disruption	0	0	0	0	0
	5 - Critical Disruption	0	0	0	0	0
Tropical Cyclones and Hurricanes	1 - No Disruption	23	10	10	10	10
	2 - Minimal Disruption	26	24	16	20	15
	3 - Moderate Disruption	6	20	27	24	28
	4 - Major Disruption	0	1	1	1	1
	5 - Critical Disruption	0	0	1	0	1
Wildfire	1 - No Disruption	29	19	15	19	15
	2 - Minimal Disruption	26	21	15	18	14
	3 - Moderate Disruption	0	15	25	18	26
	4 - Major Disruption	0	0	0	0	0
	5 - Critical Disruption	0	0	0	0	0

Tropical cyclones and hurricanes have a legacy of severe and lasting damage to U.S. regions. Four out of every five NCFs are at risk of at least minimal disruption from tropical cyclones by 2050. About half of all NCFs are at risk of moderate disruption by 2100 under both emissions scenarios. NCFs with assets and service provision along the Gulf and Atlantic Coasts are anticipated to be significantly affected.

Sea-level rise, which can temporarily or permanently inundate facilities, infrastructure assets, or key components of infrastructure systems, is expected to affect the majority of NCFs by 2050 and pose a moderate to major risk of disruption for 25 NCFs by 2100 in the high emissions scenario. These effects are due to the relatively high concentration of major urban areas and infrastructure assets in coastal zones.

Riverine and flash flooding from extreme rainfall events can cause significant infrastructure disruptions as a result of direct infrastructure damage from floodwaters or debris or as a result of loss of access to assets. Three out of every four NCFs are at risk of local to regional disruptions by 2050 in the current emissions scenario. Half of the NCFs are at a risk of moderate or greater disruption at the national scale by 2100 in the high emissions scenario. Because flooding is pervasive across much of the United States and NCF assets are concentrated along waterways or in cities prone to urban flooding, this climate driver is likely to affect all U.S. regions.

Wildfire can cause direct damage to infrastructure facilities, delay or suspend transportation and communication, and result in hazardous air quality. By 2050, under the current emissions scenario, 36 NCFs are at risk of minimal or greater disruption due to wildfire. In the high emissions scenario, 26 of the 55 NCFs are at risk of moderate disruption by 2100.

The assessments presented in this chapter offer a first-order understanding of the potential for disruption from climate change. The climate drivers were considered individually to determine direct effects on each NCF. In Chapter 4, we discuss observations on how disruption of one NCF could cascade to other NCFs, with the potential to amplify the level of disruption to a given NCF.

Chapter 4

Interdependencies

Background

The NCFs consist of a varied set of activities, assets, information, people, and distribution entities that operate as an interconnected system. By design, NCFs are also inherently critical to the security, national economy, and well-being of the United States. As a result, a disruption in one NCF can propagate, or cascade through, the system through linkages between functions. Therefore, risks to one NCF do not occur independently of risks to other NCFs. Understanding these linkages and how effects cascade is essential for identifying and characterizing all sources of risk and for resilience planning.

NCFs can be linked in various ways. One NCF might be *dependent* on another: The effect that the NCF has on another is unidirectional. In other words, one NCF might rely on another to maintain normal operations, but there is no reliance in the opposite direction. NCF pairs might also be mutually dependent: The operations of each could affect the other. This is referred to as *interdependence*. In this chapter, we refer to two directions of dependencies. An NCF is described as *upstream* when it provides critical inputs to other NCFs, while an NCF is described as *downstream* if it relies on outputs of other NCFs. Figure 4.1 provides a notional example of the types of links between NCFs.

Figure 4.1. Dependent and Interdependent Relationships Between National Critical Functions

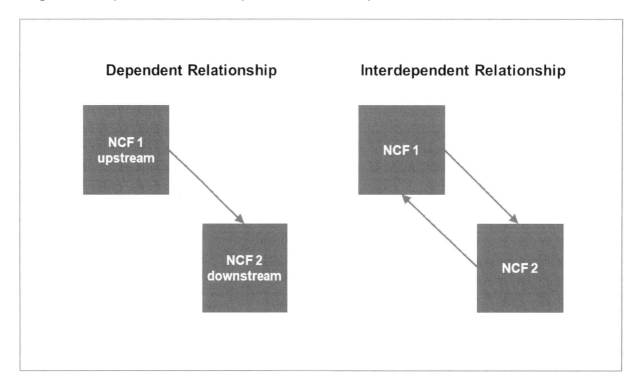

For example, as shown in Figure 4.2, *Supply Water* is upstream of *Produce and Provide Agricultural Products and Services* because it provides agricultural water to *Produce and Provide Agricultural Products and Services*. *Supply Water* is downstream of *Distribute Electricity* because it relies on electricity from *Distribute Electricity*.

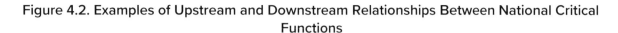

Figure 4.2. Examples of Upstream and Downstream Relationships Between National Critical Functions

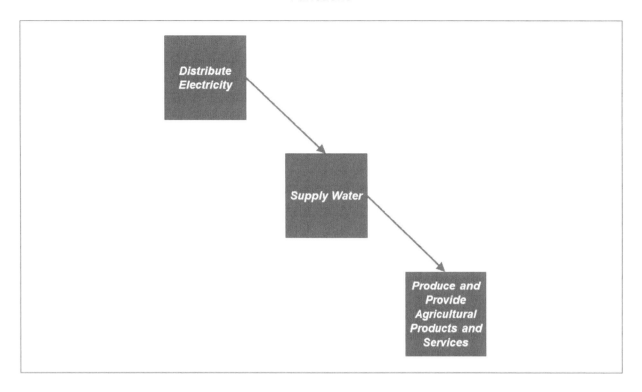

The literature describes four common types of dependencies:

- *physical*, in which the operations of one NCF rely on material outputs from another
- *cyber*, in which the operations of one NCF rely on information or data produced by another
- *geographic*, in which operations of two or more NCFs are located near one another
- *logical*, in which the operations of one NCF rely on another through policy, economic or financial inputs, or personnel (CISA, undated-a; Petit et al., 2015; Rinaldi, Peerenboom, and Kelly, 2001).

Dependencies can be strong or weak, described as *tightly* or *loosely coupled*. For tightly coupled NCFs, effects would cascade relatively quickly, or there might be few options to dampen the effects. According to Klaver, Luiijf, and Nieuwenhuijs (2011, p. 29), when infrastructures are strongly interconnected, "disturbances tend to spread more rapidly." Using indicators of strength of interdependence (also known as *coupling*), analysts can begin to evaluate the nuances of interdependence (Der Sarkissian et al., 2022; Klaver, Luiijf, and Nieuwenhuijs, 2011; Rinaldi, Peerenboom, and Kelly, 2001).

Therefore, to assess how disruptions might cascade from one NCF to another, it is necessary not only to know that a dependence or interdependence exists but also to understand the contextual and operational environments that influence how the operations of the two interact. These too will vary among NCFs, but a variety of factors can affect how disruption might cascade from one NCF to another. These include knowing how long it might take before a disruption in one influences another, whether there are alternatives or substitutes for the disrupted NCF, whether there are other buffers between the two to lessen the effects, or how quickly an NCF can recover from these cascading effects. Finally, we captured the first-order dependencies (direct dependencies between two NCFs), but effects can also cascade from seemingly unconnected NCFs through second-, third-, or *n*th-order connections (when an NCF affects another through the interdependencies of an intermediary NCF).

In the remainder of this chapter, we present the findings of an interdependence analysis using information provided by CISA on the functional dependencies between NCFs. We discuss the limitations of this approach, which are driven largely by having inadequate information on the characteristics of these dependencies for the entire system of NCFs, and we propose an approach for extending this analysis for assessing the cascading effects from climate drivers.

Interdependence Analysis

In this analysis, our goal was to understand how dependencies between NCFs can result in cascading effects from climate-induced disruptions to NCFs. We aim to identify two main groups of NCFs: (1) those that we assessed as being at lower levels of risk due to climate change but depend on a high number of NCFs assessed as being at higher levels of risk and (2) those that we assessed as being at higher levels of risk and have a high number of NCFs that depend on them. These two groupings allow us to understand which NCFs might be indirectly affected by climate change via other NCFs and which NCFs have the potential to affect a large number of NCFs, respectively. This information is important to decisionmakers because it could alter their understanding of the total underlying level of risk to individual NCFs as a result of climate change or which NCFs should be prioritized for mitigation efforts.

To conduct our analysis, we relied on information provided by CISA in its official characterization of NCF interdependencies. For each NCF, CISA identified whether the NCF was dependent on another NCF.[19] These dependencies reflect relationships in which the output of one NCF is used as an input for another, such as the relationship between *Transmit Electricity* and *Distribute Electricity*. This characterization is based largely on the functional interdependencies between NCFs and does not contain information on the characteristics of the connections between NCFs. This means we do not have an assessment of how those dependencies might change over time, geography, or different contexts, which are factors that influence how disruption of one NCF could affect another. In other words, we know there is a connection between two NCFs, but we do not have information that describes the significance or nature of the connection. As a result, we were not able to weight the importance of each identified dependence, nor were we able to characterize the severity, timing, or other aspects of any cascading effects between NCFs.

Using CISA's information and these definitions, we first totaled the number of upstream and downstream dependencies for each NCF. We then carried out two analyses on these data. For the first, we filtered the upstream dependencies for those NCFs that were assessed as being at risk of moderate disruption or greater. This gave us a table showing, for each NCF, the number of NCFs on which it depends (that is, the number of upstream NCFs) that are at risk of moderate disruption or

---

[19] A Microsoft Excel spreadsheet with a binary dependence matrix for the 55 NCFs was provided to HSOAC on September 26, 2022. It was noted that the NCF functional decomposition was ongoing at that time.

greater. For the second, we filtered the dataset to look only at NCFs at risk of moderate disruption or greater but retained all of their downstream dependencies. This allowed us to understand which of the NCFs assessed as being at risk of moderate disruption might have the potential to affect large numbers of other NCFs.

National Critical Functions That Could Face Additional Risk from Others at Moderate or Higher Risk of Disruption

Table 4.1 shows the results of our analysis of which NCFs might be indirectly affected by climate change via other NCFs. For each NCF, it presents the NCF's risk rating for 2050 under the current emissions scenario and the number of upstream NCFs at risk of moderate disruption or greater. The numbers in the table show how NCFs with the same numerical risk rating from our risk assessment (which considered only the direct effects of climate change) could likely have different exposures to climate change through the indirect effects from dependencies. This analysis shows that the ten NCFs assessed as being at no risk of disruption due to climate change are dependent on at least six other NCFs assessed as being at risk of moderate disruption or greater. Of this group, *Research and Development* has the highest number of upstream NCFs at risk of moderate disruption or greater.

Table 4.1. Number of Upstream Dependencies, by National Critical Function

NCF	Risk Rating (Current Emissions 2050)	Number of Upstream NCFs (at risk of moderate disruption or greater)
Research and Development	1	11
Provide Internet Based Content, Information, and Communication Services	1	9
Provide Internet Routing, Access, and Connection Services	1	9
Maintain Access to Medical Records	1	9
Provide Wholesale Funding	1	9
Provide Consumer and Commercial Banking Services	1	8
Provide Identity Management and Associated Trust Support Services	1	8
Operate Core Network	1	6
Provide Positioning, Navigation, and Timing Services	1	6
Provide Satellite Access Network Services	1	6
Transport Cargo and Passengers by Rail	2	15
Provide and Maintain Infrastructure	2	15
Store Fuel and Maintain Reserves	2	15

NCF	Risk Rating (Current Emissions 2050)	Number of Upstream NCFs (at risk of moderate disruption or greater)
Transport Passengers by Mass Transit	2	13
Provide Metals and Materials	2	13
Manufacture Equipment	2	12
Produce and Provide Agricultural Products and Services	2	12
Operate Government	2	11
Generate Electricity	2	11
Produce Chemicals	2	11
Protect Sensitive Information	2	10
Conduct Elections	2	9
Preserve Constitutional Rights	2	9
Provide Payment, Clearing, and Settlement Services	2	9
Educate and Train	2	8
Perform Cyber Incident Management Capabilities	2	8
Provide Capital Markets and Investment Activities	2	8
Provide Funding and Liquidity Services	2	8
Provide Insurance Services	2	8
Provide Radio Broadcast Access Network Services	2	6
Provide Materiel and Operational Support to Defense	3	19
Develop and Maintain Public Works and Services	3	17
Provide Public Safety	3	17
Manage Hazardous Materials	3	16
Prepare for and Manage Emergencies	3	16
Transport Cargo and Passengers by Air	3	15
Transport Cargo and Passengers by Vessel	3	15
Provide Medical Care	3	15
Support Community Health	3	15
Fuel Refining and Processing Fuels	3	14
Maintain Supply Chains	3	13
Transport Cargo and Passengers by Road	3	13

NCF	Risk Rating (Current Emissions 2050)	Number of Upstream NCFs (at risk of moderate disruption or greater)
Manage Wastewater	3	13
Supply Water	3	13
Exploration and Extraction of Fuels	3	12
Provide Housing	3	12
Transport Materials by Pipeline	3	11
Produce and Provide Human and Animal Food Products and Services	3	11
Enforce Law	3	10
Distribute Electricity	3	9
Provide Cable Access Network Services	3	8
Transmit Electricity	3	8
Provide Wireline Access Network Services	3	7
Provide Wireless Access Network Services	3	6
Provide Information Technology Products and Services	4	8

NOTE: 1 = no disruption or normal operations. 2 = minimal disruption. 3 = moderate disruption. 4 = major disruption. See Chapter 2 for further explanation of ratings.

Although the ten NCFs are not anticipated to be at risk of disruption from climate change, they might still face effects indirectly through other NCFs, so owner-operators responsible for these functions might want to consider resilience investments to address those dependencies. Outside of *Research and Development*, this group includes financial service NCFs, as well as NCFs related to internet-based services and operations. Owner operators might have additional equities related to climate change beyond those related to direct effects to their NCFs and, more generally, that the full level of risk to any individual NCF as a result of climate change might be understated.

Similarly, those NCFs with high numbers of upstream dependencies could be at additional risk from climate drivers, although the number of dependencies does not mean that effects will necessarily cascade. More analysis is required to determine the nature of the dependence and therefore to understand whether and how a disruption to one NCF could affect another. In Table 4.1, *Provide Materiel and Operational Support to Defense* has the highest number of upstream dependencies assessed as being at risk of moderate disruption. This suggests that resilience planning and analysis that incorporates the effects on other NCFs will be essential to lessen the potential for disruption that climate drivers could cause for normal operations for *Provide Materiel and Operational Support to Defense*. This also applies to the other NCFs that have many upstream connections to NCFs at moderate risk.

National Critical Functions That Could Affect Many Others

Table 4.2 presents the results of the second part of our interdependence analysis, which focused on identifying the NCFs that have the potential to affect many NCFs because of climate change. Table 4.2 includes the number of downstream NCFs for all NCFs assessed as being at risk of moderate disruption or greater by 2050 in the current emissions scenario and demonstrates the interconnected nature of the system. The table shows that there are four NCFs on which all other NCFs depend. Within this group of four is *Provide Information Technology Products and Services*, which was assessed as being at risk of major disruption by 2050 and at risk of critical disruption by 2100 under both emissions scenarios. An additional five NCFs also have high numbers of downstream NCFs; these include utility-related NCFs that provide essential electricity, communication, and water services to the country. Because these NCFs have the potential to affect many other NCFs and they face risk of disruption into the future, measures to reduce the risks that climate change poses to these NCFs could be integral to mitigating overall national risk. Additional information on the strength of these relationships with downstream NCFs is essential for determining resilience planning and investment priorities.

Table 4.2. National Critical Function Downstream Dependencies

NCF	Risk Rating (Current Emissions 2050)	Number of Downstream NCFs
Maintain Supply Chains	3	54
Prepare for and Manage Emergencies	3	54
Provide Medical Care	3	54
Provide Information Technology Products and Services	4	54
Distribute Electricity	3	52
Provide Wireless Access Network Services	3	50
Provide Wireline Access Network Services	3	49
Provide Cable Access Network Services	3	48
Enforce Law	3	46
Supply Water	3	39
Manage Hazardous Materials	3	30
Manage Wastewater	3	22
Provide Public Safety	3	19
Transport Materials by Pipeline	3	18
Develop and Maintain Public Works and Services	3	13
Support Community Health	3	11
Transport Cargo and Passengers by Air	3	8
Transport Cargo and Passengers by Road	3	8
Transport Cargo and Passengers by Vessel	3	8
Produce and Provide Human and Animal Food Products and Services	3	8
Provide Materiel and Operational Support to Defense	3	7
Transmit Electricity	3	4
Fuel Refining and Processing Fuels	3	3
Exploration and Extraction of Fuels	3	2
Provide Housing	3	2

NOTE: 3 = moderate disruption. 4 = major disruption.

Limitations of Our Interdependence Analysis

We used CISA's analysis of functional relationships between NCFs to determine how our understanding of risk to an NCF from climate change might change when considering its dependence on other NCFs. We identified NCFs that have large numbers of dependent relationships with higher-risk NCFs and therefore could be poised to experience greater effects from climate change than what their individual risk ratings suggest. We also identified those NCFs assessed as being at higher risk that have the potential to affect large numbers of downstream NCFs. These analyses are an important first step for understanding how effects could cascade through the system. However, there are several important limitations to this analysis.

First, the data we incorporated identify only whether a relationship exists between NCFs but do not provide the details of that relationship. For example, some NCFs might be strongly related, such that a disruption of one NCF could lead to the immediate disruption of one or more downstream NCFs. An example of this sort of relationship is *Distribute Electricity* and any NCFs that are dependent on electric power, such as *Provide Medical Care*. In contrast, other NCFs might be more loosely related to one another or might be able to continue to function for a significant time period without inputs from an upstream NCF. For example, *Provide Medical Care* might depend on *Educate and Train* but be able to continue to function reasonably well for years before effects are evident. Nor were we able to consider second- or third-order relationships among interdependent NCFs that could cause effects to ripple or cascade through the system in ways that are not readily apparent. For these reasons, we were not able to incorporate the effects of interdependencies in our numerical risk ratings, but some important distinctions are needed to effectively guide resilience planning.

Second, because NCFs are, by definition, critical to the functioning of society and operate as an interconnected system, to some extent, most NCFs are likely to be dependent on most other NCFs. For example, a complete failure of *Operate Government*, *Provide Housing*, or any number of other NCFs would eventually affect most—or possibly all—other NCFs. This demonstrates that, although identifying the interdependencies without understanding the strength of the relationship or other features of the dependence could provide clues as to how or where effects might cascade, it is not particularly useful for providing actionable analysis. Because each NCF is inherently important and because NCFs are interconnected, considering interdependencies as a part of risk ratings without an analysis of their characteristics and nuances might simply elevate the assessed level of risk to all NCFs.

Third, there are multiple forms of interdependence, many of which could be particularly relevant to climate risk. We outline several forms of interdependence in this chapter but, because of the characteristics of existing identified interdependencies between NCFs, were able to use only a high-level binary (yes/no) definition of *interdependence*. Some forms of interdependence, such as geographic interdependence, might be particularly relevant to climate risk, given the geographic nature of climate events. However, with the tools and information at hand, assessing the effect of multiple forms of interdependence simultaneously across all 55 NCFs systematically is extremely difficult.

Enhancing the Strength of Interdependence

Earlier in this chapter, we described interdependencies between NCFs, but this determined only whether a relationship exists and did not assess the strength of that relationship. According to Klaver, Luiijf, and Nieuwenhuijs (2011, p. 29), when infrastructures are strongly interconnected, "disturbances tend to spread more rapidly." Given the limitations of this analysis and the importance of understanding how interdependencies could increase risk due to climate change, more analysis is needed. Characterizing the strength of the interdependencies between NCFs could lead to a more complete understanding of the ways in which climate effects could propagate through the network of NCFs. This could also illuminate points of intervention for reducing these effects.

According to the literature, five primary indicators could form a basis for evaluating the strength of interdependence: (1) directness, (2) magnitude of the interdependence, (3) time effects, (4) buffers, and (5) adaptive capacity (Der Sarkissian et al., 2022; Klaver, Luiijf, and Nieuwenhuijs, 2011; Rinaldi, Peerenboom, and Kelly, 2001). The concepts of the directness and magnitude of a connection could provide a first-order estimate of the strength of interdependence. The inclusion of time effects is also very relevant for assessing interdependencies. Whereas a disruption to *Distribute Electricity* would cause cascading disruption to downstream NCFs with little or no time delay, a disruption to *Research and Development* might need to be sustained over months, years, or decades to degrade the operations of most downstream NCFs. Additionally, buffers might exist that extend the time to affect or reduce the magnitude of disruption, such as well-planned stockpiles of a resource or generators. Finally, adaptive capacity (including differences in recovery times and processes) might be higher for some NCFs than others (and among assets making up a single NCF).[20]

By assigning weights to these indicators, analysts can begin to quantify differences in the effects of threats and hazards between NCFs due to both direct and indirect effects. For example, by using a standardized assessment, analysts will have a vocabulary to describe differences in the ways in which *Distribute Electricity* is vulnerable to disruptions in inputs from *Transmit Electricity* compared with those from *Manufacture Equipment*. Interruptions to *Distribute Electricity*, for example, might be immediate, direct, and at a high magnitude. In the latter, disruptions might be indirect, experience time lags, and be buffered or adapted to. An NCF with many upstream dependencies on higher-risk NCFs, with a high strength of interdependence, could experience greater effects from climate drivers. This is true even if the NCF itself was assessed as being at low risk from direct climate effects. Understanding interdependencies and how risk could be amplified or muted among interconnected NCFs is necessary to fully understanding climate change risk and how to mitigate it, but achieving that understanding at the scale of all 55 NCFs requires sustained effort to gather the necessary information and improve analysis methods.

---

[20] *Adaptive capacity* in this context refers to an NCF's ability to adjust when confronted with an event that could cause damage or disrupt operations, respond to consequences such that disruptions are limited, or leverage events that present beneficial opportunities. Rinaldi, Peerenboom, and Kelly (2001) identified many factors that contribute to adaptive capacity, such as "the availability and number of substitutes for critical processes or products, workarounds and contingency plans, backup systems, training and educational programs for operational personnel, and even human ingenuity . . . ." They also identified factors that could limit or lower adaptive capacity, such as "restrictive legal and regulatory regimes, health and safety standards, social concerns, organizational policies, fixed network topologies, and the high cost of providing extensive backups and workarounds" (Rinaldi, Peerenboom, and Kelly, 2001, p. 20).

Chapter 5

Conclusion

Summary of Findings

This report presents the findings of a climate risk assessment conducted by HSOAC researchers for the National Risk Management Center to identify NCFs that are at a heightened risk because of climate change. This is the second assessment conducted by HSOAC researchers, and it updates and expands on an assessment conducted one year ago.

The Risk Management Framework

To conduct our assessment, we updated the risk management framework used in our prior effort (Miro et al., 2022) to assess the risk that climate change poses to the NCFs. As in the prior effort, the framework is intended to be compatible with and integrate into existing CISA risk management efforts and, as a result, incorporates the five-point risk rating scale from CISA's operational-level framework. Although we used the framework to support an assessment of climate-related risk to the NCFs, which was the focus of this project, the framework is also intended to provide CISA with an ongoing capability to assess climate-related risk moving forward. The risk management framework employed a four-step process:

1. Identify and characterize climate drivers.
2. Identify impact mechanisms.
3. Assess risk to each NCF.
4. Identify mitigation strategies to reduce risk.

Steps 1 through 3 are described in detail in Chapter 2 of this report. The fourth step—identify mitigation strategies—will be the subject of a forthcoming report.

In the assessment completed in 2021 (Miro et al., 2022), we focused on a subset of 27 NCFs that we identified as having the highest vulnerability. In this report, we assessed risk to all 55 NCFs for a baseline of today and two future time periods under two emissions scenarios. In many cases, we found NCFs that had been screened out of the prior year's assessment to be at risk of a moderate or greater disruption, suggesting that these NCFs should have been included in last year's assessment.

The Risk Rating Scale

The National Risk Management Center's operational-level framework defines *moderate disruption* as an NCF meeting routine operational needs in only most of the country. Using these criteria, the national-level disruption caused by a major historical weather event, such as Hurricane Katrina in

2005, would be characterized as moderate for many of the affected NCFs, such as those related to the provision of power and water, which failed at the local and regional levels as a result of effects from the storm. At a national level, these NCFs continued to function for most of the country. Accordingly, for purposes of our analysis, the risk of moderate disruption to an NCF at the national level should be regarded as highly significant and includes the potential for major disruption or failure of NCFs at a local or regional level and the potential for significant economic loss, health and safety effects, and other consequences. Designations of *major disruption*, indicating that an NCF is expected to not meet routine operational needs in most of the country, and *critical disruption*, indicating that an NCF is meeting none of its routine operational needs in most of the country, are less prevalent among the 55 NCFs, given the nature of the NCFs and the geospatial variability in climate change, although there were some instances of these ratings in our assessments.

Assessment of Climate Change's Effects on National Critical Functions

Using this risk rating scale and projected changes in the eight climate drivers identified by our analysis—drought, extreme cold, extreme heat, flooding, sea-level rise, severe storm systems, tropical cyclones and hurricanes, and wildfire—we assessed the direct risk to NCFs from climate change at present and in two future time periods (by 2050 and by 2100) and in two future emissions scenarios. The emissions scenarios represent two future trajectories of global GHG emissions that could shape the degree of warming and the resulting changes to each climate driver.

Overall, the risk ratings are higher than in the previous year's assessment, including many more NCFs assessed to be at risk of moderate disruption at certain time periods under certain scenarios. This change in ratings reflects additional guidance and the use of small-group sessions to discuss how to apply the risk rating scale at the national scale, the use of new information to guide assessments of NCF exposure and vulnerabilities to climate drivers, and the ability to refine and add depth to the analysis over time. We assessed that 45 of the 55 NCFs are at risk of at least a minimal disruption from climate change by 2050 under the current emissions scenario. This means that 45 NCFs are at risk of being affected by climate change in some way, but many will be able to meet routine operational needs. Nearly half of the NCFs are at risk of a moderate disruption or greater, which means that they might be unable to meet routine operational needs in some parts of the country because of direct effects.

Three NCFs—*Provide Information Technology Products and Services, Maintain Supply Chains*, and *Supply Water*—are at risk of major or critical disruption due to climate change by 2100. These NCFs are at risk of being disrupted to the point that they are unable to meet routine operational needs in most or all of the country at some point. We assessed that ten NCFs—primarily related to financial services and network (i.e., internet-based) operations and services—are not at risk of direct effects from climate change in either of the two time periods.

Climate change is increasing the frequency, magnitude, and duration of many of the climate drivers that pose a risk to NCFs. Although there are important regional distinctions in how and where climate drivers are expected to change, we assessed that, at the national scale, four of the eight climate drivers will have an outsized effect on NCFs.

Tropical cyclones and hurricanes have a legacy of severe and lasting damage to U.S. regions. Four out of every five NCFs are at risk of at least minimal disruption from tropical cyclones by 2050. More than half of the NCFs are at risk of moderate or greater disruption by 2100 under both emissions scenarios. NCFs with assets and service provision along the Gulf and Atlantic Coasts are likely to be significantly affected.

Sea-level rise, which can temporarily or permanently inundate facilities, infrastructure assets, or key components of infrastructure systems, is anticipated to affect the majority of NCFs by 2050 and pose a moderate to major risk of disruption for 25 NCFs by 2100 in the high emissions scenario. These effects are due to the relatively high concentration of major urban areas and infrastructure assets in coastal zones.

Riverine and flash flooding from extreme rainfall events can cause significant infrastructure disruptions as a result of direct infrastructure damage from floodwaters or debris or as a result of loss of access to assets. Three out of four NCFs are at risk of local to regional disruptions by 2050 in the current emissions scenario. Half of the NCFs are at a risk of moderate or greater disruption at the national scale by 2100 in the high emissions scenario. Because flooding is pervasive across much of the United States and because NCF assets are concentrated along waterways or in cities prone to urban flooding, this climate driver is likely to affect all U.S. regions.

Wildfire can cause direct damage to infrastructure facilities, delay or suspend transportation and communication, and result in hazardous air quality. More than half of the NCFs are at risk of minimal or greater disruption due to wildfire by 2050 under the current emissions scenario. In the high emissions scenario, 26 of the 55 NCFs are at risk of significant regional disruptions by 2100.

National Critical Functions at Risk Because of Interdependencies

The NCFs consist of an interconnected system, and every NCF relies on at least one other NCF to operate. In addition to assessing the direct effects as described earlier, we examined the functional relationships, or dependencies (unidirectional relationships),[21] between the 55 NCFs to determine how climate-induced disruption of a given NCF can cascade to another NCF. Our analysis suggests the following conclusions for 2050 under the current emissions scenario:

- NCFs with the same risk rating from our risk assessment (which considered only the direct effects of climate change) could have a greater exposure to climate change through the indirect effects from dependencies. For example, the ten NCFs assessed as being at being at no risk of disruption due to climate change are dependent on at least six other NCFs assessed as at risk of moderate disruption or greater.

---

[21] In this report, an NCF is described as upstream dependent if it provides inputs to other NCFs. An NCF is described as downstream dependent if it relies on outputs of other NCFs. *Interdependence* refers to the bidirectional relationships between two NCFs, in which the operations of one can affect the other. *Interdependencies* refers to the combination of the individual dependencies that flow upstream and downstream from one NCF to another. Functional interdependencies generally are classified into four types: physical, geographic, cyber, or logical. An NCF that is interdependent (e.g., *Distribute Electricity* is dependent on *Supply Water*, but *Supply Water* is dependent on *Distribute Electricity*) with another NCF can experience increased risk if an NCF with which it is interdependent is at risk.

- NCFs with high numbers of upstream dependencies (i.e., that rely on critical inputs from other NCFs) could be at additional risk from climate drivers, although the number of dependencies does not mean that effects will necessarily cascade. This suggests that resilience planning and analysis that incorporate the effects on other NCFs will be essential in lessening the potential for disruption to normal operations for this NCF.
- There are four NCFs at risk of moderate disruption or greater on which all other NCFs depend: *Provide Information Technology Products and Services*, *Maintain Supply Chains*, *Prepare for and Manage Emergencies*, and *Provide Medical Care*. An additional five NCFs also have high numbers of downstream NCFs. Because these NCFs have the potential to affect many other NCFs and face risk of disruption into the future, mitigation measures that reduce risk to these NCFs could be integral to mitigating overall national risk from climate change.

Limitations and Areas for Future Research

Because of the scope of the NCFs, the accelerated time period for analysis, availability of relevant research, and assumptions required to operationalize the analysis across a large and diverse variety of NCFs and their subfunctions, the application of the risk management framework was subject to some limitations. Subsequent analyses and research could address these limitations and provide additional information about climate change's effects on the NCFs.

Expand the Analysis of Cascading Disruptions

We assessed risks to the NCFs from climate change independently. Risk ratings were based on direct effects on an individual NCF without considering its dependencies on other NCFs. In our analysis, we considered NCF interdependencies but did not account for them in the numerical risk ratings. Because of the complex interdependencies among NCFs and their subfunctions, the risk of cascading disruptions among NCFs is not fully articulated in this report. Future analyses could use an iterative approach to identifying the cascading risk that climate change poses for NCFs. Such an analysis could consider more degrees of risk cascade than considered in this report. For example, our analysis used a narrow, binary definition of *dependence*. Future work could characterize NCFs at varying degrees of separation and with varying types of dependencies.

Consider a Broader Variety of Future Scenarios

In our analysis, we accounted for only a portion of the uncertainty about the future that will determine the risk that climate change presents to the NCFs. Ideally, a future analysis could account for additional uncertainty about climate change and incorporate uncertainty about the NCFs themselves, as well as societal characteristics that will influence how climate change affects the NCFs. These other factors might include population dispersal and demographics, economic conditions, and changes in technology.

The first of these factors is that climate change is uncertain, and this uncertainty grows over time. The emissions scenarios we used to capture some of this uncertainty represent only two of many plausible futures. Uncertainties about the magnitude and direction of future climate changes are greater by 2100. Although we considered two future emissions scenarios to account for this uncertainty, best practices in climate science and decisionmaking under uncertainty include assessing a broader variety of future scenarios and updating assessments as new science becomes available. Furthermore, we considered each climate driver in isolation. Developing scenarios that include potential tipping points or compounding effects of climate drivers occurring concurrently or in rapid succession would add depth to the analysis. Employing additional scenarios that include a wider variety of emissions scenarios or the compounding effects of climate drivers can help generate an understanding of how sensitive (or not) risk ratings are to future uncertainty in climate change.

In this analysis, we also assumed that NCFs follow current trends into the future. We did not account for any major population shifts, technological advancements, or significant climate adaptations. We assumed that the NCFs of today are affected by the climate of the future. To generate risk assessments that are comparable across time periods and emissions scenarios, one of the simplifying assumptions we used was that NCFs in future time periods have the same capacities and infrastructure that they have today. The approximately 80-year time period between 2021 and 2100 will see numerous changes that influence NCF operations, including changes in population and demographics, technologies, adaptations within the NCFs, governance structures, and laws and regulations. Moreover, some NCFs are likely to change much more rapidly or drastically than others, and attempts to predict these changes over such a long future time period without careful management of uncertainty are likelier to resemble science fiction than policy analysis.

Our simplifying assumption to exclude anticipated technology changes presents a key limitation of our work. Changes in technologies and the way society uses them will drastically change the future operation of the NCFs and might even change the composition of the list of NCFs. Many of today's NCFs might reasonably be excluded from a future list of NCFs, and other functions that do not yet exist might be considered NCFs in the future. Consider the example of NCFs associated with networked communications that form such a vital part of U.S. infrastructure today (e.g., *Operate Core Network*; *Provide Internet Routing, Access, and Connection Services*; and the various NCFs for access networks). The earliest implementations of these kind of networked communications came into existence only about 40 years ago, and a list of NCFs composed in 1980 would almost certainly not have included anything like them.

An understanding of how the changes within NCFs could influence future climate risk could support the development of risk mitigation strategies and climate-related policies moving forward. A more sophisticated analysis could develop scenarios around technological change, population growth, economic development, climate migration, and other factors to consider uncertainties across multiple dimensions, rather than only the two emissions scenarios we considered. This analysis might resemble the Global Trends reports, published by the National Intelligence Council every four years since 1997. These reports are intended to assess key trends and uncertainties in the subsequent 20 years to provide an analytic basis for policymakers crafting U.S. national security strategy (National Intelligence Council, 2021). Each report takes a unique approach to this task, often examining potential outcomes from interacting trends in technology, demographics, and environments, and attempting to describe a variety of future scenarios that will be useful for understanding risk, uncertainty, and potential policy levers.

Clarify the Roles of Subfunctions

Risk ratings for an NCF reflect individual ratings for the NCF's subfunctions. NCF risk ratings are therefore dependent on the definitions of these subfunctions and how they collectively characterize an NCF. By employing the subfunctions as the basis for the risk assessment, we made risk ratings dependent on how subfunctions are defined and how they collectively represent an NCF. For example, the subfunctions for *Support Community Health* focus on the systems and functions that provide community health (operations, emergency response, and communications) but not on the health of the community itself. As a result, risk to this NCF focuses on climate change's effects on these types of functions. When communicating risk, it should be made clear to stakeholders that the definitions of NCFs have been developed for specific purposes that might not align with the ways in which they are conceptualized by other communities of practice. CISA continues to review and update the subfunctions, and, as a result, our understanding of several NCFs changed during this year's assessment. As subfunctions are refined, future iterations of this risk assessment might result in further adjustments to risk ratings.

Understand Regional Disruptions

We conducted a risk assessment at the national level, which is an important limitation with significant implications for interpreting our findings. National-scale risk ratings do not fully capture regional differences in risk, which can occur because of the geographic orientation of NCFs or the geospatial nature of climate change. Climate change will not affect all areas of the United States equally. The geographic distribution of risk across the United States, which can occur because of the geographic orientation of NCFs or the geospatial nature of climate change, is therefore not fully captured by national-scale risk ratings. Future risk assessments at the regional scale could also rely on regional climate assessments or approaches that summarize and synthesize regional- to local-scale literature to ensure that the most–locally appropriate future climate information is incorporated. These analyses could help highlight which regions and NCFs might need more-regular risk monitoring and additional support for climate risk mitigation.

Identify Climate Drivers' Effects on Other Components of the Development and Maintenance of Infrastructure

The strength of the evidence on climate change's effects on NCFs varies widely across NCFs. For those with a legacy of climate effects, such as *Supply Water* and *Distribute Electricity*, a broad evidence base exists. For others, such as *Educate and Train* and *Provide Public Safety*, fewer analyses have been conducted. Future investments in research for NCFs with weaker evidence bases could support CISA in tracking climate change risk to all NCFs over time. These NCFs include *Produce Chemicals* and *Provide and Maintain Infrastructure*. Overall, there is relatively little research on extreme weather's effects on chemical production. Most of the analysis comes from reporting following extreme weather events (see, e.g., DiChristopher, 2017, on the effects of Hurricane Harvey), reports that are tangentially related to climate change (see, e.g., World Health Organization, 2018), or reports by industry stakeholder groups (see, e.g., American Chemistry Council [ACC], 2021). For *Provide and*

Maintain Infrastructure, there is sufficient literature that examines climate drivers' effect on infrastructure from an adaptation and risk management perspective. However, there is less well-documented literature on how climate drivers affect other components of development and maintenance of infrastructure, such as obtaining finances, regulation compliance, and plan and specification development.

Determine the Multilevel Consequences of National Critical Function Disruption

The consequences of NCF disruption should be factored into future assessments. The consequences of disruption to an NCF as a result of climate change might be severe at the local or regional level. Although an NCF might not be disrupted at the national level, regional disruption of an NCF can create national-level consequences, including significant threats to health and safety, economic loss, and risks to national security. A more complete analysis of the consequences of the level of disruptions projected in this response should be conducted to inform the prioritization of future risk mitigation activities.

Implications of the Findings

Prioritizing Specific National Critical Functions for Further Assessment

Our analysis suggests that CISA should consider prioritizing specific NCFs for further assessment, communication, and risk mitigation. CISA should consider prioritizing those NCFs identified as being at greatest risk for more-granular risk assessments at the local and regional levels, including further assessment of the mechanisms by which climate drivers could disrupt these NCFs. These NCFs should also be prioritized for communication and outreach to stakeholders and the general public to ensure that the risk to these NCFs—and the potential consequences—are well understood. Finally, these NCFs should be prioritized for development of mitigation strategies that address risks posed by the highest-risk drivers identified in this report. NCF owners, which could include different levels of government or private-sector critical infrastructure owners and operators, will likely need to provide resources to support implementation of mitigation strategies.

We estimate that, by 2050, 25 of the 55 NCFs will face a risk of moderate or greater disruption. Although these risks might be outside of typical planning horizons for many entities, this does not suggest that these NCFs will not suffer significant and costly disruptions in the interim. The adoption of successful risk mitigation strategies is often challenging and can take decades, particularly for strategies that involve complex physical infrastructure or individual behavior. The time to begin activities to mitigate risk to these NCFs from climate change has likely already arrived. The four drivers that we identified as posing the greatest risk of disruption to the NCFs—flooding, sea-level rise, tropical cyclones and hurricanes, and wildfire—should be prioritized for further assessment and risk mitigation activities.

Factoring the Consequences of National Critical Function Disruption into Future Assessments

Our analysis suggests that the consequences of NCF disruption should be factored into future assessments. This could be done in a variety of ways, initially drawing on research, technical literature, and event-based analyses of hazard or disaster events as they occur. However, because climate change–driven disruptions of NCFs are likely to occur on more frequently in the future, CISA should also update analysis on a continuing basis as real-world disruptions occur, to better understand the consequences of NCF disruption. Documenting and learning from these events as they occur can serve as a means to both understand the risk landscape more fully and learn from and update analyses and analytical frameworks, as needed.

The consequences of climate change disrupting an NCF could be severe at the local or regional level. Although an NCF might not be disrupted at the national level, regional disruption of an NCF can create national-level consequences, including significant threats to health and safety, economic loss, and risks to national security. A more complete analysis of the consequences of the level of disruption should be conducted to inform the prioritization of future risk mitigation activities. For example, it might be useful to assess—and communicate to stakeholders and the public—how disruptions at the scale projected in our report to *Provide Public Safety* and *Supply Water* would affect those NCFs and NCF stakeholders over time.

Communicating Risk from Climate Change and Climate Drivers

This report also illustrates the difficulty in conveying risk information about the NCFs, which, by nature, are difficult to disrupt given their national scale and inherent redundancies but which, given their criticality, can—and do—create severe life, safety, and economic effects when disrupted at the local or regional level. Furthermore, climate drivers are local or regional by nature—a flood event in Maryland does not necessarily extend to Maine, for example—and can be difficult to evaluate at the national level. Effectively communicating the risk that climate change poses to NCFs is likely to be an ongoing challenge.

Updating Assessments of Climate Change's Risk to National Critical Functions

Finally, our research suggests that CISA should continue to update assessments of the risk that climate change poses to NCFs over time. The science underlying climate projections for 2050 and 2100 will be refined over time as understanding increases of global climate and of the dynamics of how climate change affects built and natural systems. For example, the NCA5, which incorporates more-recent data and scientific information, is anticipated to be released in late 2023. At the same time, the provision of these functions could change with technological innovation, investments in adaptation or resilience measures, demographic trends, or other determinants. Data and methods for assessing complex, interrelated infrastructure systems are developing, enabling more-sophisticated analyses. CISA should also consider systematically gathering evidence of the disruptions that occur from climate drivers to document the types of disruptions, interdependencies, spatial extent, and severity of consequences to the NCFs that could inform future risk analyses. For these reasons, we recommend that these assessments be revisited and additional information be collected to inform them.

Identifying and Characterizing Climate Drivers

This appendix presents the findings of our analysis of changes in climate drivers for the two future time periods and two emissions scenarios considered in this work. As described previously, climate drivers are conceptualized as the climate-related hazards that drive climate change–induced risk to the NCFs. To assess each driver, we relied on those values or descriptions of change for each driver as published in NCA4. In some cases, changes in average values for a specific metric were available,[22] but, in many cases, NCA4 presents its own summary of how a driver is anticipated to change. When possible, we relied on quantitative changes from the underlying downscaled climate model data made available as part of NCA4 resources. These datasets rely on average values for various metrics, and we used those metrics described in NCA4 for each driver. We assigned the direction of change based on the direction of expected change—if a climate driver was projected to increase in frequency, magnitude, or duration, we assigned an increase. If a climate driver was projected to decrease in frequency, magnitude, or duration, we assigned a decrease.

In addition to the description for each climate driver, presented later in this appendix, we summarized these changes in a high-level table of the projected degree and direction of change at the national level to support CISA climate risk assessments and communication of climate changes. Each of the summary descriptions was characterized by degree of change (high, medium, low, or none) and by direction (increase or decrease). These ratings did not distinguish between frequency, magnitude, or duration and were assigned for each driver, future time period, and climate scenario.

We assigned a high degree of change if either of these conditions was true:

- Expected climate changes indicated an increase of at least 20 percent.
- Expected climate changes indicated a decrease of at least 20 percent.

We assigned a medium degree of change if either of these conditions was true:

- Expected climate changes reflected significant increases or decreases over time (at least 20 percent) but with the most-significant increases or decreases occurring in later time periods.
- Quantitative values indicated an increase or decrease of less than 20 percent.

We assigned a low degree of change if expected climate changes did not differ much from recent changes in the historical record, even if some changes were expected. The categorizations were based

---

[22] By *average value*, we mean the average value of the variety of future climate projections that could characterize future change for a given climate driver. For example, the localized constructed analog dataset includes 32 global climate models, each of which contains a projection of future change.

on NCA4 language about the degree of change qualitatively or, for climate drivers with quantitative information, quantitatively.

Table A.1 depicts the qualitative summary of national-scale projected degree of change in each of the eight climate drivers. To maintain consistency with authoritative sources that describe both historical and future changes in these drivers, changes are reported in climate drivers relative to the historical climatology utilized in the *Climate Science Special Report* volume of NCA4 (Wuebbles et al., 2017) for 1976 through 2005. As shown in this table, nearly every climate driver is expected to experience some degree of increase in intensity, magnitude, or duration by 2050. Five of the drivers— drought, extreme heat, flooding, sea-level rise, and wildfire—are projected to experience a high degree of increase by 2100. Only extreme heat is projected to see a high degree of increase by 2050 (under the high emissions scenario). Extreme cold is the only climate driver expected to decrease in the future.

Table A.1. National Projected Changes in Climate Drivers

	CURRENT EMISSIONS		HIGH EMISSIONS	
	2050	2100	2050	2100
Drought	↑↑	↑↑↑	↑↑	↑↑↑
Extreme cold	↓↓	↓↓	↓↓	↓↓↓
Extreme heat	↑↑	↑↑↑	↑↑↑	↑↑↑
Flooding	↑	↑↑	↑↑	↑↑↑
Sea-level rise	↑↑	↑↑↑	↑↑↑	↑↑↑
Severe storm systems (nontropical)	↑	↑	↑	↑
Tropical cyclones and hurricanes	↑	↑↑	↑	↑↑
Wildfire	↑↑	↑↑↑	↑↑	↑↑↑

Change in intensity, magnitude, or duration: -- None | Increase ↑ / Decrease ↓ Low | ↑↑ / ↓↓ Medium | ↑↑↑ / ↓↓↓ High

A broad consensus and scientific basis exist around future climate changes, but climate models on regional to global scales have not fully characterized all future climate changes. For each driver,

Table A.2 provides a low, medium, or high rating for confidence in the projected future trend based on the evidence basis in climate science for anticipated future changes. These definitions are modeled after those used in NCA4, and the ratings shown in Table A.2 are derived from expert assessments of confidence, where available, in NCA4.

Table A.2. Degree of Confidence in Climate Driver Trends

Climate Driver	Confidence in Future Trend
Drought	High
Extreme cold	High
Extreme heat	High
Flooding	High
Sea-level rise	High
Severe storm systems	Low
Tropical cyclones and hurricanes	Medium
Wildfire	High

We assigned a low rating if either of these conditions was true:

- The literature provided a limited understanding of physical processes.
- Inconsistencies existed among models that produced inconclusive evidence.

We assigned a medium rating if either of these conditions was true:

- The literature provided a moderate understanding of physical processes.
- There was published variation in model results.

We assigned a high rating if either of these conditions was true:

- A high degree of understanding of physical processes was exhibited in the literature.
- Consistency was evident in modeling.

As shown in Table A.2, only severe storm systems had a low degree of confidence in the prediction of the future trend.

In the rest of this appendix, future national trends in each of the eight climate drivers are described, noting differences between future time periods, emissions scenarios, and any regional nuances. Additionally, each section contains a description of the confidence in predictions of future trends.

Drought

In the near term, the declines in surface soil moisture that have occurred as average temperatures have warmed are expected to continue. In both the current and high emissions scenarios, this is expected to lead to continued incidence of drought, similar to that experienced in the past five to ten years (Easterling et al., 2017). The western United States has seen large and consistent declines in precipitation indices that track drought, meaning that drier-than-average conditions are increasing over time (U.S. Environmental Protection Agency [EPA], 2022a). In the past 20 years, roughly 20 to 70 percent of U.S. land area has experienced abnormally dry conditions or worse at any given time (EPA, 2022a). The United States lost an approximate $100 billion to $200 billion to drought-related disaster costs between 2012 and 2022 (National Centers for Environmental Information, undated-a).[23]

Changes in hydrological drought, which is governed predominantly by precipitation (including snowpack), will be uneven across the United States, with projections indicating that the western and southwestern portions of the United States are likely to experience increases in hydrological drought in both scenarios through the 21st century. Agricultural drought, which is caused by higher temperature in addition to lower precipitation, is likely to occur across the United States as soil moisture declines, with the incidence of extreme drought increasing as well. Drought incidence is likely to increase across emissions scenarios and from 2050 to 2100, which is similar to projections of hydrological drought (Wuebbles et al., 2017). The Intergovernmental Panel on Climate Change (IPCC) Sixth Assessment Report (AR6) suggests more-intense agricultural drought conditions by 2050 under the high emissions scenario (Masson-Delmotte et al., 2021).

Confidence in the projections of trends in droughts is high because there is a solid understanding of the physical processes underlying droughts and there is consistency in observations and modeling of future drought conditions under climate change. Current climate science has understood the dynamics that produce droughts relatively well, with NCA4 citing a very high confidence in the projections of changes in hydrological drought by the end of the century. For agricultural drought, global climate model representation of land surface processes and the models' ability to represent soil moisture at the root zone of crops varies, but models consistently show drying of soil moisture across the United States (Wuebbles et al., 2017).

NCA4 focuses on the underlying science describing projected changes in drought incidence for those regions that have historically experienced and are increasingly prone to drought—particularly, the western United States. Although droughts have occurred in portions of the eastern United States, these are generally less severe and much shorter in duration than those in the western United States. The future anticipated change in the eastern United States is, therefore, lower than that shown for national-scale changes (Wuebbles et al., 2017).

Extreme Cold

Unlike other climate drivers, the trend in extremely cold temperatures is expected to decline in the next century. In the past 40 years, unusually cold winter temperatures have become less frequent, and the number of record-low days has decreased (EPA, 2022a). The United States lost an approximate $1 billion to $2 billion in extreme cold–related disaster costs between 2012 and 2022 (National

---

[23] This number is derived from the National Oceanic and Atmospheric Administration's (NOAA's) listing of billion-dollar weather and climate disasters (National Centers for Environmental Information, undated-a) and includes only those disasters with estimated total costs over $1 billion.

Centers for Environmental Information, undated-a).[24] In both the current and high emissions scenarios, cold waves are expected to be generally less frequently observed than they have been in recent years (Wuebbles et al., 2017). The frequency and intensity of cold waves and low temperature extremes are expected to decrease by 2050 as temperatures warm across the United States. In the current emissions scenario, cold-wave temperatures are expected to warm by at least 4 to 6 degrees Fahrenheit on average across the United States by 2050 (Wuebbles et al., 2017). In the high emissions scenario, cold-wave temperatures are expected to warm by at least 6 to 8 degrees Fahrenheit on average across the United States by 2100 (Wuebbles et al., 2017). The current emissions scenario projects that cold-wave temperatures will warm by at least 6 to 8 degrees Fahrenheit on average across the United States by 2100 (Wuebbles et al., 2017). In the high emissions scenario, cold-wave temperatures are expected to warm by at least 10 to 14 degrees Fahrenheit by 2100 (Wuebbles et al., 2017).

Confidence in the projections of trends in extremely cold temperatures is high. A large and robust body of work describes and attributes temperature changes to climate change. Numerous datasets exist to examine specific regional to national-scale changes in extreme temperatures. Although more work has been done on high temperature extremes, the same data and models support analyses of warming cold temperatures (Wuebbles et al., 2017). Although recent extreme cold events have shown the potential for significant effects from extreme cold, future climate projections do not indicate a high likelihood of similar events in the future. In general, global climate models do not capture the Arctic polar vortex with consistency and are not in agreement on how it will shape future extreme cold events. Additionally, most experts on extreme cold agree that the general trend is that, "on average, winters are warmer and cold extremes are less likely than they were a century ago" (Lindsey, 2021).

Some regional differences in the trend in extreme cold are expected as well. Across the available metrics in NCA4 to track changes in extreme cold (such as a projected change in the number of days below freezing or the coldest five-day one-in-ten-year event), compared with national average changes in extreme cold events, the Northeast and Great Lakes are expected to see greater declines in the number of extreme cold events, and the southern half of the United States is projected to see smaller declines in the number of extreme cold events (Wuebbles et al., 2017).

Extreme Heat

In the near term, high temperature extremes and heat waves are projected to increase (Easterling et al., 2017). Both unusually hot days and hot nights have become more frequent in the past few decades (EPA, 2022a). Extreme temperatures have increased in degree and in frequency along the Gulf and Atlantic Coasts (EPA, 2022a). Major cities in the United States average six heat waves per year, compared with two per year during the 1960s (EPA, 2022a). Heat waves have also become more intense and occur during a longer portion of the year (EPA, 2022a). Research on the economic effects of extreme heat has shown that these increases have cost some states in the United States, such as Texas and California, have tens of billions of dollars since the 1990s (Callahan and Mankin, 2022).

---

[24] This number is derived from NOAA's listing of billion-dollar weather and climate disasters (National Centers for Environmental Information, undated-a) and includes only those disasters with estimated total costs over $1 billion.

High temperature extremes are expected to increase, by 2050, by about 4 degrees Fahrenheit under the current emissions scenario and by about 6 degrees under the high emissions scenario. The frequency of extreme heat days (above 90 degrees Fahrenheit) is also projected to rise by 2050. In the current emissions scenario, the number of these days is projected to increase by 20 to 40 percent; in the high emissions scenario, the increase is projected to be 40 to 60 percent across the United States (Wuebbles et al., 2017).

Extremely high temperatures and heat waves are expected to become more frequent and intense by 2100. By 2100, high temperatures are expected to increase by 6 degrees Fahrenheit across the United States in the current emissions scenario. The number of extreme heat days are also projected to rise by between 40 to 60 percent under the current emissions scenario. Under the high emissions scenario, high temperatures are expected to increase by 10 degrees or more by 2100, and the number of extreme heat days is projected to increase by about 60 to 80 percent across the country (Wuebbles et al., 2017).

Confidence in this projection of increased trend is high because there is a good understanding of the physical processes and consistency in observations and modeling of extreme heat under climate change. A large and robust body of work describes and attributes changes in temperatures to climate change. Numerous datasets exist to examine specific regional to national-scale changes in extreme temperatures (Wuebbles et al., 2017), which increases confidence in the projections of trends.

Across the available metrics in NCA4 to track changes in extreme heat (such as projected change in the number of days above 90 degrees Fahrenheit or the warmest five-day one-in-ten-year event), compared with national average changes in extreme heat events, the Northeast and Pacific Northwest are expected to warm less and the southern half of the United States is projected to warm more (Wuebbles et al., 2017). At the national scale, urban areas are likelier to experience extreme heat events because of the urban heat-island effect (Masson-Delmotte et al., 2021). Additionally, extreme heat events can be magnified in regions with high humidity, such as the southeastern and midwestern United States. The combination of heat and humidity is known as the *wet-bulb temperature*, and wet-bulb temperatures above 79 degrees Fahrenheit can affect human health (Raymond, Matthews, and Horton, 2020). Finally, heat is regional, and some regions could be more vulnerable to less extreme hot days (e.g., 90 degrees Fahrenheit instead of 105 degrees Fahrenheit) because of the prevalence or absence of air-conditioning or other existing adaptations to extreme heat.

Flooding

Flooding is expected to increase in intensity and frequency in the next century under both the current and high emissions scenarios. In both current and high emissions scenarios, observed changes in the frequency and intensity of extreme precipitation events are projected to continue in most parts of the United States, contributing to the occurrence of severe flood events, similar to what has been experienced in the past five to ten years (Easterling et al., 2017). Extreme precipitation-induced flooding has become more frequent in the past two decades, with the past 25 years seeing nine of the top ten years for extreme precipitation events (EPA, 2022a). Flooding has generally increased in magnitude across the Northeast and Midwest but decreased in the western United States. Large flood events have become more frequent in the Northeast, Pacific Northwest, and northern Great Plains

(EPA, 2022a). The United States lost an approximate $50 billion to $100 billion to flooding-related disaster costs between 2012 and 2022 (National Centers for Environmental Information, undated-a).[25]

Projections of future precipitation under the current emissions scenario show that increases in the frequency of flood-inducing extreme precipitation events (specifically, those events with a return time period of more than five years) could be 50- to 100-percent more than historical averages by the end of the century. This suggests a slightly smaller magnitude in these changes by midcentury. Similarly, under the high emissions scenario, the increase in frequency of extreme precipitation events that cause flooding could be 200 to 300 percent by the end of the century, indicating that flooding by 2050 could also be more frequent (Masson-Delmotte et al., 2021). Model projections under the current emissions scenario indicate an 8- to 10-percent increase in the size of precipitation events with a 20-year return time period, which could translate to a slight but noticeable increase in the incidence of flooding by midcentury (Wuebbles et al., 2017). In the high emissions scenario, events with return time periods of 20 years are projected to increase in frequency by a moderate amount by 2050 (Wuebbles et al., 2017).

By 2100, increases in the frequency of extreme precipitation events that cause flooding could be 50- to 100-percent more than historical averages under current emissions (Masson-Delmotte et al., 2021) and 200- to 300-percent above historical averages under the high emissions scenario. The size of 20-year return time period events is projected to increase by 10 to 13 percent under the current emissions scenario and by 16 to 21 percent under the high emissions scenario by 2100 (Wuebbles et al., 2017).

A significant volume of research of historical and recent trends in extreme precipitation and flooding supports the near-term flooding projections. NCA4 describes and cites clear trends in future precipitation extremes, and its authors attributed these changes to a strong body of evidence that ties increases in extreme precipitation to increased water vapor due to warmer temperatures (Easterling et al., 2017; Wuebbles et al., 2017). Although the evidence on general trends in precipitation is strong, limitations in this body of research do exist. In particular, global climate models' ability to capture the magnitude of future extreme precipitation events is limited, as is the forecasting of future flood occurrence as a result of changing precipitation (in part, because of uncertainties about future land use).

Finally, regional differences are expected in projected changes in flooding. Projected changes in extreme precipitation, combined with studies on changes in historical flood incidence and changing flood risk, suggest that the Northeast could experience greater increases in flooding than national averages in the future (Mallakpour and Villarini, 2015; Slater and Villarini, 2016; Wuebbles et al., 2017).

Sea-Level Rise

Coastal inundation from sea-level rise and tidal events is projected to become more frequent in the next century. U.S. coastal communities are projected to continue to experience coastal flooding from

---

[25] This number is derived from NOAA's listing of billion-dollar weather and climate disasters (National Centers for Environmental Information, undated-a) and includes only those disasters with estimated total costs over $1 billion.

rising sea levels and tidal events similar to observed events in the past five to ten years (Easterling et al., 2017). Sea-level rise along the Atlantic coastline has ranged from approximately 4 to more than 8 inches since 1960 (EPA, 2022a).

However, by 2050, global mean sea levels could rise by 1.7 to 2 feet under the current emissions scenario and by 1.7 to 3 feet under the high emissions scenario (Wuebbles et al., 2017). Under the current and high emissions scenarios, by 2100, global mean sea levels could increase, respectively, by 2.5 to 4.5 feet and 2.5 to 5.8 feet (Wuebbles et al., 2017). During these time periods, sea-level rise will affect U.S. coastlines, resulting in inundation of facilities within affected areas, coastal erosion and permanent inundation, heightened wave impacts, coastal flooding, and saltwater intrusion into coastal freshwater (surface water and groundwater) (Wuebbles et al., 2017).

There is a high degree of confidence in sea-level rise projections. Scientists are more readily able to detect and project sea-level rise than they are other climate drivers. According to NCA4, this is due, in part, to the fact that "the trend signal for sea level change tends to be large relative to natural variability" (Wuebbles et al., 2017, Section 12.2).

Some regional differences in sea-level rise are likely. According to NCA4, although rates of change in the Northeast are projected to be slightly higher than along the southern Atlantic coastline, estimated differences between the two are not likely to be significant. However, projected relative sea-level change is amplified in the Gulf Coast compared with that along the rest of the Atlantic coastline and slightly lower in the Pacific Northwest (Wuebbles et al., 2017).

Severe Storm Systems

Convective storm systems (including thunderstorms and tornadoes), extreme winter storms, and atmospheric rivers are expected to change only slightly from historical climatology in the next century. In the short term, the incidence of severe thunderstorms, tornadoes, hail, strong wind events, and other nontropical extreme weather is similar to observations for the past five to ten years. In the near term, potential increases in the frequency and intensity of nor'easters and other extreme winter storms are similar to those seen between 2008 and 2018 (Easterling et al., 2017). The United States is estimated to have lost more than $200 billion from severe storm–related disaster costs between 2012 and 2022.[26]

The 2050 and 2100 projections are also similar in the current and high emissions scenarios. An increase in the frequency of the environments that produce severe thunderstorms is likely throughout the mid- to late 21st century. Global climate models also suggest the possibility of increases in winter storm frequency and intensity in the same time period. The frequency and severity of atmospheric rivers are likely to increase throughout the 21st century with warming global temperatures (Wuebbles et al., 2017). In the high emissions scenario, some estimates suggest that a large increase in the number of days with atmospheric rivers could occur by the end of the 21st century (Gao et al., 2015; Wuebbles et al., 2017).

---

[26] This number is derived from NOAA's listing of billion-dollar weather and climate disasters (National Centers for Environmental Information, undated-a) and includes only those disasters with estimated total costs over $1 billion.

Additionally, the confidence in projections of the future trend of severe storm systems is the lowest of all the climate drivers. There is a limited understanding of physical processes and significant inconsistencies among models predicting changes in severe storm systems. Limitations in both the consistency and accuracy of historical data and the ability of global and regional climate models to capture the dynamics that produce severe storms contribute to lower confidence in the projections of changes of convective storms. However, a small body of work supports similar findings in future trends. The complexity of the dynamics that lead to winter storm events, combined with a limited understanding of the effects that greater arctic warming has on winter weather across the United States, limits the conclusiveness of future projections for winter storms. Studies on atmospheric rivers, however, are more conclusive (Wuebbles et al., 2017).

Each of the subcategories of storm types within this category—winter storms, convective storms, and atmospheric rivers—has a regional footprint, with atmospheric rivers occurring in the western United States; severe winter storms throughout the Midwest, intermountain west, and Northeast; and convective storms occurring throughout the United States, although less frequently along the continental Pacific coastline. Even with these regional differences, national projections of aggregate changes in severe storm systems are appropriate for the majority of regions. However, projected changes are anticipated to be elevated for convective storms in the midwestern and southern United States by 2100 (Diffenbaugh, Scherer, and Trapp, 2013; Wuebbles et al., 2017).

Tropical Cyclones and Hurricanes

In the near term, no change in hurricane activity is expected to be observed by 2030. These storms will continue to affect the eastern United States, and hurricane seasons are expected to be similar to those of the past five to ten years (Easterling et al., 2017). An average Atlantic hurricane season sees 14 named storms and seven hurricanes, three of which are major hurricanes (category 3, 4, or 5) (NOAA, 2021). The average number of hurricanes has not changed much in the past 100 years, although some evidence shows increases in their intensity (EPA, 2022a). The United States is estimated to have lost over $500 billion in tropical cyclone–related disaster costs between 2012 and 2022.[27]

Tropical cyclone and hurricane activity is projected to increase in the next century. In the medium and long terms, the projected changes in global mean tropical cyclone and hurricane wind speeds and precipitation rates will likely increase in both current and high emissions scenarios. Hurricane frequency will likely remain the same as seen in the historical record in the current emissions scenario. However, hurricane frequency could increase under the high emissions scenario. Increases in storm intensity and frequency are likelier to be seen by 2100, with more-moderate changes expected by 2050 (Stocker et al., 2013; Wuebbles et al., 2017).

However, confidence in projections of this trend is lower than confidence in projections of the changes of most other climate drivers. The literature reveals a moderate understanding of physical processes and variation in model results of future storms. The IPCC, based on available science and

---

[27] This number is derived from NOAA's listing of billion-dollar weather and climate disasters (National Centers for Environmental Information, undated-a) and includes only those disasters with estimated total costs over $1 billion.

examination of AR6 projections (Masson-Delmotte et al., 2021), suggests a medium to high confidence rating for projections of changes in global tropical cyclone rainfall rates and intensities. Other studies, and NCA4, have confirmed a medium confidence in projections of changes in the frequency and intensity of global mean tropical cyclones (Knutson, Sirutis, et al., 2015; Wuebbles et al., 2017). For the Atlantic basin, research shows that uncertainty in these future changes in storm frequency and intensity is higher than at the global scale (Knutson, Camargo, et al., 2020).

The regions that will be most affected by changes in hurricane activity will remain those that have been most affected historically. Specifically, future simulations of Saffir–Simpson category 4 and 5 hurricanes show the majority of hurricane landfalls and hurricane tracks affecting the Atlantic coastline and the Gulf Coast. Remnants of tropical cyclones can make their way farther inland and contribute to significant flooding and storm-related damage (Villarini et al., 2014). However, research on anticipated changes in tropical remnants into the future, as well as their plausible regions of effect, is limited.

Wildfire

In the near term, the conditions likely to trigger wildfire, including low soil moisture, declines in precipitation, and warming temperatures, which have contributed to the increased incidence of very large wildfires since the 1980s, are expected to continue. The extent of burned area from wildfires in the past 40 years has increased, with fires burning more land in the western United States (EPA, 2022a). In 2021, damage from wildfire surpassed $5 billion, which was the seventh consecutive year with national wildfire damage exceeding $2 billion (Masters, 2022). The United States lost an approximate $50 billion to $100 billion in wildfire-related disaster costs between 2012 and 2022.[28]

Significant increases in the potential for large wildfires are likely across the western United States through the end of the 21st century. These increases are anticipated to be greater by 2100 and increase between the current and high emissions scenarios. Projections of future wildfires in the Southeast and Alaska also suggest increases (Stavros, McKenzie, and Larkin, 2014; Wuebbles et al., 2017). AR6 suggests an increase in the incidence of conditions that can lead to wildfire by 2050 under the high emissions scenario (Masson-Delmotte et al., 2021).

Confidence in the projections of an increasing trend in wildfire is high because of sufficient understanding of the physical processes involved in wildfire and consistency in observations and modeling of how wildfire will progress under climate change. The climatic factors that contribute to wildfire risk—drought, temperature, and vapor pressure—are well-understood climatic variables. Their relationship to wildfire risk is also well understood. A substantial body of evidence supports an increased risk of wildfire occurrence in the future, noting variability, however, across local ecosystems and land management practices that are difficult to fully consider in future projections (Wuebbles et al., 2017).

Some regional differences in wildfire risk are expected as climate change progresses through the century. NCA4 focuses on the underlying science describing projected changes in wildfire incidence

---

[28] This number is derived from NOAA's listing of billion-dollar weather and climate disasters (National Centers for Environmental Information, undated-a) and includes only those disasters with estimated total costs over $1 billion.

for those regions that have historically experienced and are prone to wildfire—the western United States and Alaska. Although wildfires have occurred in portions of the eastern United States, these are generally less severe than those in the West. The future anticipated change in the eastern United States is, therefore, lower than that of the western United States (Wuebbles et al., 2017).

Appendix B

National Critical Function Risk Assessments

This appendix includes the full risk assessments that characterize and describe the risk that each climate driver poses to each NCF at the subfunction level. We also describe the strength of evidence that supports each risk assessment, as well as the potential for cascading effects onto each NCF. Note that many of these assessments draw from the language used in the team's previous report (Miro et al., 2022), with some omissions and minor adjustments. The earlier report assessed 27 high-priority NCFs rather than the full set of 55.

Given the number of NCFs, we have provided a list of NCFs, by category, with the page numbers to aid in finding an NCF of interest. The NCFs are grouped into five categories and clustered within those categories by similar NCFs. For example, *Generate Electricity* is included directly before *Transmit Electricity*.

Health, Food, Housing, and Education

Maintain Access to Medical Records

Table B.1. Scorecard for *Maintain Access to Medical Records*

	National Risk Assessment				
	Baseline	Current Emissions		High Emissions	
	Present	2050	2100	2050	2100
Maintain Access to Medical Records	1	1	1	1	1

Description: "Maintain, use, and share actionable data (including personally-identifiable information and personal health information such as care history) effectively, appropriately, bi-directionally and in a timely fashion, for patient care, billing, and operational and clinical research" (CISA, 2020, p. 4).

Highlights

- We do not anticipate direct effects of climate change on access to medical records, particularly given the shift toward electronic medical records.
- This NCF could be subject to cascading risk from other upstream NCFs, such as *Transmit Electricity* or *Provide Internet Routing, Access, and Connection Services.*

Synopsis

We do not foresee significant direct effects of climate change on *Maintain Access to Medical Records*. Although, in past years, climate driver–induced damage to physical infrastructure in hospitals or health care facilities has prevented or delayed access to paper medical records, the shift to digital records protects medical records from some effects of climate change. As more records are maintained digitally, there is greater opportunity to protect patient medical records through redundancy and backups and to avoid the destruction of paper medical records, as occurred in Hurricane Katrina (Gettinger, 2017).

Data from 2019 through 2021 show that 86 percent of nonfederal general acute-care hospitals have adopted 2015 edition certified electronic health records (EHRs) (Office of the National Coordinator for Health Information Technology, 2022),[29] although other types of hospitals, such as rehabilitation and specialty hospitals, had much lower rates of adoption. Because some facilities still have not adopted EHRs, there is still risk to this NCF on a local or regional level, although we assessed there to be no disruption on a national scale in either of the time frames or emissions scenarios.

This assessment is based on what we know about medical records today, but the field is changing as new technology is introduced, so the future could look very different. Access to medical records is increasingly digital, and, in the future, more health information exchange is anticipated. Any or a combination of a variety of factors could affect access to medical records: all records being controlled

---

[29] For more information about this certification, see Office of the National Coordinator for Health Information Technology, 2023.

by a single institution, downtime procedures being eliminated, or records being stored in a new way. For now, our assessment is of a minimal direct effect from the climate drivers under consideration.

Subfunctions

Neither subfunction for this NCF—*Provide Access to Medical Records* (a subfunction of *Provide Medical Care*) or *Application of Protect Sensitive Information Skeleton*—is likely to be at risk from climate change. Although access to electronic records is subject to interruptions in electricity and information exchange, which can lead to delays in providing care (E. Larsen et al., 2019), these are climate change's cascading, rather than direct, effects on this NCF. There might also be increase in demand for access to medical records due to the increase in disease associated with changes in the climate drivers, which could indirectly lead to problems in accessing medical records if a hospital is overwhelmed, but this is not a direct effect.

Table B.2. Subfunction Risk for *Maintain Access to Medical Records*

	Baseline	Current Emissions		High Emissions	
Subfunction	Present	2050	2100	2050	2100
Provide Access to Medical Records (subfunction of *Provide Medical Care*)	1	1	1	1	1
Application of Protect Sensitive Information Skeleton	1	1	1	1	1

NOTE: 1 = no disruption or normal operations. See Chapter 2 for further explanation of ratings.

Climate Drivers

We do not anticipate direct effects to this NCF on a national scale from any of the eight climate drivers.

Table B.3. Risk from Climate Drivers for *Maintain Access to Medical Records*

	Baseline	Current Emissions		High Emissions	
Climate Driver	Present	2050	2100	2050	2100
Drought	1	1	1	1	1
Extreme cold	1	1	1	1	1
Extreme heat	1	1	1	1	1
Flooding	1	1	1	1	1
Sea-level rise	1	1	1	1	1
Severe storm systems	1	1	1	1	1
Tropical cyclones and hurricanes	1	1	1	1	1
Wildfire	1	1	1	1	1

NOTE: 1 = no disruption or normal operations. See Chapter 2 for further explanation of ratings.

Impact Mechanisms

The primary impact mechanism affecting this NCF would be physical damage to medical records or the infrastructure in which the records are maintained. As noted above, the increased use of EHRs reduces the likelihood of impact on a national scale.

Cascading Risk

We assessed the primary risk to this NCF to derive not from direct effects but from cascading risk. In particular, access to electronic medical records could be impeded by effects on *Generate Electricity*, *Transmit Electricity*, and *Distribute Electricity*. Climate drivers, including extreme heat, flooding, and wildfire can affect the electrical grid (see, for example, *Transmit Electricity*). To the extent that these drivers disrupt any of the electricity-related NCFs, EHR access could be affected. However, loss of electricity at one site would likely be a local event, and health care facilities often have mitigation strategies for power shortages using generators, which can mitigate the effect of an outage, although these generators might be used primarily for critical patient care areas, such as intensive-care units and surgery and not for other uses (Federal Emergency Management Agency [FEMA] and Assistant Secretary for Preparedness and Response, 2019).

This NCF could also be affected by cascading risk to such NCFs as *Provide Internet Routing, Access, and Connection Services* and *Provide Wireless Access Network Services*, either of which could affect access to electronic communications. Disruption of internet access would affect electronic communications (e.g., email) and other kinds of data exchange and would affect access to records that are stored remotely. However, hospitals often store copies of their EHR data locally, and interruptions of internet do not greatly affect access to medical records, either at a local level or more broadly.

As data exchange increases over the longer term, there could be an increase in cascading risk due to interruptions to electricity and internet access. This effect is likely worsened in later time periods and in the high emissions scenario.

This NCF also depends on the *Protect Sensitive Information* NCF.

Two NCFs also depend on *Maintain Access to Medical Records*. These are *Provide Medical Care*, because tracking a patient's medical history and treatments given is important for ensuring patient safety, and *Research and Development*, because lack of access to medical records could impede the work of medical, biomedical, or medical technology researchers who are analyzing clinical data.

Strength of Evidence

The strength of evidence on climate change's effect on *Maintain Access to Medical Records* is generally low. Although there is evidence that lack of access to EHRs can be disruptive to medical care (E. Larsen et al., 2019; Rahurkar, Vest, and Menachemi, 2015), research on the effect of climate change (e.g., disruptions due to lack of electricity and limited data exchange) in the context of health records is relatively sparse. However, the shift to electronic medical records suggests that there is less risk of destruction of paper medical records than occurred in Hurricane Katrina (Gettinger, 2017).

Table B.4. Strength of Evidence for *Maintain Access to Medical Records*

Strength of Evidence	
Medium	Flooding; tropical cyclones and hurricanes
Low	Drought; extreme cold; extreme heat; sea-level rise; severe storm systems; wildfire

Risk Rating Interpretations

Table B.5 provides examples of how to interpret the risk ratings for this NCF.

Table B.5. Example Interpretations of Risk Ratings for *Maintain Access to Medical Records*

Risk Rating	NCF Example
1: No disruption or normal operations	Medical record access continues to function as it does today, with increasing frequency of data exchange, but this includes all kinds of data exchange, both push and pull. Complete access is not generally available outside a given health system. During normal operations, there are occasional interruptions to medical record access because of technical and other issues unrelated to climate change, so the baseline is not 100 percent uptime.
2: Minimal disruption	More-frequent electricity outages would lead to somewhat more-frequent downtimes, which would limit access to local data or exchange of records.
3: Moderate disruption	Frequent or longer electricity outages would lead to more regular downtime, which would limit access to records and historical treatment information.
4: Major disruption	Electricity would be lacking in hospitals regularly, very often with no access to records or historical treatment information. Health care facilities would need to rely on paper records as a backup.
5: Critical disruption	Regular lack of electricity in hospitals could lead to permanent use of paper records and inhibited patient data exchange.

Provide Medical Care

Table B.6. Scorecard for *Provide Medical Care*

	National Risk Assessment				
	Baseline	Current Emissions		High Emissions	
	Present	2050	2100	2050	2100
Provide Medical Care	2	3	3	3	3

Description: "Ensure the provision of healthcare services" (CISA, 2020, p. 5).

Highlights

- *Provide Medical Care* is at risk of moderate disruption from climate change by 2050 because of the threat of physical damage that flooding, sea-level rise, severe storm systems, tropical cyclones and hurricanes, and wildfire pose to health care facilities.
- Physical damage compromises the ability to *Provide Operational Facilities*, a key subfunction of this NCF.
- Climate change also increases the demand for medical care. In addition to the drivers above, which cause injuries and disease, drought, extreme cold, and extreme heat can affect human health.
- Although this is not explicitly included in the ratings above, operational facilities are extremely dependent on having water and power. Those risks are covered in other NCFs.

Synopsis

Climate change poses a risk to *Provide Medical Care* in two ways. The first involves the threat of physical damage that tropical cyclones and hurricanes, severe storm systems, and flooding pose to health care facilities (Guenther and Balbus, 2014). Wildfire also poses a threat of physical damage and, in certain parts of the country, is the greatest climate threat to health care facilities (Adelaine et al., 2017). In addition to actual damage, the threat of harm posed by flooding, severe storm systems, tropical cyclones and hurricanes, or wildfire could be enough to warrant evacuation of facilities, thus causing disruption in *Provide Medical Care* (Office of Inspector General, 2014). Extreme weather events can also cut off utilities and prevent people from getting to work (Calma, 2021; Division of Foodborne, Waterborne, and Environmental Diseases, 2019; Montgomery and Romero, 2021).

The second way in which climate change poses a risk to *Provide Medical Care* involves the potential for an increase in injury and disease, which, in turn, causes an increase in the demand for health care. An increase in the number and severity of storms and floods can lead to injury and disease. Extreme cold can also cause injuries and deaths due to traffic accidents and hypothermia. Extreme heat has been shown to result in an increase in hospital visits due to heart, lung, and kidney problems, as well as cause problems with fetal health and increasing premature births (Johnson et al., 2019; Konkel, 2019; L. Turner et al., 2012). Warmer temperatures enable the spread of mosquitoes and ticks into regions that had previously been too cold for them, thereby increasing the incidence of vector-borne illnesses (Thomson and Stanberry, 2022). Warmer temperatures also increase the incidence of water-related illnesses, such as those caused by toxic algal blooms (Trainer et al., 2020). Climate change is also expected to increase ozone levels and airborne allergens (Fann, Nolte, et al., 2021). Wildfire adversely affects air quality through smoke and increases in levels of particulate matter, thereby increasing the incidence of respiratory illness (Liu et al., 2015). Drought increases the frequency and severity of dust storms that can lead to an increase in the occurrence of respiratory diseases (United Nations Environment Programme, World Meteorological Organization, and United Nations Convention to Combat Desertification, 2016). Drought is also expected to affect the availability, safety, and nutrient content of food, and that will have an effect on health (da Silva et al., 2011). Disasters can also take a toll on people's mental health. Thus, climate change is expected to increase demand for medical care, causing a strain on health care providers and institutions (H. Anderson et al., 2017; Carlton et al., 2016; K. Levy et al., 2016; Reidmiller et al., 2018).

Currently, the *Provide Medical Care* function experiences some disruption from climate change. Although this disruption would be considered major in the affected areas, from a national perspective, it would be considered minimal because the function would be generally meeting routine operational needs. By 2050, increases in the frequency, intensity, or duration are expected for all the climate drivers except for extreme cold. This will lead to an increase in the likelihood of physical damage, as well as an increase in demand for this NCF, which could result in a moderate disruption to this NCF and the inability to meet routine operational needs in some parts of the country. Although the frequency, intensity, or duration of climate drivers will increase further in 2100, as well as in the high emissions scenario, we did not assess that the disruption would rise to a severe level such that the NCF would be unable to meet needs in most of the country.

Key Considerations for *Provide Medical Care*

Providing medical care requires having operational facilities, which means not only that a facility be free of severe damage but also that it have functioning utilities, particularly power and water. In addition, providing medical care requires that staff can come to work. This requires not only the physical ability to make it to the facility amid damage to roads or mass transit but also that the staff themselves are physically and mentally healthy and are not occupied with taking care of their own homes and families.

Subfunctions

Among the subfunctions, *Provide Operational Facilities* is the one most at risk of disruption because of the risk of physical damage. This subfunction could face disruption in areas affected by specific climate events, rendering it unable to meet needs and therefore presenting a risk of moderate disruption. If facilities are operating, the next subfunction at risk is *Provide Patient Care*, which will experience an increase in demand due to an increase in the incidence of disease caused by the climate drivers. However, because the increase is expected to be gradual over time, allowing health care institutions the opportunity to increase hiring and capacity, the subfunction should be able to meet operational needs. However, injuries caused by disasters created by climate drivers, as well as illnesses caused by extreme heat, could strain patient care resources on a short-term basis. The remaining subfunctions, which include *Teaching*, *Research and Development*, and various *Business and Regulatory Functions*, should not be affected by climate drivers, assuming that facilities are operational and that staff can get to work. Although mortuary services could face an increase in demand, we did not assess that it would be at a level that would cause disruption.

Table B.7. Subfunction Risk for *Provide Medical Care*

Subfunction	Baseline Present	Current Emissions 2050	Current Emissions 2100	High Emissions 2050	High Emissions 2100
Provide Operational Facilities	2	3	3	3	3
Provide Patient Care	2	3	3	3	3

NOTE: 2 = minimal disruption. 3 = moderate disruption. See Chapter 2 for further explanation of ratings.

Climate Drivers

All climate drivers except for drought and extreme cold are anticipated to affect the ability to *Provide Medical Care*. Chief among the concerns will be drivers with the potential to threaten or cause physical damage to facilities, including flooding, sea-level rise, severe storm systems, tropical cyclones and hurricanes, and wildfire. When such events occur, they can disrupt the NCF in the affected parts of the country. These climate drivers will also increase the demand for medical care, producing a strain on the system.

Table B.8. Risk from Climate Drivers for *Provide Medical Care*

Climate Driver	Baseline Present	Current Emissions 2050	Current Emissions 2100	High Emissions 2050	High Emissions 2100
Drought	1	1	1	1	1
Extreme cold	1	1	1	1	1
Extreme heat	2	3	3	3	3
Flooding	2	3	3	3	3
Sea-level rise	2	3	3	3	3
Severe storm systems	2	3	3	3	3
Tropical cyclones and hurricanes	2	3	3	3	3
Wildfire	2	3	3	3	3

NOTE: 1 = no disruption or normal operations. 2 = minimal disruption. 3 = moderate disruption. See Chapter 2 for further explanation of ratings.

Impact Mechanisms

The impact mechanism of most concern is physical damage to health care facilities because this could render the facilities inoperable, to say nothing of the injuries that could be caused to personnel and patients in the health care facilities. This could be caused by flooding, sea-level rise, severe storm systems, tropical cyclones and hurricanes, or wildfire. Even the threat of physical damage could cause a facility to order a preemptive evacuation and thus render it inoperable.

The other impact mechanism is demand changes. The climate drivers listed above that can cause physical damage to health care facilities can also damage homes, businesses, and other facilities in the

67

community, causing injuries as well as disease. The other drivers—drought, extreme cold, and extreme heat—also cause increases in demand for medical care.

Cascading Risk

Provide Medical Care is dependent on the following NCFs:

- *Distribute Electricity*
- *Maintain Supply Chains*
- *Supply Water*
- *Manage Wastewater*
- *Transport Cargo and Passengers by Road*
- *Transport Passengers by Mass Transit*
- *Maintain Access to Medical Records*
- *Support Community Health.*

Health care facilities require utilities to function. Electricity is needed to operate medical equipment and to *Maintain Access to Medical Records,* as well as to operate the buildings themselves. Although hospitals have backup generators, those generators cannot function indefinitely and require, at minimum, resupply of fuel. Health care facilities also require clean water for sanitation, cooking, and medical procedures; wastewater must be managed as well.

Because much medical care requires an in-person component, in addition to functioning facilities, *Provide Medical Care* requires trained personnel who can get to their jobs. This requires the ability to *Transport Cargo and Passengers by Road.* The transportation function, along with the maintenance of supply chains, is also needed so that health care facilities can be resupplied with medications, medical supplies, personal protective equipment, and other items. Maintaining access to medical records is a dependence because tracking a patient's medical history and treatments given is so important for ensuring patient safety. Hospitals can find short-term work-arounds when their systems are down, and they regularly treat patients without having full medical histories. Finally, *Support Community Health,* as conducted by public health departments, is crucial to *Provide Medical Care,* through disease surveillance, laboratory testing, coordination across health care agencies, and facilitation of medical supply chains.

Strength of Evidence

Experience from recent events has demonstrated health care facilities' vulnerability to extreme weather (Guenther and Balbus, 2014). Furthermore, there is an extensive literature on climate change's effects on people's health (H. Anderson et al., 2017; Reidmiller et al., 2018), which could increase demand for medical care. However, there appear to be limitations in the research on the magnitude of the health effects, the number of people affected, and the size of the burden imposed on the health care system. Although there is a relatively robust literature on climate drivers' effects on the subfunction *Provide Operational Facilities,* there is less evidence on climate change's effects on some of the other subfunctions of medical care, such as *Teaching, Research and Development,* and *Business and Regulatory Functions.*

Table B.9. Strength of Evidence for *Provide Medical Care*

Strength of Evidence	
Medium	Flooding; sea-level rise; severe storm systems; tropical cyclones and hurricanes; wildfire
Low	Drought; extreme cold; extreme heat

Risk Rating Interpretations

Table B.10 provides examples of how to interpret the risk ratings for this NCF.

Table B.10. Example Interpretations of Risk Ratings for *Provide Medical Care*

Risk Rating	NCF Example
1: No disruption or normal operations	Medical care continues to function as it does today, with time periods of delays or diversions due to capacity problems stemming from climate events.
2: Minimal disruption	A few hospitals each year are evacuated or suffer damage, rendering them inoperable for some time, but patients can be cared for in nearby facilities.
3: Moderate disruption	Many hospitals each year are rendered inoperable by damage from climate events, including multiple hospitals within a region causing a shortage of health care in that area.
4: Major disruption	Multiple regions of the country experience a shortage of health care stemming from large numbers of hospitals in each region being rendered inoperable for large amounts of time.
5: Critical disruption	Nationwide, morbidity and mortality increase because health care is unavailable.

Support Community Health

Table B.11. Scorecard for *Support Community Health*

	National Risk Assessment				
	Baseline	**Current Emissions**		**High Emissions**	
	Present	**2050**	**2100**	**2050**	**2100**
Support Community Health	2	3	3	3	3

Description: "Conduct epidemiologic surveillance, environmental health, migrant and shelter operations, food establishment inspections, and other community-based public health activities" (CISA, 2020, p. 6).

Highlights

- *Support Community Health* was assessed to be at risk of moderate disruption from climate change by 2050 because the increase in the occurrence of diseases stemming from all climate drivers will increase the demand for disease surveillance, community outreach, disease control, and other public health efforts.
- This increase in demand will strain the capacity of health departments, many of which are already struggling to meet current needs independent of climate change.

Synopsis

Support Community Health is anticipated to be at risk of moderate disruption from climate change as each of the climate drivers causes or exacerbates disease, affecting the health of the population, and thus increasing demand for this NCF.

Climate change's effect on community health is both significant and varied. Direct threats to human health come from temperature extremes and severe storms, while indirect threats result from an increased range for infectious-disease vectors, degradations in air quality, and persistent stress and distress resulting from emergent events and extreme heat (Balbus et al., 2016). These exposure pathways are likely to affect distinct populations in different ways based on a host of vulnerability factors, including sensitivity and susceptibility to the harm and the capacity for adaptation.

The adverse public health effect of climate drivers is clear. Flooding, severe storm systems, and tropical cyclones and hurricanes can result in injury (H. Anderson et al., 2017). Extreme heat is responsible for both increasing the incidence of heat-related illness and exacerbating existing chronic respiratory, cardiovascular, and renal disease (Konkel, 2019). Elevated temperatures are changing the natural range for mosquitoes and ticks, so vector-borne illnesses could become more common (Thomson and Stanberry, 2022). Increased frequency and extent of wildfires, drought, and changing wind patterns result in poor air quality, including increased ozone, allergens, and fine particulate matter, all of which can both exacerbate existing chronic respiratory conditions and increase incidence of newly diagnosed respiratory disease (Fann et al., 2016). Finally, the mental health effect of persistent stress due to weather events and pollution is well documented (Cuijpers et al., 2023).

By 2050, there will be increases in the frequency, intensity, or duration of events associated with all of the climate drivers except extreme cold. As a result, the *Support Community Health* function will experience increased demand for both chronic and infectious-disease surveillance, primary-care services, pest mitigation, health education, mental health services, and emergency shelter operations, which we assessed will put the function at risk of moderate disruption.

The COVID-19 pandemic revealed myriad weaknesses in community health functions, including insufficient funding and workforce to meet the surveillance and emergency response demands of an ongoing public health crisis in addition to routine operational needs (DeSalvo et al., 2021).

Key Considerations for *Support Community Health*

COVID-19 has shown that the national, state, and local public health agencies that collectively provide the *Support Community Health* NCF have been chronically underfunded and in weak condition. Although climate drivers are not expected to cause increases in disease as large as COVID-19, the fragility of public health infrastructure means that the NCF could struggle to cope with any increase in its workload. Also, like all NCFs, *Support Community Health* requires workers who are able to come to work. This includes not only the ability to physically arrive at workplaces but also that they be physically and mentally healthy and not occupied taking care of their families and their homes amid disasters caused by climate drivers.

Subfunctions

The subfunctions at greatest risk for this NCF are those for which demand would be expected to increase as the result of increased disease caused by climate change:

- Demand for *Track Community Health* is expected to increase in order to promote early detection of climate-related threats and the development of effective countermeasures.
- Demand for *Communicate Health Information* would increase as public health agencies educate people on what actions to take to protect themselves.
- Demand for *Support Emergency Response Capability* would increase as the number of events causing emergencies and disasters increases. These events could include drought, extreme heat, flooding, tropical cyclones and hurricanes, and wildfire.
- Although *Administer Public Health System* is primarily an administrative and regulatory function, some of the sub-subfunctions would experience increases in demand, including the control of communicable disease, provision of immunizations, and review and approval of drugs.
- Climate change could affect *Enable Health Supply Chain* because flooding, severe storm systems, and wildfire have the capacity to disrupt manufacturing facilities, as well as the transportation of goods, through the actual or threatened destruction of infrastructure. However, this subfunction focuses on the more general establishment of the industrial base and relations among different elements of the supply chain. Still, among the sub-subfunctions are the production, storage, and transportation of pharmaceuticals and supplies, items that could be necessary for coping with diseases brought on by climate change.
- Demand for *Promote Secure Operations* would not be expected to increase because it concerns the promotion of physical and cybersecurity, functions that would not be affected by changes in climate.

Table B.12. Subfunction Risk for *Support Community Health*

Subfunction	Baseline Present	Current Emissions 2050	Current Emissions 2100	High Emissions 2050	High Emissions 2100
Track Community Health	2	3	3	3	3
Communicate Health Information	2	3	3	3	3
Support Emergency Response Capability	2	3	3	3	3
Administer Public Health System	2	3	3	3	3
Promote Secure Operations	1	1	1	1	1
Enable Health Supply Chain	2	3	3	3	3

NOTE: 1 = no disruption or normal operations. 2 = minimal disruption. 3 = moderate disruption. See Chapter 2 for further explanation of ratings.

Climate Drivers

Nearly all climate drivers will cause a risk of at least a moderate disruption of the *Support Community Health* function. Climate drivers and have both direct and indirect effects on community health, including increased incidence and prevalence of respiratory and heat-related illness; exacerbations of chronic disease; increases in food-, water-, and vector-related disease; and mental health stressors. The one exception is extreme cold. Extreme cold does create community health problems, particularly in the elderly and in economically vulnerable populations, which would warrant public health action. However, the overall projection is that extreme cold events will not increase over current levels, so the effect on the operating of this NCF is not expected to change.

Table B.13. Risk from Climate Drivers for *Support Community Health*

Climate Driver	Baseline Present	Current Emissions 2050	Current Emissions 2100	High Emissions 2050	High Emissions 2100
Drought	2	3	3	3	3
Extreme cold	1	1	1	1	1
Extreme heat	2	3	3	3	3
Flooding	2	3	3	3	3
Sea-level rise	2	3	3	3	3
Severe storm systems	2	3	3	3	3
Tropical cyclones and hurricanes	2	3	3	3	3
Wildfire	2	3	3	3	3

NOTE: 1 = no disruption or normal operations. 2 = minimal disruption. 3 = moderate disruption. See Chapter 2 for further explanation of ratings.

Impact Mechanisms

The impact mechanism of greatest concern is changes in demand for community health services, including tracking and communicating health information, emergency response, administration, and ensuring effective supply chains. All the climate drivers can cause increases in disease, resulting in an increase in the demand for medical care. As noted above, although extreme cold also causes an increase in the demand for medical care, we assessed that the frequency of cold events will decrease over time.

It is conceivable that an extreme climate event will cause physical damage to the facilities used by health departments. However, we assessed that health department functions are more decentralized and more easily reconstituted than those of, say, hospitals, which are critical to *Provide Medical Care*. Therefore, we have not assessed physical damage to be a major impact mechanism.

Cascading Risk

Support Community Health is dependent on the following NCFs:

- *Distribute Electricity*
- *Transport Cargo and Passengers by Road*
- *Transport Passengers by Mass Transit*
- *Provide Medical Care*
- *Operate Government.*

The key functions of *Support Community Health* include disease surveillance, lab testing, communicating with the public, coordinating among health care entities, and enabling the health supply chain. These functions require operational facilities with working utilities, especially electricity. They also require personnel who can arrive at their jobs and therefore need functioning transportation. Although the functions of public health departments are distinct from those of *Provide Medical Care*, they are nonetheless intertwined, with much of the data required by public health departments, such as incidence of disease and hospital bed capacity, collected by health care providers and institutions. Finally, because public health departments serve a governmental function, an operating government is required to support their work.

Strength of Evidence

Evidence of climate change's effect on human health is well established, and unprecedented changes in the frequency or severity of many natural hazards are expected to increase the need for community outreach efforts, including providing mental health support (H. Anderson et al., 2017; Carlton et al., 2016; Dodgen et al., 2016; K. Levy et al., 2016; Reidmiller et al., 2018). It is logical to draw the conclusion that an increase in negative health outcomes due to climate change will cause a corresponding increase in the need to *Track Community Health*, *Communicate Health Information*, and *Support Emergency Response*. However, evidence of the magnitude of that effect in terms of increased workload of public health departments is not well documented. There is also a lack of research on whether any causal relationship exists between climate change and the other administrative and coordination subfunctions of *Support Community Health*.

Table B.14. Strength of Evidence for *Support Community Health*

	Strength of Evidence
Medium	Drought; extreme cold; extreme heat; flooding; sea-level rise; severe storm systems; tropical cyclones and hurricanes; wildfire

Risk Rating Interpretations

Table B.15 provides examples of how to interpret the risk ratings for this NCF.

Table B.15. Example Interpretations of Risk Ratings for *Support Community Health*

Risk Rating	NCF Example
1: No disruption or normal operations	Health departments function as they do today, with surges in workload due to outbreaks of disease connected to climate drivers.
2: Minimal disruption	Health departments experience a sustained increase in workload due to diseases connected to climate change, necessitating occasional redeployment of staff effort.
3: Moderate disruption	Health departments in some parts of the country are frequently overwhelmed responding to emergency situations and must suspend other services temporarily.
4: Major disruption	Health departments in most of the country are so overwhelmed responding to emergency situations that nearly all other services must be suspended.
5: Critical disruption	Health departments throughout the country are so overwhelmed responding to emergency situations that nearly all other services must be suspended.

Produce and Provide Agricultural Products and Services

Table B.16. Scorecard for *Produce and Provide Agricultural Products and Services*

	National Risk Assessment				
	Baseline	Current Emissions		High Emissions	
Produce and Provide Agricultural Products and Services	**Present**	**2050**	**2100**	**2050**	**2100**
	2	2	3	3	3

Description: "Grow and harvest plant and animal commodities (including crops, livestock, dairy, aquaculture, and timber) and produce inputs required to support agricultural production (such as fertilizers, pesticides, animal food, crop seeds, and veterinary services)" (CISA, 2020, p. 6).

Highlights

- We assessed that, by 2100, *Produce and Provide Agricultural Products and Services* would be at risk of moderate disruption due to the increasing frequency or severity of climate drivers.
- Increased frequency or severity of severe storm systems and tropical cyclones and hurricanes poses risks of minimal to moderate disruption because of expected damage to agricultural infrastructure and disruptions to critical-input supply chains.
- Despite efforts to minimize disruptions to agricultural production and operations from climate drivers, increasing incidence of extreme heat events and other effects from intensifying climate change is expected to reduce future agricultural yields.
- The *Maintain Infrastructure* subfunction will be the most affected across scenarios because physical damage and disruption to infrastructure will affect agriculture production and growth.
- Agricultural operations in rural areas are likelier than those in other areas to be affected, particularly in the mid-Atlantic, Southeast, and Midwest regions of the country.

Synopsis

The *Produce and Provide Agricultural Products and Services* NCF was assessed as being at risk of at least minimal disruption by 2050 and moderate disruption by 2100 because of climate change. Climate change can induce disruptions to weather patterns and temperature and increase the frequency and severity of natural hazards, which make food production and agricultural management more difficult for producers (S. Wang et al., 2018). Severe storm systems and tropical cyclones and hurricanes are expected to cause significant damage and disruption to agricultural infrastructure and operations. Hurricanes and severe storms can damage or ruin crops and timber (American Farm Bureau Federation, 2021; Lonas, 2021), destroy agricultural production and storage facilities ("Hurricane Ida," 2021), and disrupt production of critical inputs, such as fertilizers, pesticides, and equipment (Huffstutter and Weinraub, 2021). Drought, extreme cold, and extreme heat make it difficult for producers to manage crops and animals. A substantial increase in the number of heat-stress days, for example, will negatively affect the health of livestock, decreasing cattle production (S. Miller et al., 2017). Extreme temperatures (both heat and cold) and severe storm systems in particular increase risk to the agricultural workforce (de Lima et al., 2021; Pal, Patel, and Banik, 2021) because they lead to unhealthy and dangerous working conditions.

Producers have worked to minimize disruptions to agricultural production and operations from climate drivers, but we assessed that, as climate change intensifies, there would be a risk of moderate disruption from climate by 2050 in the high emissions scenario and by 2100 in the current and high emissions scenarios to this NCF. Currently, commodity markets have been able to manage disruptions and low crop harvests, buttressed by stockpiles and the relative infrequency with which these climate drivers affect growing seasons. Average yields of many commodity crops (for example, corn, soybean, wheat, rice, sorghum, cotton, oats, and silage) decline once the growing environment exceeds certain maximum temperature thresholds (in conjunction with rising atmospheric carbon dioxide levels), so long-term increases in temperature, which can result in increased incidence of extreme heat, are expected to reduce future yields under both irrigated and dryland production conditions (Marshall et al., 2015). However, even with efforts to reduce GHG emissions and introduce more-sustainable farming practices designed to reduce risk, by 2100, instability in agriculture will increase, as will the inability to reliably produce sufficient yields.

We assessed *Produce and Provide Agricultural Products and Services* to be at risk of minimal disruption from climate change and is expected to stay at that level of risk through 2050 in the current emissions scenario. Although climate change will affect agricultural infrastructure and operations through all climate drivers, the NCF is expected to meet routine operational needs. The NCF is expected to be at risk of moderate disruption from climate by 2050 in the high emissions scenario and by 2100 in the current and high emissions scenarios. These ratings are due chiefly to an expected increase in the frequency or severity of most climate drivers: drought, extreme heat, flooding, tropical cyclones and hurricanes, severe storm systems, and wildfire. We assessed that physical damage to infrastructure, disruption to the production of critical inputs and resources (e.g., fertilizers, pesticides, equipment), and workforce shortages from these climate drivers will cause moderate disruption, resulting in the inability to meet operational needs for producing and providing agricultural products and services in some parts of the country.

Key Considerations for *Produce and Provide Agricultural Products and Services*

Produce and Provide Agricultural Products and Services relies on water resources and management, making it vulnerable to changes in *Supply Water*. Disruption and contamination of water supply and damage to water supply infrastructure due to climate change reduce the amount and quality of the water available for agricultural uses. Drought can deplete groundwater aquifers used for crop irrigation, for example, while floodwaters can pollute water with soil, waste, salt, or other contaminants. Agricultural producers could also pose risks to *Supply Water* through increased demand, particularly during time periods of drought and extreme heat.

Subfunctions

Most of the subfunctions of this NCF—*Purchase Materials, Store Agricultural Materials, Maintain Infrastructure, Manage Animals, Manage Vegetation, Practice Sustainability*, and *Produce Raw Agricultural Commodities*—are expected to be at risk from physical damage and disruption from natural hazards affected by climate change. We have assessed the risk to increase in several scenarios for *Maintain Infrastructure, Manage Animals, Manage Vegetation*, and *Produce Raw Agricultural Commodities*. Infrastructure will be at increasing risk of physical damage (e.g., the destruction of agricultural production and storage facilities ["Hurricane Ida," 2021]), as well as at increasing risk of disruption to the supply of critical inputs (e.g., fertilizers, and pesticides [Huffstutter and Weinraub,

2021]) due to flooding, sea-level rise, severe storm systems, hurricanes, and wildfire (Gowda et al., 2018). Animal management will be at increasing risk from extreme heat, flooding, and severe storm systems (Food and Agriculture Organization of the United Nations, 2016; Rojas-Downing et al., 2017). The variability and unpredictability of climate and temperature fluctuation affect producers' ability to *Manage Vegetation* and *Produce Raw Agricultural Commodities* (Parker et al., 2020).

Table B.17. Subfunction Risk for *Produce and Provide Agricultural Products and Services*

Subfunction	Baseline	Current Emissions		High Emissions	
	Present	2050	2100	2050	2100
Purchase Materials	2	2	2	2	3
Store Agricultural Materials	2	2	2	2	2
Maintain Infrastructure	2	2	3	3	3
Regulate Plant and Animal Management	1	1	1	1	1
Manage Animals	2	2	3	2	3
Manage Vegetation	2	2	3	2	3
Conduct Biotechnological Research	1	1	1	1	1
Practice Sustainability	2	2	2	2	3
Produce Raw Agricultural Commodities	2	2	3	2	3

NOTE: 1 = no disruption or normal operations. 2 = minimal disruption. 3 = moderate disruption. See Chapter 2 for further explanation of ratings.

Climate Drivers

All climate drivers are expected to pose a risk of at least minimal disruption to *Produce and Provide Agricultural Products and Services*. Drought, extreme heat, severe storm systems, tropical cyclones and hurricanes, and wildfire are primary drivers with potential to cause physical damage and injury, as well as disruption to the production of critical inputs to agricultural operations. The mid-Atlantic, Southeast, and Midwest regions will face higher exposure to risks from extreme heat, flooding, and severe storm systems. Extreme cold and sea-level rise are expected to affect the NCF, but we do not anticipate an increase in risk for this NCF from these drivers over time. Although fisheries will experience some disruption from sea-level rise, most agricultural operations are not located in affected areas (Stoltz, Shivlani, and Glazer, 2021). Episodes of extreme cold are not expected to occur more frequently over current levels.

Table B.18. Risk from Climate Drivers for *Produce and Provide Agricultural Products and Services*

Climate Driver	Baseline Present	Current Emissions 2050	Current Emissions 2100	High Emissions 2050	High Emissions 2100
Drought	2	2	3	2	3
Extreme cold	2	2	2	2	2
Extreme heat	2	2	3	2	3
Flooding	2	2	2	2	3
Sea-level rise	2	2	2	2	2
Severe storm systems	2	2	3	3	3
Tropical cyclones and hurricanes	2	2	3	3	3
Wildfire	2	2	3	2	3

NOTE: 2 = minimal disruption. 3 = moderate disruption. See Chapter 2 for further explanation of ratings.

Impact Mechanisms

The impact mechanism of most concern is physical damage or disruption, which can damage or destroy the physical infrastructure crucial to the production and operation of agricultural goods and services. Physical damage can result from any of the climate drivers. Lack of resources is also an impact mechanism of concern because each of the climate drivers has the potential to interrupt the supply chains for agricultural inputs. Extreme temperatures could also cause workforce shortages because workers would be unable to work in extreme heat or cold.

Cascading Risk

Produce and Provide Agricultural Products and Services is dependent on the following NCFs:

- *Supply Water*
- *Manage Wastewater*
- *Distribute Electricity*
- *Transport Cargo and Passengers by Road*
- *Produce and Provide Human and Animal Food Products and Services*
- *Maintain Supply Chains.*

Produce and Provide Agricultural Products and Services relies on water resources and management to sustain crops and animals and protect water quality from contamination. Agricultural producers also require electricity to operate efficiently—in particular, to *Manage Animals, Store Agricultural Materials,* and *Maintain Infrastructure.* Transportation and supply chain infrastructure remain necessary to ensure the movement of food products off farms to processing facilities and markets for the food system to operate efficiently and properly.

Strength of Evidence

There is an extensive literature on climate change effects on agricultural and livestock systems, including assessment reports written by the U.S. Department of Agriculture (USDA) (Office of the Chief Economist, undated; USDA, 2021) and Reidmiller et al. (2018) and decades of peer-reviewed research (Praveen and Sharma, 2019). In addition, climate variability's consequences for different types of agricultural and livestock systems have been well documented (S. Miller et al., 2017; S. Wang et al., 2018), including multiyear drought events (Woloszyn et al., 2021) and extremes of heat (Food and Agriculture Organization of the United Nations, 2016; Parker et al., 2020), flooding (Shirzaei et al., 2021), and hurricanes (Economic Research Service, 2022; Sharma et al., 2021; Wiener, Álvarez-Berríos, and Lindsey, 2020) that destroy crops and kill livestock. However, there appears to be a gap in the literature on the effect that extreme cold temperatures have on agricultural production. Although there is a robust literature on the effects of tropical cyclones and hurricanes, there is less research that engages with the effects of nontropical severe storm systems on agriculture specifically. For wildfire, the academic literature engages with effects on timber (M. Jones et al., 2020); studies on wildfire's effects on other agricultural products are less developed.

Table B.19. Strength of Evidence for *Produce and Provide Agricultural Products and Services*

Strength of Evidence	
High	Drought; extreme heat; flooding; tropical cyclones and hurricanes; sea-level rise
Medium	Wildfire
Low	Extreme cold; severe storm systems

Risk Rating Interpretations

Table B.20 provides examples of how to interpret the risk ratings for this NCF.

Table B.20. Example Interpretations of Risk Ratings for *Produce and Provide Agricultural Products and Services*

Risk Rating	NCF Example
1: No disruption or normal operations	Agricultural production and operations continue to function as they do today, with occasional delays due to weather and operational problems.
2: Minimal disruption	Agricultural production in areas affected by hurricanes, flooding, extreme heat, and drought experience longer time periods of disruption due to the increased frequency and severity of these events, but production and operations in other areas can meet national needs.
3: Moderate disruption	Agricultural operations in areas affected by persistent drought, flooding, or sea-level rise might relocate or close; damage and destruction of crops and livestock and disruption to input supply chains become more frequent because of extreme heat, hurricanes, and severe storm systems.
4: Major disruption	Agricultural production in several regions is no longer viable, which drives up commodity prices and increases food insecurity.
5: Critical disruption	Agricultural production becomes so limited that the country experiences widespread food insecurity.

Produce and Provide Human and Animal Food Products and Services

Table B.21. Scorecard for *Produce and Provide Human and Animal Food Products and Services*

	National Risk Assessment				
	Baseline	Current Emissions		High Emissions	
Produce and Provide Human and Animal Food Products and Services	Present	2050	2100	2050	2100
	2	3	3	3	3

Description: "Produce food products from raw agricultural commodities and provide to final consumers (including processing, packaging and production, product storage as well as retail and food service)" (CISA, 2020, p. 6).

Highlights

- By 2050, *Produce and Provide Human and Animal Food Products and Services* was assessed as being at risk of moderate disruption due to increasing frequency or severity of climate drivers.
- A risk of moderate disruption could result in decreasing production from lack of inputs and frequent damage to food processing, storage, and distribution facilities from flooding, sea-level rise, tropical cyclones and hurricanes, and wildfire.
- The concentration of food processing, storage, and distribution facilities in regions likely to be affected by more-frequent and severer climate drivers increases the exposure of the function.
- The ability to process, store, and *Distribute Food Products* will be affected across scenarios because physical damage to infrastructure will cause disruption to operations.
- Risk will be unevenly distributed, with coastal and larger food processing, storage, and distribution facilities assessed as being at risk of moderate disruption.

Synopsis

Produce and Provide Human and Animal Food Products and Services was assessed as at risk of moderate disruption by 2050 by climate change. Natural hazards affected by climate change, such as flooding, severe storm systems, and tropical cyclones and hurricanes (U.S. Food and Drug Administration, 2022); sea-level rise and wildfire (Chatain et al., 2021); and drought (P. Rao, Sholes, and Cresko, 2019; National Integrated Drought Information System, undated), damage and disrupt food processing, storage, and distribution facilities. The workforce is at risk of the effects of extreme heat as millions of U.S. workers are exposed to heat in their workplaces, including in food production and service settings, such as manufacturing and warehousing facilities and commercial bakeries and kitchens (Occupational Safety and Health Administration, undated; Park, Pankratz, and Behrer, 2021), which can lead to potential labor shortages (Al-Farah et al., 2022; Gowda et al., 2018).

Together, the effects of climate change are expected to damage and disrupt food processing, storage, and distribution infrastructure and operations. Certain events, such as extreme heat and flooding, will make it more difficult to execute the food safety system because of the increased risk of spoilage (Brown et al., 2015; Naumova, 2022) and contamination (U.S. Climate Resilience Toolkit, 2021; EPA, 2017).

Produce and Provide Human and Animal Food Products and Services was assessed as being at risk of minimal disruption from climate change at present because of the effects of the increasing frequency or severity of severe storm systems and tropical cyclones and hurricanes. In June 2022, for example, severe storms and flooding led to a prolonged closure a major baby formula factory, halting production (Beach, 2022; T. Murphy and Ungar, 2022). The NCF was assessed as experiencing increasing risk of moderate disruption from climate change by 2050 in both emissions scenarios. These ratings are due chiefly to the anticipating effects of sea-level rise and an expected increase in the frequency and severity of flooding, severe storm systems, and tropical cyclones and hurricanes. Drought and extreme heat will also pose risks of moderate disruption to the NCF by 2100 in the high emissions scenario. We assessed that physical damage to infrastructure, disruption of supply inputs and resources, and workforce shortages from these climate drivers will cause moderate disruption, resulting in the inability to meet operational needs for *Produce and Provide Human and Animal Food Products and Services* in some parts of the country.

Key Considerations for *Produce and Provide Human and Animal Food Products and Services*

Produce and Provide Human and Animal Food Products and Services relies on plant and animal commodities as inputs, making it vulnerable to changes in *Produce and Provide Agricultural Products and Services*. Disruption and damage to crops, livestock, and fisheries due to climate drivers reduce the amount and quality of the agricultural commodities available for processing and distribution. For example, a substantial increase in the number of extreme heat events can cause heat stress in plants and animals, decreasing crop yields and quality and leading to livestock illness and death.

Subfunctions

Most of the subfunctions of this NCF—*Execute Food Safety System, Distribute Food Products, Maintain Infrastructure, Process Food Products*, and *Store Food*—are expected to be at risk of physical damage and disruption from natural hazards affected by climate change. We have assessed the risk to increase in the two emissions scenarios for *Execute Food Safety System, Distribute Food Products, Maintain Infrastructure, Process Food Products*, and *Store Food. Execute Food Safety System* has been assessed to be at risk due to extreme heat and flooding (Tirado et al., 2010; U.S. Food and Drug Administration, 2022), which can cause spoilage of food products (Brown et al., 2015; Naumova, 2022) and expose food products to heavy metal, chemical, bacterial, and mold contamination from floodwaters (EPA, 2017; U.S. Climate Resilience Toolkit, 2021). Food processing, storage, and distribution facilities and infrastructure will be at increasing risk of physical damage and disruption of inputs due to drought (Woloszyn et al., 2021), extreme heat (Parker et al., 2020), flooding (American Farm Bureau Federation, 2021; Shirzaei et al., 2021), sea-level rise (Mendhall et al., 2020), hurricanes and severe storms (Huffstutter and Weinraub, 2021; Lonas, 2021), and wildfire (M. Jones et al., 2020). Physical damage, particularly damage to facilities that serve large agricultural operations, can interrupt food supply chains (EPA, 2017) when processing and distribution are halted or facilities are closed for repairs.

Table B.22. Subfunction Risk for *Produce and Provide Human and Animal Food Products and Services*

Subfunction	Baseline Present	Current Emissions 2050	Current Emissions 2100	High Emissions 2050	High Emissions 2100
Execute Food Safety System	1	2	2	2	3
Distribute Food Products	2	3	3	3	3
Maintain Infrastructure	2	3	3	3	3
Conduct Biotechnological Research	1	1	1	1	1
Process Food Products	1	2	3	3	3
Store Food	1	2	2	3	3

NOTE: 1 = no disruption or normal operations. 2 = minimal disruption. 3 = moderate disruption. See Chapter 2 for further explanation of ratings.

Climate Drivers

Almost all climate drivers are expected to pose a risk of at least minimal disruption to *Produce and Provide Human and Animal Food Products and Services*. Drought, flooding, sea-level rise, severe storm systems, tropical cyclones and hurricanes, and wildfire are primary drivers with potential to cause physical damage and disruption to food processing, storage, and distribution facilities and infrastructure. Extreme heat can increase potential heat stress and harm workers' health, leading to labor shortages. Although episodes of extreme cold can cause damage to the agricultural inputs to this function, extreme cold is not expected to occur more frequently than it does currently and was not assessed as a key driver of risk to this NCF.

Table B.23. Risk from Climate Drivers for *Produce and Provide Human and Animal Food Products and Services*

Climate Driver	Baseline Present	Current Emissions 2050	Current Emissions 2100	High Emissions 2050	High Emissions 2100
Drought	1	2	2	2	2
Extreme cold	1	1	1	1	1
Extreme heat	1	2	2	2	3
Flooding	1	2	3	3	3
Sea-level rise	1	2	3	3	3
Severe storm systems	2	2	2	2	3
Tropical cyclones and hurricanes	2	3	3	3	3
Wildfire	1	2	2	2	3

NOTE: 1 = no disruption or normal operations. 2 = minimal disruption. 3 = moderate disruption. See Chapter 2 for further explanation of ratings.

Impact Mechanisms

The impact mechanism of most concern is physical damage or disruption, which can damage or destroy the processing, storage, and distribution facilities, as well as other physical infrastructure crucial to the production of human and animal food. Physical damage can result from flooding, sea-level rise, severe storm systems, tropical cyclones and hurricanes, and wildfire. This function could also experience disruption to input supply chains from drought, extreme heat, flooding, sea-level rise, severe storm systems, tropical cyclones and hurricanes, and wildfire. Extreme heat can also cause workforce shortages because workers would be unable to work in such conditions.

Cascading Risk

Produce and Provide Human and Animal Food Products and Services is dependent on the following NCFs:

- *Distribute Electricity*
- *Transport Cargo and Passengers by Road*
- *Produce and Provide Agricultural Products and Services*
- *Maintain Supply Chains*
- *Transport Cargo and Passengers by Rail.*

Without effective *Distribute Electricity*, livestock and plant crops cannot be processed efficiently. Once crops are harvested, each step of getting them to the consumer, including storage, processing, and transportation, requires a functioning electric grid to reduce spoilage. Food processing occurs downstream of food production and thus has less immediate and direct exposure but is still vulnerable to climate change. The increasing frequency and severity of drought (S. Wang et al., 2018), extreme heat (Newburger, 2019), flooding (Gowda et al., 2018), and severe storm systems and tropical cyclones and hurricanes (Polansek, 2020; Wilde, 2021) will damage crops and livestock, leading to a decrease in agricultural inputs for processing and distribution. For the food system to operate efficiently and properly, transportation and supply chain infrastructure remain necessary to ensure the movement of food products from farms to processing facilities and markets. Of those, road and rail networks remain the most crucial for transporting processed food items between regions.

Strength of Evidence

More research has focused on the effects that climate change and extreme weather events have on agricultural production (*Produce and Provide Agricultural Products and Services*) than on their effects on the production of food products and services. Literature focusing on the *Produce and Provide Human and Animal Food Products and Services* portion of the food system includes research by USDA (Canning et al., 2020; Rehkamp, Canning, and Birney, 2021), NCA4 (Gowda et al., 2018), and a large body of peer-reviewed academic literature (Antle and Capalbo, 2010; Niles et al., 2018; Vermeulen, Campbell, and Ingram, 2012). Research on infrastructure and supply chains is often not food system–specific and considers the broader manufacturing and transportation communities, including sector-specific assessments (Chatain et al., 2021).

Table B.24. Strength of Evidence for *Produce and Provide Human and Animal Food Products and Services*

	Strength of Evidence
Medium	Flooding; extreme heat; sea-level rise; severe storm systems; tropical cyclones and hurricanes; wildfire
Low	Drought; extreme cold

Risk Rating Interpretations

Table B.25 provides examples of how to interpret the risk ratings for this NCF.

Table B.25. Example Interpretations of Risk Ratings for *Produce and Provide Human and Animal Food Products and Services*

Risk Rating	NCF Example
1: No disruption or normal operations	Food processing, storage, and distribution facilities and the food safety system continue to function as they do today, with occasional disruptions due to natural hazards affected by climate change and resulting operational problems.
2: Minimal disruption	Food processing facilities in regions affected by increases in severe storm systems experience longer time periods in which they cannot operate safely, but agricultural inputs can be processed in other regional facilities, or temporary delays do not prevent national needs from being met.
3: Moderate disruption	Food production operations in areas affected by sea-level rise, flooding, severe storm systems, or hurricanes relocate or close; more-frequent heat waves reduce worker availability because of poor working conditions.
4: Major disruption	Most food production facilities in several regions are no longer viable, which drives up commodity prices and increases food insecurity.
5: Critical disruption	Food production becomes so limited that the country experiences widespread food insecurity.

85

Provide Housing

Table B.26. Scorecard for *Provide Housing*

	National Risk Assessment				
	Baseline	**Current Emissions**		**High Emissions**	
	Present	**2050**	**2100**	**2050**	**2100**
Provide Housing	2	3	3	3	3

Description: "Construct and/or provide safe and secure permanent or temporary shelter for people (includes physical construction and emergency sheltering)" (CISA, 2020, p. 6).

Highlights

- *Provide Housing* was assessed as being at risk of moderate disruption from flooding, tropical cyclones and hurricanes, and wildfire by 2050 and through 2100 in both the current and high emissions scenarios.
- The subfunction *Provide Permanent Housing* is at risk from an increase in demand that strains material supply chains and construction labor, resulting in higher material costs, labor shortages, and delays to reconstruction efforts. Increased regulations and costs associated with high-risk areas could also affect affordability and delay construction of new permanent housing.
- The subfunctions *Provide Short-Term Shelter or Housing* and *Provide Long-Term Housing* are at risk from a strain on federal and state, local, tribal, and territorial (SLTT) government resources due to increasing demand from severer or more-frequent climate drivers.
- Climate drivers affect *Provide Housing* regionally: flooding in the Northeast and Midwest; tropical cyclones and hurricanes along the Atlantic and Gulf Coasts; and wildfire in the West.

Synopsis

Provide Housing refers to the construction or provision of safe and secure shelter. It was assessed as being at risk of moderate disruption by 2050 and 2100 in both the current and high emissions scenarios. The risk of moderate disruption is driven by an increase in demand for repair and restoration of the housing stock due to more-intense or -frequent flooding, tropical cyclones and hurricanes, and wildfire. The physical damage caused by these events will increase the demand for reconstruction of permanent housing and the provision of emergency shelter. This unprecedented demand could result in changes to housing market dynamics that cause delays or price increases for repair and reconstruction activities. In addition, new regulatory constraints on the construction of permanent housing in high-risk areas could delay construction or decrease housing affordability and availability overall. For example, after Hurricane Michael, a survey of Florida homeowners revealed that repairing homes took an average of 18 months, ranging from just over four months for those needing minimal repairs to close to two years for those with severe damage or complete destruction (Sweeney, Wiley, and Kousky, 2022). The physical damage will also strain federal and SLTT government resources that provide temporary shelter and long-term housing in the aftermath of increasingly frequent or severe climate drivers.

For the baseline scenario, *Provide Housing* was assessed as being at risk of minimal disruption because changes in extreme heat, flooding, sea-level rise, and wildfire and stronger tropical cyclones and hurricanes are already apparent (Reidmiller et al., 2018). Although this NCF is still expected to meet routine operational needs, systemic issues persist. For example, more-vulnerable, low-income populations continue to live in housing in areas at greater risk of physical damage due to climate drivers, exacerbating the potential for federal and SLTT governments to have to increasingly provide resources, such as shelter and temporary housing, at previously unexperienced levels in the wake of natural disasters. According to data from NOAA's National Centers for Environmental Information, in the past five years, there were more natural disasters (including drought, extreme heat, flooding, severe storm systems, tropical cyclones and hurricanes, and wildfire) in the United States that exceeded $1 billion in damage than ever before, with average costs tripling even after adjustments for inflation (Smith and Vila, 2020; National Centers for Environmental Information, undated-a). By the end of the century, in both the current and high emissions scenarios, there is a risk of moderate disruption to *Provide Housing* from more-intense flooding, tropical cyclones and hurricanes, and wildfire.

Subfunctions

The need for housing increases after natural disasters (FEMA, 2020c). The subfunction *Provide Permanent Housing* is expected to be at risk from unprecedented demand increases to repair and restore the physical damage to housing for individuals and families caused by climate drivers. High-risk areas include the wildland–urban interface and flood-prone and coastal regions. The increase in demand for reconstruction and repair resulting from this destruction could change market dynamics and adversely affect public and private initiatives to plan, fund, construct, and maintain permanent housing, including affordable housing. *Provide Housing* would be disrupted by the need to remove debris, increases in construction costs (both labor and materials) driven by the unusually high demand, and permitting restrictions and higher insurance premiums designed to dissuade rebuilding in high-risk areas. In an analysis of disaster-related home improvement spending, researchers found that spending was higher for up to three years after a major disaster and longer for rental property (Kermit Baker and Hermann, 2017). For these reasons, the cost to reconstruct permanent housing will increase and experience construction delays that could be more substantial for rental housing (Coria, 2022; Hermann, 2018; Schuetz, 2022; Spader, 2015; Wilson, Tate, and Emrich, 2021).

The subfunctions *Provide Short-Term Shelter or Housing* and *Provide Long-Term Housing* could also become strained as federal and SLTT governments continue to meet the increasing need for short- and long-term emergency shelter and housing, respectively, as a direct result of the increased severity and frequency of climate drivers. For example, FEMA awarded nearly $6 billion in grants to 2 million natural disaster survivors via its Individuals and Households Program for repairs, losses, rental assistance, and temporary housing from 2016 to 2018 to supplement SLTT assistance. This did not include 1.7 million survivors deemed ineligible because of issues implementing the program (Gómez, 2020). As this number grows because of more-frequent and severer climate drivers as well as policy changes, this could place a substantial burden on federal and SLTT government budgets (Blom, 2021; DeGood, 2020), potentially resulting in tax increases or higher debt burdens to offset spending.

Table B.27. Subfunction Risk for *Provide Housing*

	Baseline	Current Emissions		High Emissions	
Subfunction	Present	2050	2100	2050	2100
Provide Permanent Housing	2	3	3	3	3
Provide Long-Term Housing	2	3	3	3	3
Provide Short-Term Shelter or Housing	2	3	3	3	3

NOTE: 2 = minimal disruption. 3 = moderate disruption. See Chapter 2 for further explanation of ratings.

Climate Drivers

The majority of climate drivers could pose a risk of minimal to moderate disruption to *Provide Housing* through 2050 and 2100 in both the current and high emissions scenarios. Only one climate driver, extreme cold, is not anticipated to increase over current levels nationally. It is therefore assessed as being at no risk of disruption to *Provide Housing*. However, black swan events, such as the deep freeze in Texas in 2021, could occur and cause disruption at the local or regional level.

Drought and extreme heat were assessed as being at a minimal risk of disruption. Either could adversely affect *Provide Housing* but not to the extent of other climate drivers. For example, issues related to construction worker safety due to extreme temperatures, or water constraints, or the inability to construct foundations in dry soil could lead to delays and change orders when constructing permanent housing (Hegeman, 2019). Additionally, extreme heat could require changes to building practices, such as the inclusion of air-conditioning in all housing in certain regions of the United States that could cause temporary disruption (Mann and Schuetz, 2022).

Sea-level rise and severe storm systems were assessed as being at a minimal risk of disruption. In contrast with drought and extreme heat, there is less confidence in changes to the frequency and intensity of storm systems through 2050 and 2100. Moreover, the physical damage from severe thunderstorms, hail, and snow is less acute than from other climate drivers with higher wind speeds, such as tropical cyclones and hurricanes, so the demand for repairs and reconstruction will increase modestly (FEMA, 2020c). Sea-level rise will affect a large proportion of the housing stock, but these effects will be gradual, allowing material suppliers and construction companies to adjust supplies and operations over time. By 2100, sea levels along the contiguous U.S. coast could increase by 5.8 feet on average in the high emissions scenario, affecting nearly 2 percent of all housing nationally (K. Rao, 2017), gradually increasing the demand for new construction away from coastal areas. Some of the states the most at risk are Delaware, Florida, Hawaii, Louisiana, New Jersey, New York, North Carolina, South Carolina, and Texas (First Street Foundation, 2020; K. Rao, 2017).

Flooding was assessed as being at a moderate risk of disruption. In the Northeast and Midwest, increased precipitation from climate change will cause more-significant riverine and rainfall flooding. For example, Pittsburgh, Pennsylvania, has experienced a significant increase in flooding since 2013 (Reidmiller et al., 2018); in West Virginia, nearly a quarter of properties currently face a substantial risk of flooding (First Street Foundation, 2020). By the end of the century, precipitation could increase by 20 percent in these regions in the high emissions scenario, resulting in severe physical damage.

Tropical cyclones and hurricanes were assessed as being at a moderate risk of disruption. Tropical cyclones and hurricanes are forecast to grow in intensity along the Atlantic and Gulf Coasts through 2050 and 2100, damaging homes not only along the coast but even further inland. However, predictions of increases in storm frequency are less conclusive (Colbert, 2022).

Wildfire was assessed as being at a moderate risk of disruption. Since 1990, approximately 60 percent of new homes have been built in the wildland–urban interface, primarily because of urban expansion in Arizona, California, Colorado, Texas, and other western states (Richards, 2019). Approximately one in three houses is now in the wildland–urban interface, which is where wildfire poses the greatest risk, in part because of the difficulties in containing it and the availability of vegetation for fuel (Radeloff, Helmers, et al., 2018). Reconstruction in the wake of recent wildfire has been impeded by labor and material shortages, confusion about debris removal, and new restrictions from home insurers (Popescu, 2018). The growth in housing in the wildland–urban interface is expected to coincide with moderate to large increases in wildfire through 2050 and the end of the century (Reidmiller et al., 2018).

Table B.28. Risk from Climate Drivers for *Provide Housing*

Climate Driver	Baseline Present	Current Emissions 2050	Current Emissions 2100	High Emissions 2050	High Emissions 2100
Drought	1	2	2	2	2
Extreme cold	1	1	1	1	1
Extreme heat	1	2	2	2	2
Flooding	2	3	3	3	3
Sea-level rise	2	2	2	2	2
Severe storm systems	2	2	2	2	2
Tropical cyclones and hurricanes	2	3	3	3	3
Wildfire	2	3	3	3	3

NOTE: 1 = no disruption or normal operations. 2 = minimal disruption. 3 = moderate disruption. See Chapter 2 for further explanation of ratings.

Impact Mechanisms

Tropical cyclones and hurricanes and wildfire will lead to more damage regionally along the Atlantic and Gulf Coasts and in the wildland–urban interface. An increase in demand for permanent housing created by losses in flood zones, along coastal regions, and in the wildland–urban interface will likely occur. This will affect housing market dynamics and potentially impede the construction of permanent housing when demand reaches new heights. Displaced families and individuals will increase demand and further strain federal and SLTT governments that seek to provide temporary shelter and long-term housing in the wake of increasingly frequent and severe natural disasters from climate drivers.

Cascading Risk

Provide Housing could experience cascading upstream effects from disruptions to the following NCFs:

- *Distribute Electricity*
- *Transmit Electricity*
- *Develop and Maintain Public Works and Services*
- *Operate Government*
- *Prepare for and Manage Emergencies*
- *Provide and Maintain Infrastructure*
- *Provide Capital Markets and Investment Activities*
- *Provide Consumer and Commercial Banking Services*
- *Generate Electricity*
- *Supply Water.*

These NCFs directly affect the ability of public and private entities to plan, fund, construct, sustain, maintain, and secure permanent and affordable housing for individuals and families. Additionally, disruptions to these NCFs could limit federal and SLTT governments' ability to provide temporary shelter and short- or long-term housing to individuals and families affected by natural disasters.

Strength of Evidence

The strength of evidence of climate change's effect on *Provide Housing* falls in the medium to high range. Evidence of the effect of individual climate drivers is based largely on historical data collected from FEMA and NCA4. Costly flooding has increased in recent years. With a higher percentage of residential properties at risk, sea-level rise could affect nearly 40 percent of the U.S. population living in coastal areas through 2100, and tropical cyclones and hurricanes have contributed to $1 trillion in economic damage in the past 40 years with the severity of such windstorms projected to increase (FEMA, undated). Many of the costliest wildfires have occurred since 2015; for example, the Marshall Fire in Colorado in 2021 destroyed more than 1,000 homes (FEMA, undated). The historical record suggests that extreme cold and severe storm systems, such as those bringing heavy snow have left housing largely intact, although these climate drivers present a greater threat to electrical and water infrastructure, which pose cascading risks to *Provide Housing* (FEMA, undated). However, the frequency and intensity of extreme cold is forecast to decrease through 2050 and 2100 (Reidmiller et al., 2018). Although drought and extreme heat have been responsible for an increase in fatalities in recent years and could increase demand for residential air-conditioning, the evidence for disrupting *Provide Housing* is less well established (FEMA, undated).

Table B.29. Strength of Evidence for *Provide Housing*

	Strength of Evidence
High	Flooding; sea-level rise; tropical cyclones and hurricanes; wildfire
Medium	Extreme cold; severe storm systems
Low	Drought; extreme heat

Risk Rating Interpretations

Table B.30 provides examples of how to interpret the risk ratings for this NCF.

Table B.30. Example Interpretations of Risk Ratings for *Provide Housing*

Risk Rating	NCF Example
1: No disruption or normal operations	The climate driver does not disrupt *Provide Housing*. The NCF is still expected to meet routine, operational needs, albeit with existing deficiencies. For example, extreme cold is not forecast to rise above existing levels that would result in sustained physical damage, resource shortages, or changes to demand for *Provide Housing*.
2: Minimal disruption	The climate driver disrupts *Provide Housing* in a minimal way; for example, extreme heat impedes and delays the construction of permanent housing and requires additional resources, such as air-conditioning and weatherization in housing in some regions of the United States.
3: Moderate disruption	The climate driver disrupts *Provide Housing* in a moderate way; for example, more-frequent and -intense wildfires destroy housing and lead to shortages and an increase in demand for permanent housing while increasing the need for temporary shelter and long-term housing in the West, straining public and private entities that plan, fund, construct, and maintain the NCF.
4: Major disruption	The climate driver disrupts *Provide Housing* in a major way; for example, flooding increases by more than forecast, damaging more housing than expected along inland U.S. rivers across regions and resulting in an immediate demand for permanent housing, short-term shelter, and long-term housing at a multiregional level.
5: Critical disruption	The climate driver disrupts *Provide Housing* in a critical way; for example, sea-level rise accelerates more quickly and to a greater extent than forecast, damaging more housing than expected along the contiguous U.S. coast and resulting in an immediate demand for permanent housing, short-term shelter, and long-term housing at a national level.

Educate and Train

Table B.31. Scorecard for *Educate and Train*

	National Risk Assessment				
	Baseline	**Current Emissions**		**High Emissions**	
	Present	**2050**	**2100**	**2050**	**2100**
Educate and Train	2	2	3	2	3

Description: "Provide education and workforce training including PreK–12 [prekindergarten through 12th grade], community college, university, and graduate education, technical schools, apprenticeships, non-formal education, and on-the-job training" (CISA, 2020, p. 4).

Highlights

- The physical infrastructure, necessary resources, and workforce capacity of the *Educate and Train* NCF were assessed to be at risk of moderate disruption by extreme heat, sea-level rise, and wildfire by late century, particularly in coastal areas, as well as FEMA regions 4, 6, 9, and 10.[a]
- Acute shocks to the NCF due to the increasing incidence and severity of extreme heat and wildfire put the provision of education at risk across multiple regions by late century.
- The location of education institutions and inadequate cooling and filtration contribute to this vulnerability.
- Analysis indicates a risk of moderate disruption by late century to the *Educate and Train* subfunctions of *Provide Formal Education* and *Provide Workforce Training Skeleton* due to extreme heat, sea-level rise, and wildfire through effects on physical infrastructure, necessary resources, and workforce capacity.
- Other drivers are likely to cause acute, short-term disruption throughout the century but were not assessed to put the NCF at risk of moderate or major disruption.

[a] FEMA region 4 is Alabama, Florida, Georgia, Kentucky, Mississippi, North Carolina, South Carolina, and Tennessee. FEMA region 6 is Arkansas, Louisiana, New Mexico, Oklahoma, and Texas. Region 9 is Arizona, California, Hawaii, Nevada, Guam, American Samoa, Commonwealth of Northern Mariana Islands, Republic of Marshall Islands, and Federated States of Micronesia. Region 10 is Alaska, Idaho, Oregon, and Washington. See FEMA, 2022.

Synopsis

Increasing incidence and severity of weather events associated with several climate drivers pose a threat to the physical infrastructure and staffing of educational institutions and programs. Demand for education and training is not predicted to increase overall because of climate change. However, extreme weather events could displace students, causing an increase in demand for education and training in the areas to which they relocate (and proportional decrease in demand in their home areas, recently affected by acute weather events; see Özek, 2021).

Previous natural disasters and broader climate trends have begun to demonstrate the vulnerabilities of the *Educate and Train* NCF. Extreme weather and natural disasters are the impetus for more than 90 percent of unplanned school closures in the United States (Wong et al., 2014). Acute weather events can damage or destroy an educational institution's physical plant and other key infrastructure, which can prevent school from being held, keep students and staff from attending, and force migration or displacement (e.g., Pfefferbaum et al., 2016; E. Schulze et al., 2020). Extreme heat and wildfire have already contributed to increasing incidence of short-term K–12 school closures across the country due to poor cooling and ventilation in school buildings and transportation (e.g., Melillo, 2022; Rebecca Miller and Hui, 2022).

Disruption to *Educate and Train* due to climate change is minimal at present. We assessed that, by late century, there would be a risk of moderate disruption to the *Educate and Train* subfunctions of *Provide Formal Education* and *Provide Workforce Training Skeleton* due to extreme heat, sea-level rise, and wildfire as drivers affecting physical infrastructure, necessary resources, and workforce capacity.

Key Considerations for *Educate and Train*

The *Educate and Train* NCF is plagued by disproportionate distribution of resources because K–12 public schools are largely locally funded and administered. Communities with fewer resources are more vulnerable not only to the broad effects of extreme weather events but also to these same events posing a higher-than-average risk of disrupting education and training for community members. People with disabilities constitute another particularly vulnerable group being served by the operation of this NCF.

Subfunctions

Both of the subfunctions for this NCF—*Provide Formal Education* and *Provide Workforce Training Skeleton*—are expected to be at risk due to the likelihood of physical damage or disruption, lack of resources, and workforce shortages associated with changes in the climate drivers. Especially of concern are the effects of extreme heat, sea-level rise, and wildfire.

Subfunction risks were determined to be different in the current time period because of differences in the governance of educational institutions by subfunction. K–12 public schools make up the bulk of educational institutions whose operation is captured by the *Provide Formal Education* subfunction, and these institutions, as providers of compulsory elementary and secondary education, are more closely overseen by local and state government authorities than the majority of workforce training opportunities are (captured under *Provide Workforce Training Skeleton*). Consequently, when extreme weather events or other emergencies occur, there is often clear shutdown guidance for formal educational institutions and less-clear guidance for workforce training programs.

Table B.32. Subfunction Risk for *Educate and Train*

Subfunction	Baseline Present	Current Emissions 2050	Current Emissions 2100	High Emissions 2050	High Emissions 2100
Provide Formal Education	2	2	3	2	3
Provide Workforce Training Skeleton	1	2	3	2	3

NOTE: 1 = no disruption or normal operations. 2 = minimal disruption. 3 = moderate disruption. See Chapter 2 for further explanation of ratings.

Climate Drivers

By late century, extreme heat, sea-level rise, and wildfire will put the *Provide Formal Education* and *Provide Workforce Training Skeleton* subfunctions at risk of moderate disruption. Sea-level rise, which is projected to increase in intensity and magnitude from present to 2050 and again to 2100 under both emissions scenarios, is likely to cause increasingly disruptive physical damage to the physical plant of education and training facilities and other key resources. Increased sea-level rise could also contribute to workforce shortages in coastal areas, both during acute events and because of long-term changes in residential patterns, particularly as sea-level rise increases in intensity and flooding of coastal areas becomes more widespread.

Extreme heat has already begun to disrupt this NCF for short time periods in multiple regions because of lack of resources: School and training cancellation due to extreme heat is increasing in frequency because many schools and school transportation vehicles (i.e., school buses) lack adequate heating, ventilation, and air-conditioning capacity to keep students and staff safe. As with sea-level rise, the intensity and duration of extreme heat are projected to rise from present to 2050 and again to 2100 under both emissions scenarios. Longer time periods of extreme heat will lead to greater disruption of formal education and workforce training due to lack of resources.

Similarly, wildfire disruptions are increasingly common in FEMA regions 9 and 10 because of a lack of resources (i.e., adequate air filtration) and the risk of physical damage and disruption. As wildfire increases in frequency and severity, as is projected from present through 2100 under both emissions scenarios, there is likely to be increasing incidence of localized or regional workforce shortages due to the need for staff to relocate or meet their family needs during acute events.

Table B.33. Risk from Climate Drivers for *Educate and Train*

Climate Driver	Baseline Present	Current Emissions 2050	Current Emissions 2100	High Emissions 2050	High Emissions 2100
Drought	1	1	1	1	1
Extreme cold	1	1	1	1	1
Extreme heat	2	2	3	2	3
Flooding	2	2	2	2	2
Sea-level rise	1	2	3	2	3
Severe storm systems	1	1	2	1	2
Tropical cyclones and hurricanes	1	2	2	2	2
Wildfire	2	2	3	2	3

NOTE: 1 = no disruption or normal operations. 2 = minimal disruption. 3 = moderate disruption. See Chapter 2 for further explanation of ratings.

Impact Mechanisms

The increase in frequency and severity of extreme weather events and coastal flooding due to sea-level rise have the potential to cause damage to the physical plant necessary for the provision of the NCF, as well as access to the personnel and other resources needed for NCF operation, which, in turn, increases the risk of school cancellations and degradation of education and training quality.

Without upgrades to existing heating, ventilation, and air-conditioning systems, **extreme heat** and **wildfire** will also cause more short-term school closures, interrupting educational processes.

There has been some evidence that extreme weather events can shift demand for public education from affected communities to less affected communities within an evacuation radius (e.g., Özek, 2021) as families resettle for short- to longer-term time periods and seek to reenroll students in school as quickly as possible.

Research documenting the effects that extreme weather events have on students (referred to later as *secondary effects on students*) are illustrative of other impact mechanisms and the overall vulnerability of the *Educate and Train* NCF to extreme weather events. For instance, there is growing evidence that extreme heat and the stress of natural disasters could both impede student learning and negatively affect academic performance (Laurito, 2022; Park, Goodman, et al., 2020). These effects could be particularly pronounced for people with disabilities (Frederick et al., 2018).

Cascading Risk

Educate and Train is dependent on the following NCFs:

- *Generate Electricity*
- *Distribute Electricity*
- *Transmit Electricity*
- *Develop and Maintain Public Works and Services*
- *Manage Wastewater*
- *Supply Water.*

For this NCF to operate effectively (as presently conceived), it must be supported by public utilities and infrastructure. Major or critical disruption to the above NCFs would lead to disruption of *Educate and Train*.

Strength of Evidence

There is mounting evidence of climate change's effects on the *Educate and Train* NCF and the secondary effects on students themselves. The effect of acute events, such as flooding, severe storm systems, tropical cyclones and hurricanes, and wildfire, on schools and students is well established. In addition, there is strong evidence of the vulnerability of the physical plant of coastal educational institutions to extreme weather events (Pfefferbaum et al., 2016; E. Schulze et al., 2020), which are expected to increase in frequency or severity because of climate change. However, evidence on the effects of drought and extreme cold is less well established.

Table B.34. Strength of Evidence for *Educate and Train*

	Strength of Evidence
High	Extreme heat; severe storm systems; tropical cyclones and hurricanes; wildfire
Medium	Flooding; sea-level rise
Low	Drought; extreme cold

Risk Rating Interpretations

Table B.35 provides examples of how to interpret the risk ratings for this NCF.

Table B.35. Example Interpretations of Risk Ratings for *Educate and Train*

Risk Rating	NCF Example
1: No disruption or normal operations	*Educate and Train* continues to function as it does today, with occasional school or training closures due to weather events and related physical plant deficiencies.
2: Minimal disruption	Schools in areas affected by extreme heat, sea-level rise, and wildfire experience longer time periods in which they cannot operate safely, but students can pursue education remotely or enroll in alternative educational programs and institutions.
3: Moderate disruption	Some educational programs and institutions close temporarily or permanently because of extreme heat, sea-level rise, or wildfire; cancellations due to extreme heat and wildfire (and the damage they cause to educational infrastructure) become far more frequent.
4: Major disruption	Most education and training provision is not viable because of extreme acute and long-term conditions; a significant number of educational institutions and training programs close permanently.
5: Critical disruption	Public education essentially ceases; alternatives might be available through private providers.

Research and Development

Table B.36. Scorecard for *Research and Development*

	National Risk Assessment				
	Baseline	Current Emissions		High Emissions	
	Present	2050	2100	2050	2100
Research and Development	1	1	2	1	2

Description: "Conduct basic research, innovate, test, and introduce new products and services or improve existing products and services" (CISA, 2020, p. 7).

Highlights

- The *Research and Development* NCF is likely to be at risk of minimal disruption by late century because of effects on physical infrastructure, necessary resources, and workforce capacity.
- Analysis indicates that, by late century, extreme heat, flooding, sea-level rise, severe storm systems, tropical cyclones and hurricanes, and wildfire will lead to a risk of localized disruption to the subfunctions of *Standalone Function* and *Supporting Skeleton* due to effects on physical infrastructure, necessary resources, and workforce capacity.
- *Research and Development* capacity has traditionally been concentrated in the United States, but some research areas in other countries—particularly China—have been growing at a rapid rate. Continued U.S. competitiveness across fields and industries is dependent on building *Research and Development* resilience to the climate drivers.

Synopsis

Anticipated increases in the frequency and severity of extreme weather events, including extreme heat, flooding, sea-level rise, severe storm systems, tropical cyclones and hurricanes, and wildfire, pose a minimal but rising risk to *Research and Development* because of the potential for damage to physical infrastructure, disruption to or inability of necessary resources, and reduced workforce capacity for basic research, applied research, and development across all *Research and Development* performance sectors.[30] This NCF consists of stand-alone *Research and Development* functions and supporting functions,[31] each of which is at minimal but rising risk of disruption to normal operation by late century.

---

[30] Basic research is experimental or theoretical work undertaken to acquire new knowledge without any particular application or use. Applied research is investigation undertaken to acquire new knowledge that has a practical aim or objective. Development is systematic work that draws on research to produce additional knowledge or new products or processes or to improve existing products or processes. See Moris (2018) for additional information.

Three primary sectors undertake U.S. research and development: private enterprise, higher education institutions, and the federal government. The business sector's share of the *Research and Development* enterprise has grown since 2010, but the federal government remains an important source of support for all *Research and Development*–performing sectors and is the largest funder of basic research (Sargent et al., 2022).

[31] The stand-alone functions relate to the management of research; the maintenance of the intellectual property environment; the management of information-sharing; the maintenance of research facilities; and the maintenance of a supply of researchers. The supporting *Research and Development* skeleton consists of database maintenance; monitoring of advances; management of resources, including personnel, needed materials, and manufacturing capacity; and the management of *Research and Development* prioritization. See "Subfunctions" for additional discussion.

All types of *Research and Development*, whether basic research, applied research, or development (Sargent et al., 2022), have some vulnerability to climate change, but different types of research are likely to have different exposure to climate drivers and, consequently, different mechanisms of impact. For instance, whereas basic research, conducted largely in labs and building settings, is likeliest to be affected by physical damage to infrastructure, research done in applied field settings can be affected by the difficulty of operating and maintaining necessary equipment in the field during and after extreme weather events. For example, field experiments could be disrupted if technical equipment (such as sensors) or environmental elements (such as tree seedlings in forest research) are damaged by extreme heat, drought, or severe storm systems to which they are respectively sensitive (e.g., Blazier, 2021; Zong et al., 2021). In the summer of 2022, record-breaking heat disrupted ecological and biological research being conducted by a scholar at the University of California, Los Angeles, modifying the scope, design, and costs of planned experiments related to the feeding and mating behaviors of native and invasive species (Robitzski, 2022). Research done in the field can also be particularly vulnerable to extreme weather conditions in terms of the potential reduction of worker productivity and increases in absenteeism (e.g., Somanathan et al., 2021) and increasing threats to worker safety, affecting workforce supply and performance. A plant ecologist at the University of California, Los Angeles, postponed and redesigned planned field research because of risks to human field safety as a result of wildfire risk (Robitzski, 2022).

By late century in both the current and high emissions scenarios, research activities in areas affected by extreme weather events are likely to experience greater disruption to resources and research activities and longer time periods in which they cannot operate safely or effectively. Nonetheless, *Research and Development* is likely to be at risk of minimal disruption because personnel are likely to be able to restart research efforts, continue their efforts remotely, or move their research apparatus to other institutions. By contrast, by 2050, there is unlikely to be additional disruption to normal operations: *Research and Development* is likely to continue to function as it does today, with occasional facility closures, experiment failures, and data collection delays due to extreme weather events and related physical plant deficiencies.

Key Considerations for *Research and Development*

Research and Development is an over $700 billion enterprise in the United States, undertaken by an interlocking fabric of private enterprise, higher education, and the federal government. The United States has been losing market share of global research and development in recent decades, largely to China, whose research and development annual rate of growth is more than double that of the United States (Boroush and Guci, 2022). United States' *Research and Development* productivity has been a key driver in the country's global competitiveness through the centuries. Although there appears to be minimal risk of disruption to the overall *Research and Development* enterprise due to extreme weather, *Research and Development* is critically dependent on other key NCFs (discussed in "Cascading Risk") and so could be at increased risk itself should those NCFs be disrupted.

Subfunctions

Both subfunctions of this NCF are likely to suffer minimal disruption due to effects on physical infrastructure, necessary resources, and workforce capacity. For example, planning and facilitation of

Research and Development efforts could be delayed or made impossible by extreme weather events either directly by disrupting research efforts themselves (e.g., disruption of maintenance of research facilities) or indirectly by causing disruption to the administrative coordination of *Research and Development* activities (e.g., contract review and execution, resource allocation). Extreme weather events can also pose a risk to necessary intellectual property protection (i.e., disrupt the management of the patent office) or disrupt access to key resources and supplies (i.e., management portfolio of *Research and Development* activities).

Table B.37. Subfunction Risk for *Research and Development*

	Baseline	Current Emissions		High Emissions	
Subfunction	Present	2050	2100	2050	2100
Standalone Function	1	1	2	1	2
Skeleton Function	1	1	2	1	2

NOTE: 1 = no disruption or normal operations. 2 = minimal disruption. See Chapter 2 for further explanation of ratings.

Climate Drivers

Nearly all climate drivers will create risk for *Research and Development* by late century. Extreme heat, flooding, sea-level rise, severe storm systems, tropical cyclones and hurricanes, and wildfire have all shown their potential to disrupt the NCF. For example, in 2001, Tropical Storm Allison caused catastrophic flooding at the medical school at the University of Texas Health Science Center, with total losses estimated at more than $205 million (Committee on Strengthening the Disaster Resilience of Academic Research Communities, 2017; B. Goodwin and Donaho, 2010). More than 3,200 faculty, staff, and students were displaced for a month, and more than 1 million square feet of space for teaching, research, support, and animal care were out of service for longer. Flooding destroyed years of stored research data and drowned thousands of animals within the Texas Medical Center complex.

Extreme weather events present both opportunities and risks for field research (Cornwall, 2022). If necessary equipment is not damaged by the weather event, an event can yield rich data to inform scientific understanding of the environmental or agricultural effects of acute weather phenomena (e.g., Rey, Corbett, and Mulligan, 2020, examined the effects of hurricanes in a back-barrier estuary). However, if key equipment or facilities are destroyed, extreme weather events disrupt or destroy carefully planned research projects that are often years in the making. For example, the University of Kentucky Research and Education Center suffered considerable damage from the December 2021 cluster tornado event in the mid-South (McClellan, 2021).

There is little evidence that drought and extreme cold will be disruptive to *Research and Development*. The overall projection is that extreme cold events are not expected to increase over current levels, and drought will likely affect the NCF only in very extreme circumstances in which the water supply experiences major or critical disruptions. For instance, in a hypothetical extreme circumstance in which drought results in a major or critical disruption to the provision of water, zoological research that requires water for the preservation of endangered species would be affected.

Table B.38. Risk from Climate Drivers for *Research and Development*

Climate Driver	Baseline	Current Emissions		High Emissions	
	Present	2050	2100	2050	2100
Drought	1	1	1	1	1
Extreme cold	1	1	1	1	1
Extreme heat	1	1	2	1	2
Flooding	1	1	2	1	2
Sea-level rise	1	1	2	1	2
Severe storm systems	1	1	2	1	2
Tropical cyclones and hurricanes	1	1	2	1	2
Wildfire	1	1	2	1	2

NOTE: 1 = no disruption or normal operations. 2 = minimal disruption. See Chapter 2 for further explanation of ratings.

Impact Mechanisms

Research and Development is at minimal risk of disruption from climate change from physical damage, lack of resources, and workforce shortages. Facilities and key equipment necessary for research could be damaged by acute events. It might become difficult to procure necessary supplies to conduct research. For instance, flooding could disrupt the shipping and receipt of crucial resources needed for the ongoing operation of laboratory research or might destroy critical instruments, such as water sensors, that have been placed in the field for environmental or agricultural research. After extreme weather events, research personnel might be displaced and unable to carry out critical, time-sensitive activities for their research projects, resulting in workforce shortages that disrupt *Research and Development* efforts.

Climate events and conditions could also make it difficult for research personnel to commit to the continuity necessary to successfully conduct their research efforts. For instance, researchers might not be able to physically reach their laboratory or field settings to complete key tasks and measurements because of evacuations and road closures brought on by flooding, tropical cyclones and hurricanes, and wildfire.

Cascading Risk

Research and Development is dependent on the following NCFs:

- *Generate Electricity*
- *Distribute Electricity*
- *Transmit Electricity*
- *Develop and Maintain Public Works and Services*
- *Supply Water*
- *Provide Wireless Access Network Services*
- *Provide Wireline Access Network Services*

100

- *Operate Government.*

For this NCF to operate effectively (as presently conceived), it must be supported by public utilities and infrastructure, including electricity, water, and internet access, in order to support the facilitation of research, as well as management and coordination of *Research and Development*. Research undertaken in laboratory, office, clinical, and field settings is often critically dependent on reliable power, access to water, and wireless access to the internet for the operation of key equipment and the transmission and storage of data. Major or critical disruption to the above NCFs would lead to disruption of *Research and Development*. Furthermore, given the role of the federal government in funding research efforts and the training of research personnel across many sectors and fields, *Operate Government* is a key NCF to the management and coordination of *Research and Development*, the maintenance of the intellectual property environment, the funding and maintenance of research universities and federally funded research and development centers, and the maintenance of the supply of researchers.

Notably, research in some fields is also dependent on additional NCFs, such as *Maintain Access to Medical Records* and *Protect Sensitive Information*. For instance, a disruption to the *Maintain Access to Medical Records* NCF could undermine the ability of medical, biomedical, or medical technology researchers to obtain and analyze clinical data needed for their research projects.

Strength of Evidence

Although there is abundant evidence of climate change's effect on natural hazards, there is limited evidence on natural disasters' effect on *Research and Development*. There is growing evidence of the effect that acute and prolonged emergencies have on *Research and Development*; the existing literature base consists largely of discipline- and event-specific research that does not comprehensively consider the risk to the *Research and Development* enterprise due to extreme weather events. News reporting is complemented by a growing research literature in examining the implications of extreme weather events for the destruction of critical research infrastructure, damage to equipment and resources, and overall disruption of ongoing projects across settings and disciplines. When taken together, these sources provide a window into the risk to *Research and Development* of extreme weather events.

One exception to this trend is a report published by the National Academies of Sciences, Engineering, and Medicine that profiled the ways in which a series of natural disasters affected the biomedical research enterprise, sponsors, institutions, and research personnel and how the development of knowledge affected biomedical research (Committee on Strengthening the Disaster Resilience of Academic Research Communities, 2017). The report provides a comprehensive, multifaceted analysis of a diverse array of examples of *Research and Development* disruption due to natural disasters, albeit in only one discipline. The report concludes that biomedical *Research and Development* is at risk of disruption and loss due to natural disasters and that there is a need to strengthen the disaster resilience of biomedical *Research and Development*. Although the report did not focus solely on hurricanes, wildfire, severe storm systems, extreme cold, extreme heat, drought, flooding, or sea-level rise, its insights about the potential for damage to research facilities and research progress are well documented and could apply to other domains in the *Research and Development* enterprise.

These sources offer important perspective on the potential effect that the increasing severity and frequency of acute weather events and related sustained effects to key resources could have on *Research and Development*: Lack of access to facilities and resources disrupts research and diverts attention to other professional needs, such as administrative management, response to crisis, or increased family responsibilities and, consequently, has the potential to affect *Research and Development* workforce capacity. The existing evidence does not suggest that disruptions to research will appreciably affect the researcher training pipeline (L. Johnston, Wilson, and MacKenzie, 2016) to doctoral completion but that they could affect career progress and trajectories (Committee on Strengthening the Disaster Resilience of Academic Research Communities, 2017).

There has been some ethical discussion among researchers regarding the appropriateness of continuing research in the face of crisis, particularly for researchers who are also trained in clinical medicine or other fields that might contribute to disaster response (e.g., House, Marete, and Meslin, 2016).

Table B.39. Strength of Evidence for *Research and Development*

Strength of Evidence	
Medium	Tropical cyclones and hurricanes; wildfire
Low	Drought; extreme cold; extreme heat; flooding; sea-level rise; severe storm systems

Risk Rating Interpretations

Table B.40 provides examples of how to interpret the risk ratings for this NCF.

Table B.40. Example Interpretations of Risk Ratings for *Research and Development*

Risk Rating	NCF Example
1: No disruption or normal operations	*Research and Development* continues to function as it does today, with occasional facility closures, data collection delays, or administrative disruptions due to weather events and related physical plant deficiencies.
2: Minimal disruption	Research facilities and projects in areas affected by climate drivers experience longer time periods in which they cannot operate safely or have research efforts ruined by extreme weather events, but research personnel can continue their efforts remotely or restart efforts. *Research and Development* management, intellectual property management, and *Research and Development* resource management experience longer time periods of disruption but can continue and support the ongoing operation of the *Research and Development* apparatus.
3: Moderate disruption	Some research facilities close temporarily or permanently because of consistent disruption and damage due to climate drivers; facility closures, ruined experiments, and research and administrative delays become far more frequent, limiting *Research and Development* progress and global competitiveness.
4: Major disruption	Most *Research and Development* is not viable due to extreme acute and long-term conditions; a significant number of research facilities close permanently, substantially curtailing *Research and Development* efforts across the country.
5: Critical disruption	*Research and Development* essentially ceases.

Law, Government, and National Security

Enforce Law

Table B.41. Scorecard for *Enforce Law*

	National Risk Assessment				
	Baseline	**Current Emissions**		**High Emissions**	
	Present	**2050**	**2100**	**2050**	**2100**
Enforce Law	2	3	3	3	3

Description "Operate Federal, State, local, tribal, territorial, and private sector assets, networks, and systems that contribute to enforcing laws, conducting criminal investigations, collecting evidence, apprehending suspects, operating the judicial system, and ensuring custody and rehabilitation of offenders" (CISA, 2020, p. 4).

Highlights

- *Enforce Law* was assessed as being at risk of minimal disruption for the baseline time period and at risk of moderate disruption for 2050 and 2100 under both current and high emissions scenarios because of increases in demand for this NCF brought on by increases in the frequency, intensity, or duration of natural hazards, which will likely contribute to a growth in crime, although the effect of specific drivers is not well established.
- The justice system is interconnected, creating some risk of disruption across all of *Enforce Law*'s subfunctions, but the *Contribute to Public Safety* subfunction is expected to experience the most stress.
- As a function distributed throughout the country, *Enforce Law* faces similar risks throughout the United States.

Synopsis

Enforce Law was assessed as being at risk of moderate disruption from climate change in both 2050 and 2100 for both the current and high emissions scenarios. An increase in the number of natural hazards due to climate change is expected to increase demand for law enforcement and, more broadly, for law enforcement support to other public safety agencies. Some climate drivers, such as extreme heat, are associated with increases in violent and property crime (Abbott, 2008; Goin, Rudolph, and Ahern, 2017), and there is a literature demonstrating relationships between weather and crime (Lauritsen and White, 2014) and between weather and disorder (Ranson, 2014).

Research into the relationship between climate drivers and crime, although limited, indicates that climate change is likely to increase crime through a variety of social, individual, and environmental mechanisms. For example, as climate change leads to increases in violent and property crime (Ranson, 2014), this increase in crime rates would likely lead to increased demand for multiple subfunctions of *Enforce Law*, including increased demand for investigative time, prosecutions, and correctional services. Indeed, because the different aspects of the justice system are interconnected, all the subfunctions of the NCF face risk of disruption. In particular, the *Contribute to Public Safety* subfunction already faces challenges in the western United States due to wildfire, and we assessed that risks will grow in 2050 and 2100. Because the *Enforce Law* NCF is distributed throughout the country, the risk is similar across regions of the United States.

Subfunctions

The subfunctions of *Enforce Law* are expected to face some disruption in 2050 and 2100 due to the increased demand for law enforcement activities in response to an increase in natural hazards due to climate change. Although there is limited evidence directly linked to individual subfunctions, the literature indicates that natural hazards contribute to increases in different types of crime (e.g., Agnew, 2012; Ranson, 2014), which, in turn, will likely increase demand for the subfunctions associated with *Enforce Law*.

We assessed every subfunction except for *Contribute to Public Safety* to be at risk of minimal disruption in 2050 and 2100 due to increases in demand. For example, an increase in violent and property crimes will increase demand for the *Conduct Investigations*, *Charge Alleged Suspects*, *Prosecute Crimes*, and *Operate Judicial System* subfunctions. High demand for these subfunctions could leave some states and localities with less responsive law enforcement and judicial organizations that are unable to meet demand. *Operate Detention Systems* and *Operate Corrections System* are also at risk because rising crime could result in the need for more detention and correctional facilities (Committee on Causes and Consequences of High Rates of Incarceration, 2014). These facilities can also face some risk from infrastructure concerns, such as lack of ventilation and air-conditioning, which affects the incarcerated (A. Jones, 2019; Skarha et al., 2020). However, this effect has not been well studied (Skarha et al., 2020).

Contribute to Public Safety faces a dynamic similar to those of the other subfunctions: Increases in crime will lead to more demand for public safety. However, this subfunction was assessed to be at minimal disruption in the present scenario because of the demand already placed on public safety workforces, such as firefighters, who face disruption due to natural hazards affected by climate change (Quinton, 2021), and law enforcement agencies providing support.

Table B.42. Subfunction Risk for *Enforce Law*

Subfunction	Baseline Present	Current Emissions 2050	Current Emissions 2100	High Emissions 2050	High Emissions 2100
Contribute to Public Safety	2	3	3	3	3
Conduct Investigations	1	2	2	2	2
Charge Alleged Suspects	1	2	2	2	2
Prosecute Crimes	1	2	2	2	2
Operate Judicial System	1	2	2	2	2
Operate Corrections System	1	2	2	2	2
Operate Detention Systems	1	2	2	2	2

NOTE: 1 = no disruption or normal operations. 2 = minimal disruption. 3 = moderate disruption. See Chapter 2 for further explanation of ratings.

Climate Drivers

Many of the climate drivers affect *Enforce Law*. For example, there is some evidence that drought could contribute to an increase in crime—particularly property crimes—although this topic is not well evaluated, and the study we identified focused solely on California (Goin, Rudolph, and Ahern, 2017). Historical research indicates that, at least in certain sociopolitical and economic environments, natural hazards, such as drought, have been associated with the growth of criminal organizations (Acemoglu, De Feo, and De Luca, 2017). Some studies have shown that hotter weather—and extreme heat—are correlated with an increase in violent crime and property crime; less clear is whether there is a causal relationship and what the magnitude of that effect is (Abbott, 2008; Goin, Rudolph, and Ahern, 2017). Wildfire risk is projected to increase, which could also affect this NCF. The United States has faced record-wildfire seasons in recent years, and longer and more-intense wildfire seasons are expected in the future (Wehner et al., 2018). This driver has increased the need for law enforcement agencies to assist with search and rescue and evacuation, causing threats to personnel safety and the need to divert staff to fulfill these duties (Estill, 2018). Extreme cold has been found to have a positive effect on crime, potentially lowering the incidence of property crime, for example (Butke and Sheridan, 2010). However, the overall projection is that the number of annual extreme cold events is not expected to increase over current levels, so the effect on this NCF is not expected to change.

There is little evidence in the literature on the effect of the other climate drivers—flooding, sea-level rise, severe storm systems, and tropical cyclones and hurricanes—on crime and *Enforce Law*. However, there is a literature demonstrating general relationships between weather and crime (Lauritsen and White, 2014) and between weather and disorder (Ranson, 2014). Evidence shows that weather can contribute to increases in crime (Brunsdon et al., 2009). Climate change could increase the burden of certain categories of crime (e.g., immigration and human trafficking offenses and offenses related to fraud after natural disasters) (Levenson, 2021; "Local Prosecution in the Era of Climate Change," 2022).

Furthermore, the increased risk from these climate drivers might require more-frequent evacuation of cities, as well as closures of courts and correctional facilities, which could affect crime rates. The literature suggests that, in general, increases in the frequency, duration, or severity of the climate drivers will increase the risk of disruption to this NCF.

Table B.43. Risk from Climate Drivers for *Enforce Law*

Climate Driver	Baseline Present	Current Emissions 2050	Current Emissions 2100	High Emissions 2050	High Emissions 2100
Drought	1	1	1	1	1
Extreme cold	1	1	1	1	1
Extreme heat	1	2	2	2	2
Flooding	1	2	2	2	2
Sea-level rise	1	2	2	2	2
Severe storm systems	1	2	2	2	2
Tropical cyclones and hurricanes	1	2	2	2	2
Wildfire	2	3	3	3	3

NOTE: 1 = no disruption or normal operations. 2 = minimal disruption. 3 = moderate disruption. See Chapter 2 for further explanation of ratings.

Impact Mechanisms

Past analyses have suggested that changes in the climate drivers will increase demand for law enforcement because of projected increases in violent crime (e.g., murder, rape, assault) and property crime (e.g., burglary, larceny) (Ranson, 2014), as well as opportunities for crime (Agnew, 2012).

Cascading Risk

Enforce Law has a significant overlap in workforce and mission with *Provide Public Safety*; a failure of one would cause the failure of the other. Upstream functions on which *Enforce Law* depends include *Operate Government*, *Preserve Constitutional Rights*, and *Prepare for and Manage Emergencies*, given that, for this NCF to operate effectively, it must be supported by a capable and fully functioning government that protects civil liberties and can support efforts prior to and during emergencies. Furthermore, *Enforce Law* requires *Transport Cargo and Passengers by Road* because the NCF requires law enforcement personnel to be able to travel to their jobs and to wherever crimes have occurred or are threatened to occur. *Enforce Law* also requires functioning correctional and detention facilities for alleged and convicted criminals and therefore is dependent on *Distribute Electricity*. Finally, *Enforce Law* requires the ability to coordinate and communicate and is dependent on NCFs providing those functions: *Provide Wireless Access Network Services*, *Provide Wireline Access Network Services*, and *Provide Satellite Access Network Services*.

Strength of Evidence

The strength of evidence of climate change's effect on this NCF falls in the medium and low categories. There is a well-established body of research examining the relationship between weather and crime (which has cascading effects through the criminal justice system), the seasonality of crime (with the summer typically witnessing more violent crime) (Lauritsen and White, 2014), and law enforcement's relationships and demands as they relate to natural hazards (Birkland and Schneider, 2007; Goin, Rudolph, and Ahern, 2017). The effect that some of the specific climate drivers have on crime is less well established.

Table B.44. Strength of Evidence for *Enforce Law*

Strength of Evidence	
Medium	Extreme cold; extreme heat
Low	Drought; flooding; sea-level rise; severe storm systems; tropical cyclones and hurricanes; wildfire

Risk Rating Interpretations

Table B.45 provides examples of how to interpret the risk ratings for this NCF.

Table B.45. Example Interpretations of Risk Ratings for *Enforce Law*

Risk Rating	NCF Example
1: No disruption or normal operations	Law enforcement organizations continue to function as they do today, with occasional delays due to capacity constraints, weather, and operational problems.
2: Minimal disruption	Law enforcement organizations in some states and localities experience time periods in which they face operational challenges, such as slower response times and staff shortages, but they can still fulfill core operational responsibilities. Law enforcement organizations are able to shift resources within regions to assist jurisdictions experiencing climate change–related public safety challenges.
3: Moderate disruption	Law enforcement organizations face operational challenges at the regional level, such as slower response times and staff shortages, but they can still typically fulfill core operational responsibilities. Staff shortages are severer and sustained, and response times are significantly slowed. Law enforcement organizations can shift resources from region to region to assist regions experiencing climate change–related public safety challenges.
4: Major disruption	Law enforcement organizations throughout the country and at the national or federal level face serious operational challenges that significantly affect their ability to fulfill core operational responsibilities and provide public safety services. These challenges include severe workforce shortages and damage to physical infrastructure (such as police stations and correctional facilities).
5: Critical disruption	Law enforcement organizations throughout the country and at the national or federal level cannot fulfill core operational responsibilities; across the country, laws are no longer systematically enforced.

Prepare for and Manage Emergencies

Table B.46. Scorecard for *Prepare for and Manage Emergencies*

	National Risk Assessment				
	Baseline	Current Emissions		High Emissions	
	Present	2050	2100	2050	2100
Prepare for and Manage Emergencies	2	3	3	3	3

Description: "Organize and manage resources and responsibilities for dealing with all aspects of emergencies (prevent, protect, mitigate, respond, and recover), to be resilient to and reduce the harmful effects of all hazards" (CISA, 2020, p. 5).

Highlights

- *Prepare for and Manage Emergencies* was assessed as being at risk of moderate disruption under the current and high emissions scenarios for 2050 and 2100 due to increases in the frequency, intensity, or duration of nearly all of the climate drivers—which will lead to increasing demand that strains the NCF.
- This NCF will experience the effect of climate drivers through increased demand for emergency management services and through a parallel decrease in the availability of government actors at all levels to supply those, thanks to workforce shortages.
- Chief among the climate drivers contributing to increasing demand will be those with the potential to cause widespread damage and injury, such as flooding, hurricanes, severe storm systems, and wildfire; extreme cold is not expected to have much effect on this NCF because the frequency of such events is not expected to increase.
- As the frequency of extreme weather events increases, demand for nearly all of the subfunctions will increase; the one subfunction for which there is limited evidence of effect is *Prevent Attacks*.

Synopsis

This NCF was assessed as being at risk of minimal disruption for the baseline. For the 2050 and 2100 current and high emissions scenarios, projections show increases in the frequency, intensity, or duration of nearly every climate driver. These increases will likely result in significant demand increases for this NCF and contribute to its inability to meet operational needs in some parts of the country. However, because the localized or regional nature of most emergent responses and the wide proliferation of mutual-aid agreements in most among SLTT partners, such shortcomings manifesting simultaneously across the whole country is an unlikely scenario—especially as experience working across jurisdictional boundaries spreads (Waugh, 2007). For these reasons, this NCF was assessed as being at risk of moderate disruption in coming decades.

Climate change is expected to increase the frequency and severity of natural hazards that it affects, which will, in turn, increase demand for *Prepare for Emergencies* and to manage response and recovery when disasters occur (Reidmiller et al., 2018). Flooding and severe storm systems have the capacity to cause widespread damage and destruction to homes, businesses, and other facilities, which, during acute moments, can tax the emergency response and management systems with emergent fire, search and rescue, and medical responses. Extreme cold can lead to blizzards and other winter storms— creating a need for snow clearance and assistance for vulnerable community members whose utilities or travel have been disrupted (Bolinger et al., 2022). Extreme heat can lead to an increase in heat-

109

related illnesses, including heart, lung, and kidney problems, increasing the need for acute emergency medical response during extreme heat events and for increased support from emergency management, such as planning and coordination with public health agencies (Seltenrich, 2015). Wildfire can threaten homes, businesses, and other facilities, necessitating fire suppression, evacuation, and rescue. Drought can lead to an increase in the number and severity of dust storms.

Together, the effects of climate change are expected to increase demand for emergency management services for extended time periods because certain seasonal natural hazards—such as wildfire—will likely occur outside of their traditional seasons and multiple hazards occur concurrently. Furthermore, workforce shortages will likely also contribute to the risk of disruption. For example, wildland firefighting already faces workforce shortages, as has the emergency management field more generally—and these challenges will likely increase with the increased risk of climate change (C. Johnson, 2022; Wigglesworth, 2022).

Subfunctions

The demand for nearly every subfunction is expected to increase over time as the frequency of natural hazards affected by climate change increases. This will cover the spectrum for *Prepare for Emergencies, Manage Emergencies, Protect Against Threats and Hazards, Mitigate Impacts of Threats and Hazards, Respond to Incidents,* and *Recover from Incidents* (FEMA, 2020a; Silverman et al., 2010). The one subfunction for which there is not enough evidence to project an increase in demand resulting from climate change is *Prevent Attacks.*

As the frequency and severity of climate-related emergencies increase, so will the need to prepare and plan for emergencies. Preparation typically involves the engagement of communities, organizations, corporations, and government entities. In the context of natural hazards, protection can also take the form of physically hardening facilities to make them less vulnerable to damage or to having their functions degraded or disrupted. As the frequency and severity of emergencies increase, so will the need for such preparation and planning, as well as execution of hardening strategies. There is a small literature examining hazard mitigation plans that has found that the quality of these state plans could be improved (Berke, Smith, and Lyles, 2012; Lyles, Berke, and Smith, 2014) and that, although local plans generally discuss climate change, it is not a major focus (Stults, 2017). This literature has also shown that some states lack the capacity to support local community efforts to develop and implement their own mitigation plans (Smith and Vila, 2020).

An increase in the frequency and intensity of climate events is expected to lead to a corresponding increase in the need for emergency management to coordinate and manage response and recovery. This need is expected to be especially strong for climate drivers with the capacity to cause widespread damage, such as flooding, severe storm systems, tropical cyclones and hurricanes, and wildfire. Even weather events that are not expected to cause high levels of physical damage, such as extreme heat, could increase demand for emergency management to coordinate agencies' outreach and sheltering efforts.

Demand for this NCF is also expected to be high after a weather incident because managing recovery and rebuilding after a disaster requires engagement with the community, rebuilding in a way that mitigates future hazards, and careful management of reconstruction grants (Central Office for Recovery, Reconstruction and Resiliency, 2018).

To mitigate these increased demands, emergency management and response entities at each level of government will face a need to increase investments in staffing and equipment. The emergency management community in the United States is also likely to continue advancing lessons learned in past disasters, including record-breaking hurricane and wildfire seasons along with the COVID-19 pandemic. Such lessons include increasing mutual-aid partnerships with neighboring jurisdictions, a tactic that smooths out capability differences and builds capacity that could help ensure that incidents that can overwhelm local responders are kept isolated (Dzigbede, Gehl, and Willoughby, 2020).

Table B.47. Subfunction Risk for *Prepare for and Manage Emergencies*

Subfunction	Baseline Present	Current Emissions 2050	Current Emissions 2100	High Emissions 2050	High Emissions 2100
Prepare for Emergencies	1	2	2	2	2
Manage Emergencies	2	3	3	3	3
Prevent Attacks	1	1	1	1	1
Protect Against Threats and Hazards	1	2	2	2	2
Mitigate Impacts of Threats and Hazards	1	2	2	2	2
Respond to Incidents	2	3	3	3	3
Recover from Incidents	2	3	3	3	3

NOTE: 1 = no disruption or normal operations. 2 = minimal disruption. 3 = moderate disruption. See Chapter 2 for further explanation of ratings.

Climate Drivers

Each of the drivers has some effect on the *Prepare for and Manage Emergencies* NCF. For example, drought can lead to an increase in the frequency of dust storms (Reed and Nugent, 2018). However, drought (and associated dust storms) is largely a regional event in the United States and unlikely to contribute to a national disruption of this NCF. Extreme cold can contribute to blizzards and storms, which can disrupt travel, utilities, and create a need for snow clearance and assistance for vulnerable community members (Bolinger et al., 2022). However, the overall projection is that extreme cold events are not expected to increase over current levels, so their effect on this NCF is not expected to change. Extreme heat also creates health effects that could require emergency management support, such as planning and coordination with public health agencies (Seltenrich, 2015).

Flooding, sea-level rise, severe storm systems, and tropical cyclones and hurricanes have the capacity to cause widespread damage or destruction to homes, businesses, infrastructure, and natural habitats. This, in turn, can create a need for planning, coordination, debris clearance, repair, and other actions supporting disaster preparedness, mitigation, response, and recovery (Luther, 2008). The United States has faced record-wildfire seasons in recent years, and longer and more-intense wildfire seasons are expected in the future (Wuebbles et al., 2017). Wildfire can have effects similar to those of the other drivers—such as widespread damage—that create a need for emergency management activities that include planning, search and rescue, and repair and rebuilding.

111

Table B.48. Risk from Climate Drivers for *Prepare for and Manage Emergencies*

Climate Driver	Baseline	Current Emissions		High Emissions	
	Present	2050	2100	2050	2100
Drought	1	1	1	1	1
Extreme cold	1	1	1	1	1
Extreme heat	1	2	2	2	2
Flooding	2	3	3	3	3
Sea-level rise	2	3	3	3	3
Severe storm systems	2	3	3	3	3
Tropical cyclones and hurricanes	2	3	3	3	3
Wildfire	2	3	3	3	3

NOTE: 1 = no disruption or normal operations. 2 = minimal disruption. 3 = moderate disruption. See Chapter 2 for further explanation of ratings.

Impact Mechanisms

The primary pathways through which this NCF will experience the effects of climate drivers are through increased demand for emergency management services and through a parallel decrease in the availability of government actors at all levels to supply those because of workforce shortages.

In the case of the former, drivers that increase the prevalence or severity of severe weather events will demand an increasing amount of emergency management attention for longer time periods. In seasons during which multiple severe weather events are possible, such as the Gulf Coast hurricane season, this increased demand will strain emergency management resources when conditions demand multiple extended responses. Additionally, certain seasonal natural hazards can be expected to occur outside the bounds of their traditional seasons, with unseasonal rainfall bringing flood risk at unfamiliar times or hot and dry conditions increasing wildfire risk in winter.

Compounding this demand-side pressure are supply-side effects on the NCF as the result of climate drivers. Some drivers, such as extreme heat, make work in managing emergencies—notably, firefighting—a riskier affair. In the case of wildland fire, for example, this increased risk is already combining with longer fire seasons, low pay, and other factors to cause shortfalls in key seasonal firefighting hiring (Wigglesworth, 2022). Similarly, emergency management workforces have already faced shortages, such as during the 2017 hurricane and wildfire seasons (C. Johnson, 2022), and these challenges are likely to grow as climate risks increase.

Cascading Risk

Prepare for and Manage Emergencies has overlap with the *Provide Public Safety* NCF, and a failure in one could contribute to failure in the other. Upstream functions on which *Prepare for and Manage Emergencies* relies include *Operate Government* because this NCF requires a functioning public sector to operate effectively. Communication infrastructure is critical to this NCF because it supports communication and coordination among stakeholders and the public. As a result, *Provide Wireless*

112

Access Network Services, Provide Wireline Access Network Services, and *Provide Satellite Access Network Services* are required, as is *Distribute Electricity.* The ability to transport by road is necessary to ensure that personnel can get to their workplaces and enable those personnel to get to where their services are needed; as a result, this NCF relies on *Transport Cargo and Passengers by Road. Provide Positioning, Navigation, and Timing Services* also provides upstream functions, supporting transportation, response, and recovery activities.

Strength of Evidence

The strength of evidence documenting how climate drivers affect the *Prepare for and Manage Emergencies* NCF was assessed as medium. Individual studies have documented the effects of different climate drivers and generally demonstrate that increases in the frequency, intensity, or duration of these drivers will contribute to increases in their negative effects, such as damage to infrastructure. However, the magnitude of climate change's effect on the increased workload of emergency management is not well documented, nor is their effect on specific subfunctions. It therefore requires an application of logic to predict likely consequences that such increased frequency, intensity, or duration of drivers will bring for the country's ability to prepare for and manage emergencies.

Table B.49. Strength of Evidence for *Prepare for and Manage Emergencies*

Strength of Evidence	
Medium	Drought; extreme cold; extreme heat; flooding; sea-level rise; severe storm systems; tropical cyclones and hurricanes; wildfire

Risk Rating Interpretations

Table B.50 provides examples of how to interpret the risk ratings for this NCF.

Table B.50. Example Interpretations of Risk Ratings for *Prepare for and Manage Emergencies*

Risk Rating	NCF Example
1: No disruption or normal operations	Emergency management organizations at the local, state, regional, and national or federal levels continue to function as they do today, with occasional challenges due to weather and operational problems restricted to one response type.
2: Minimal disruption	Emergency management organizations in some states and localities experience time periods in which they face operational challenges, such as slower response times and staff shortages, but they can still fulfill core operational responsibilities.
3: Moderate disruption	Emergency management organizations face operational challenges, such as slower response times and staff shortages, at the regional level, but they can still fulfill core operational responsibilities.
4: Major disruption	Emergency management organizations throughout the country and at the national or federal level face serious operational challenges that significantly affect their ability to provide emergency management services. These challenges include severe workforce shortages, budget constraints, and widespread disruptions stemming from the ongoing management of multiple incidents.
5: Critical disruption	Emergency management organizations throughout the country and at the national or federal level cannot fulfill core operational responsibilities.

Provide Public Safety

Table B.51. Scorecard for *Provide Public Safety*

	National Risk Assessment				
	Baseline	Current Emissions		High Emissions	
	Present	2050	2100	2050	2100
Provide Public Safety	2	3	3	3	3

Description: "Provide public services—to include police, fire, and emergency medical services [EMS]—to ensure the safety and security of communities, businesses and populations" (CISA, 2020, p. 6).

Highlights

- By 2050, *Provide Public Safety* was assessed as being at risk of moderate disruption from climate change in both the current and high emissions scenarios as the frequency, intensity, or duration of nearly every climate driver is expected to increase, leading to an increase in the demand for this NCF.
- Workforce shortages currently pose challenges for wildland firefighters; similar issues could create difficulties for other workforces included in this NCF, such as public health support workers and emergency managers.
- All the subfunctions for this NCF are expected to be at risk due to the increase in demand associated with natural disasters; however, risk to the *Provide Fire and Rescue Services* subfunction is expected to increase as the likelihood of wildland firefighter workforce shortages grows.
- Wildfire risk is severest in the western United States; as a result, the *Provide Fire and Rescue Services* subfunction faces the greatest risk of disruption there among all regions in the country.

Synopsis

Provide Public Safety is vulnerable to climate change because of expected increases in the frequency, intensity, or duration of the climate drivers that will put increased pressure on the NCF. Flooding, sea-level rise, severe storm systems, and tropical cyclones and hurricanes have the capacity to cause widespread damage or destruction to homes, businesses, and other facilities. Extreme cold, particularly when paired with winter storms, can cause injuries and deaths from traffic accidents, as well as from hypothermia, particularly among the poor and elderly, who often have inadequate heat. Extreme heat can cause an increase in heat-related illnesses, including heart, lung, and kidney problems, as well as issues with fetal health. Some studies have also shown that hotter weather is correlated with an increase in crime and terrorist attacks; less clear are whether there is a causal relationship and what the magnitude of that effect is (Abbott, 2008; Goin, Rudolph, and Ahern, 2017).[32] High energy usage associated with heat waves can cause blackouts and damage to power transmission facilities. Wildfire can threaten homes, businesses, and other facilities, necessitating fire suppression, evacuation, and rescue while smoke from those fires causes additional harm, including potentially acting as an infectious-disease vector (Kobziar and Thompson, 2020). Drought can lead to an increase in the frequency of dust storms, which, in turn, can increase respiratory diseases.

Together, the effects of climate change are expected to affect demand for this NCF and its workforce. Specifically, climate change will likely increase the need for law enforcement to maintain

---

[32] See also the discussion of *Enforce Law* for more information about the relationship between crime, instability, and climate.

order, fire services to suppress fires and provide rescue, and EMS to provide prehospital care—service providers who will also need to help conduct evacuations and support displaced people. During large disasters, emergency management personnel will need to coordinate with public health, while restoration of damaged utilities will require support from public works (Calma, 2021). The effects of climate change could also contribute to workforce shortages. For example, the United States currently faces a shortage of wildland firefighters, and this shortage could grow as the threat of wildfire increases (A. Phillips, 2021; Quinton, 2021; Safo, 2022).

Provide Public Safety was assessed as being at risk of moderate disruption by 2050 from climate change in both the current and high emissions scenarios. These ratings are due, in part, to an expected increase in the frequency, intensity, or duration of wildfires. Public safety needs in this area are already not being fully met in some parts of the country because of the current wildfire situation (as discussed in "Subfunctions," next). By 2100, the frequency, intensity, and duration of nearly all of the climate drivers are expected to increase, leading to an increase in the demand for this NCF, which we assessed as at risk of moderate disruption under the current and high emissions scenarios, resulting in the inability to meet operational needs in some parts of the country.

Subfunctions

All the subfunctions for this NCF—*Contribute to Law Enforcement, Provide Fire and Rescue Services, Provide EMS, Provide Public Works Support, Provide Emergency Management,* and *Support Public Health Services*—are expected to be at risk due to the increase in demand associated with natural disasters. The ability to combat wildfire is of particular concern. In the 2021 fire season, firefighting capacity often faced challenges due to workforce shortages, resulting in slower suppression and, consequently, larger fires (A. Phillips, 2021; Quinton, 2021). Risk to this subfunction is expected to increase to moderate disruption as the likelihood of wildland firefighter workforce shortages grows in response to the effects of climate change. The public health and emergency management workforces face similar risks, which affects the *Support Public Health Services* and *Provide Emergency Management* subfunctions (Currie, 2022; Kintziger et al., 2021).

Table B.52. Subfunction Risk for *Provide Public Safety*

Subfunction	Baseline Present	Current Emissions 2050	Current Emissions 2100	High Emissions 2050	High Emissions 2100
Contribute to Law Enforcement	2	3	3	3	3
Provide Fire and Rescue Services	2	3	3	3	3
Provide EMS	2	3	3	3	3
Provide Public Works Support	2	3	3	3	3
Provide Emergency Management	2	3	3	3	3
Support Public Health Services	1	2	2	2	2

NOTE: 1 = no disruption or normal operations. 2 = minimal disruption. 3 = moderate disruption. See Chapter 2 for further explanation of ratings.

Climate Drivers

Each of the climate drivers affects the *Provide Public Safety* NCF. For example, there is some evidence that drought can contribute to increases in property crime, although this is not well studied, and the identified study was on only California (Goin, Rudolph, and Ahern, 2017). Drought can also lead to an increase in the frequency of dust storms, which can increase respiratory diseases (Reed and Nugent, 2018). However, drought (and associated dust storms) is largely a regional event in the United States and unlikely to contribute to a national disruption of this NCF. Extreme cold has also been found to have some effect on crime, potentially depressing property crime (Butke and Sheridan, 2010), and it has a host of health effects that might require public health and EMS support (Seltenrich, 2015). However, the overall projection is that extreme cold events are not expected to increase over current levels, so their effect on this NCF is not expected to change. Extreme heat also creates health effects that might require public health and EMS interventions (Seltenrich, 2015), and some studies have shown that hotter weather is correlated with an increase in violent and property crime. The causal relationship and the magnitude of that effect is less clear, however (Abbott, 2008; Goin, Rudolph, and Ahern, 2017).

The literature demonstrates a relationship between weather and crime (Lauritsen and White, 2014) and between weather and disorder (Ranson, 2014). Consequently, as climate change contributes to an increase in the frequency, duration, or severity of the remaining climate drivers— flooding, sea-level rise, severe storm systems, and tropical cyclones and hurricanes—these drivers will likely contribute to increases in these social ills. Furthermore, among flooding, severe storm systems, and tropical cyclones and hurricanes, exacerbated by sea-level rise, each has the capacity to cause widespread damage or destruction to homes, businesses, infrastructure, and natural habitats. This, in turn, can create a need for fire suppression, debris clearance, repair, and other actions supporting disaster response and recovery, public works, and firefighting (Luther, 2008). Furthermore, these drivers can contribute to injuries, sickness, and other health issues (B. Waddell, 2021). Wildfire can cause similar challenges—such as widespread damage, injury, and sickness. Furthermore, the United States has faced record-wildfire seasons in recent years, and longer and more-intense wildfire seasons are expected in the future (Wuebbles et al., 2017). This risk has also increased the need for law enforcement agencies to assist with search and rescue and evacuation, causing threats to personnel safety and the need to divert staff to fulfill these duties (Estill, 2018).

Table B.53. Risk from Climate Drivers for *Provide Public Safety*

Climate Driver	Baseline	Current Emissions		High Emissions	
	Present	2050	2100	2050	2100
Drought	1	1	1	1	1
Extreme cold	1	1	1	1	1
Extreme heat	1	2	2	2	2
Flooding	2	3	3	3	3
Sea-level rise	2	3	3	3	3
Severe storm systems	2	3	3	3	3
Tropical cyclones and hurricanes	2	3	3	3	3
Wildfire	2	3	3	3	3

NOTE: 1 = no disruption or normal operations. 2 = minimal disruption. 3 = moderate disruption. See Chapter 2 for further explanation of ratings.

Impact Mechanisms

Climate drivers that drive increased incidence and severity of natural hazard–related incidents will exert pressure on the *Provide Public Safety* NCF. The increased resource needs of those incidents— notably, flooding, severe storm systems, and wildfire—will increase the need for law enforcement to maintain order, fire services to suppress fires and provide rescue, and EMS to provide prehospital care—service providers who will also need to help conduct evacuations and support displaced people. Additionally, large disasters tied to these natural hazards will bring increased strain because lengthy response and recovery times will strain the abilities of agencies at all levels to coordinate with public health and provide public works support to coordinate restoration of damaged utilities.

Further complicating the outlook for this NCF through 2050 and beyond are the extant workforce issues for wildland firefighters, public health support workers, and emergency managers (Currie, 2022; Kintziger et al., 2021; Wigglesworth, 2022). Similar issues could be seen in the workforces responsible for other subfunctions as climate drivers intensify.

Cascading Risk

Provide Public Safety has significant overlap in workforce and mission with the *Enforce Law* and *Prepare for and Manage Emergencies* NCFs. A failure in one could contribute to failure in another. Upstream functions on which the *Provide Public Safety* NCF depends include *Operate Government* and *Preserve Constitutional Rights*. For this NCF to operate effectively, it must also be supported by a capable and fully functioning government that protects civil liberties. Communication infrastructure is critical to the dispatch of public safety services. Therefore, *Provide Wireless Access Network Services*, *Provide Wireline Access Network Services*, and *Provide Satellite Access Network Services* are required, as are *Distribute Electricity* and *Provide Positioning, Navigation, and Timing Services*. The ability to transport by road is necessary not only to ensure that personnel can get to their workplaces but also to allow those personnel to get to where their services are needed; as a result, this NCF relies on

Transport Cargo and Passengers by Road. Finally, given water's role in suppressing fires, the *Supply Water* NCF is required to support this NCF. *Provide Public Safety* also requires functioning correctional and detention facilities for alleged and convicted criminals and therefore requires *Distribute Electricity*. Finally, *Provide Public Safety* requires the ability to coordinate and communicate and is dependent on NCFs providing those functions: *Provide Wireless Access Network Services*, *Provide Wireline Access Network Services*, and *Provide Satellite Access Network Services*.

Strength of Evidence

Evidence of climate change's effects on natural disasters—and their potential to cause widespread damage or destruction to entire communities—is well established and has been recognized by numerous government agencies (H. Anderson et al., 2017; Reidmiller et al., 2018). There is an extensive literature on natural disasters' effect on the demand for law enforcement and other public safety personnel during and after such events. Although the literature shows a correlation between extreme heat and increased crime rates and terrorist attacks, which suggests that there could be an increase in demand for public safety during heat waves, evidence is mixed on the magnitude of this effect and whether there is a causal relationship. On the other hand, evidence suggests that there is already a shortage of wildland firefighters in wildfire-prone areas (A. Phillips, 2021; Quinton, 2021; Safo, 2022). Although the need for public safety services has been shown to increase during natural disasters, such as major floods and wildfires, the magnitude of climate change's effect on the increased workload of public safety services is less well documented.

Table B.54. Strength of Evidence for *Provide Public Safety*

Strength of Evidence	
High	Extreme cold; extreme heat; flooding; tropical cyclones and hurricanes; wildfire
Medium	Sea-level rise; severe storm systems
Low	Drought

Risk Rating Interpretations

Table B.55 provides examples of how to interpret the risk ratings for this NCF.

119

Table B.55. Example Interpretations of Risk Ratings for *Provide Public Safety*

Risk Rating	NCF Example
1: No disruption or normal operations	Public safety services continue to function as they do today, with occasional challenges due to weather and operational problems restricted to one response type.
2: Minimal disruption	Public safety organizations in some states and localities experience time periods in which they face operational challenges, such as slower response times and staff shortages, but can still fulfill core operational responsibilities.
3: Moderate disruption	Public safety organizations face operational challenges at the regional level, such as slower response times and staff shortages, but they can still fulfill most core operational responsibilities.
4: Major disruption	Public safety organizations throughout the country and at the national or federal level face serious operational challenges that significantly affect their ability to provide public safety services. These challenges include severe workforce shortages, budget constraints, and widespread disruptions stemming from the ongoing management of multiple incidents.
5: Critical disruption	Public safety organizations throughout the country and at the national or federal level cannot fulfill core operational responsibilities.

Provide Materiel and Operational Support to Defense

Table B.56. Scorecard for *Provide Materiel and Operational Support to Defense*

	National Risk Assessment				
	Baseline	**Current Emissions**		**High Emissions**	
Provide Materiel and Operational Support to Defense	**Present**	**2050**	**2100**	**2050**	**2100**
	2	3	3	3	3

Description: "Develop, produce, and sustain defense systems and components and provide support to defense operations" (CISA, 2020, p. 6).

Highlights

- The risk rating for *Provide Materiel and Operational Support to Defense* was assessed to be moderate based on the projected growth in number and intensity of climate drivers that could affect military installations across the country.
- Drought is the most prevalent climate driver affecting installations at present, but we assessed that extreme heat, flooding, tropical cyclones and hurricanes, and wildfire will also present risk that could cause physical damage to base infrastructure and present unsafe conditions for operations. Increased prevalence of climate drivers will also generate greater demand on the function from domestic entities.
- Vulnerabilities to climate drivers arise from installation locations in many areas subject to increased risk, but those vulnerabilities will vary depending on the type of facility (e.g., Navy base versus Army installation).
- All seven subfunctions are vulnerable to climate drivers given that these subfunctions are geographically distributed across the country.
- Military installations in coastal areas and in the West, Southwest, and Southeast are particularly vulnerable given climate drivers' likely effects in these regions by 2050 and 2100.

Synopsis

Climate drivers will have varied effects on the NCF depending on the geographic distribution of subfunctions and DoD installations, but, in aggregate, this function was assessed to be at moderate risk of disruption in 2050 and 2100. DoD operates more than 4,000 sites across more than 500 installations in the United States (DoD, undated-a; DoD, undated-b). Resident units, functions, and agencies at DoD installations carry out different subfunctions of the NCF, although many installations will have primary residents that focus on one or more core missions, such as airlift or logistics.

DoD routinely provides support to domestic disaster response efforts through defense support of civil authorities and overseas through humanitarian assistance/disaster relief operations. As climate effects lead to more occurrences and increase the severity of climate drivers, such as tropical cyclones, demands for DoD support are expected to increase as well. To support relief operations in the aftermath of Hurricane Ian in 2022, 5,000 National Guard members were mobilized (Sheely, 2022). DoD can be called on to provide strategic and tactical airlift, maritime domain awareness for search and rescue, medical support, and other humanitarian relief in the aftermath of large natural hazards (Kreisher, 2005; Shaw, 2013).

Increasingly frequent and severe natural hazards affected by climate change are also expected to cause physical disruption and damage and to affect worker availability on DoD installations. Wildfire can limit access to bases for personnel, and hazardous air quality can limit the activities in which personnel can safely engage, particularly in outdoor areas. Extreme heat can result in damage to infrastructure necessary for deployment and sustainment of operational forces. Railroad tracks can buckle or warp in extreme heat, especially when coupled with very large and heavy loads (e.g., an Abrams tank weighs approximately 55 tons with full armor).[33] Roads and airport runways are also affected by extreme heat ("UK Royal Air Force Halts Flights at Brize Norton Base Due to Heatwave," 2022). Extreme heat can limit air operations because aircraft lift (particularly for larger transport aircraft) is limited by less-dense air, requiring more power to take off, longer runways, or reduced weight (Coffel, Thompson, and Horton, 2017; Heise, 2021).

These effects result in an assessment of risk of moderate disruption to *Provide Materiel and Operational Support to Defense* in 2050 and in 2100 under both current and high emissions scenarios. The broad distribution of DoD installations and subfunctions across those installations mitigates greater risk to the NCF.

Key Considerations for *Provide Materiel and Operational Support to Defense*

Although this analysis focused on natural hazards' effects in the United States, instances of climate drivers globally will likely also place greater demands on the NCF because the United States provides humanitarian relief and other support to other countries affected by natural hazards. DoD has provided military capabilities (e.g., search and rescue, medical support, strategic and tactical airlift, and surveillance and reconnaissance) for large-scale natural hazards, such as the 2004 Indian Ocean tsunami, the 2010 earthquake in Haiti, and the 2011 Fukushima earthquake in Japan.

Subfunctions

Because of the distributed nature of subfunctions across DoD installations, all seven subfunctions are subject to risk from multiple types of climate drivers—particularly, drought, extreme heat, flooding, severe storm systems, and tropical cyclones and hurricanes. Deployment and distribution at air- and seaports of embarkation will be affected. Supply, maintenance, and logistics services can be disrupted by physical damage to facilities and workforce shortages at installations because these functions are spread across many installations (e.g., some maintenance is conducted at base level, while more-complex maintenance occurs at centralized facilities). Commercially sourced logistics and other support are particularly susceptible to effects on the workforce that are not subject to the same response requirements as military personnel, as well as damage to transportation infrastructure that can occur (e.g., roads buckling in extreme heat, areas made impassable by flooding). Joint health services are provided through military hospitals and health clinics in 38 states and the District of Columbia, exposing these facilities to multiple climate drivers (TRICARE, 2022).

---

[33] Rail operations on military bases operate under different business models: government owned, government operated; government owned, contractor operated; and privatized operations. This analysis is specific to the first two models; off-base and privatized rail operations are addressed in "Cascading Risk." For more on rail operations on Army installations, see Pint et al., 2017.

Table B.57. Subfunction Risk for *Provide Materiel and Operational Support to Defense*

Subfunction	Baseline Present	Current Emissions 2050	Current Emissions 2100	High Emissions 2050	High Emissions 2100
Conduct Deployment and Distribution	2	3	3	3	3
Supply the Force	2	3	3	3	3
Provide Maintenance Operations	2	3	3	3	3
Provide Logistics Services	2	3	3	3	3
Provide Operational Contract Support	2	3	3	3	3
Provide Engineering Support	2	3	3	3	3
Provide Joint Health Services	2	3	3	3	3

NOTE: 2 = minimal disruption. 3 = moderate disruption. See Chapter 2 for further explanation of ratings.

Climate Drivers

Nearly every climate driver will have at least a moderate effect on the function. Drought, severe storm systems, tropical cyclones and hurricanes, and wildfire have the potential to threaten or cause physical damage and displace or otherwise affect workers. Demands for military support to civil authorities and overseas are expected to increase as climate events occur more often and with greater severity. Sea-level rise is expected to affect naval facilities and will require mitigations or moving facilities. Extreme cold is one climate driver that is not expected to present a risk to the function based on current and high emissions scenarios in 2050 and 2100.

From what we glean from historical examples and existing analyses in this space, climate-driver effects could affect this NCF in the following ways:

- Langley Air Force Base in Virginia and MacDill Air Force Base in Florida are particularly susceptible to flooding in emissions scenarios, with more than 75 percent of the installations in flood hazard zones at present (Narayanan, Lostumbo, et al., 2021). Flooding can affect the usability of runways and access to the base for critical personnel, such as when one-third of Offutt Air Force Base experienced severe flooding in March 2019 (Losey, 2020). Recovery at Offutt took three years, partly because the Air Force pursued improvements in the design of facilities (Garcia, 2022). Coastal and riverine flooding is also a dominant hazard for Navy installations, particularly on the Atlantic Coast, in the Gulf of Mexico, and for overseas locations (Pinson et al., 2021, p. 34).
- Sea-level rise is a particular concern for the Hampton Roads area in Virginia, home to 15 installations, including Naval Station Norfolk, the largest naval base in the world, with more than 80,000 active-duty personnel assigned.
- Severe storm systems and tropical cyclones and hurricanes can cause physical damage to installations, potentially in the billions of dollars. Previous analysis shows that Lackland, Eglin, and Robins Air Force Base and Patrick Space Force Base could experience particularly high damage in a category 5 hurricane (Narayanan, Lostumbo, et al., 2021).

- Wildfire risk is relatively low for most Air Force installations, but several important facilities, including Vandenberg Space Force Base and Eglin, Beale, Mountain Home, and Moody Air Force Bases, are in areas subject to extreme risk (Narayanan, Lostumbo, et al., 2021). Joint Base Lewis-McChord (home of I Corps and the 62nd Airlift Wing) in Washington State is also susceptible to wildfire.
- Extreme cold was not assessed to be a key driver of risk to the function.
- Extreme heat will affect installations in the West and Southeast, which can affect the availability of workforce and lead to physical damage to infrastructure, such as roads and runways.
- Drought is the dominant hazard expected to affect Army installations (Pinson et al., 2021, p. 32). Drought conditions also increase the probability of wildfire sparked by accident and can affect readiness if installations have to limit or cancel training events or take extra precautions during operations (Pinson et al., 2021, p. 50).

Table B.58. Risk from Climate Drivers for *Provide Materiel and Operational Support to Defense*

Climate Driver	Baseline	Current Emissions		High Emissions	
	Present	2050	2100	2050	2100
Drought	2	3	3	3	3
Extreme cold	1	1	1	1	1
Extreme heat	2	3	3	3	3
Flooding	2	3	3	3	3
Sea-level rise	2	3	3	3	3
Severe storm systems	2	3	3	3	3
Tropical cyclones and hurricanes	2	3	3	3	3
Wildfire	2	3	3	3	3

NOTE: 1 = no disruption or normal operations. 2 = minimal disruption. 3 = moderate disruption. See Chapter 2 for further explanation of ratings.

Impact Mechanisms

Physical damage and disruption, workforce shortages, and demand changes are the impact mechanisms expected to affect the function. Physical damage to facilities and installations can cost billions of dollars and can affect weapon systems and other military capabilities that have long lead times to replace or repair. For example, Hurricane Michael in 2018 damaged more than 400 buildings and led to the displacement of F-22 aircraft from Tyndall Air Force Base to Eglin Air Force Base in Florida (Reeves, 2019). Although military personnel and civilians might try to press on with normal operations, working in extreme heat can lead to heat exhaustion and heat stroke (Armed Forces Health Surveillance Branch, 2021). In 2019, the majority of heat illnesses reported in the Army occurred at installations in the Southeast (Fort Moore [formerly Fort Benning], Georgia; Fort Liberty [formerly Fort Bragg], North Carolina; and Fort Campbell, Kentucky), which tend to experience days of high heat coupled with humidity (U.S. Army Public Health Center, undated).

Cascading Risk

The upstream functions that affect this NCF are *Distribute Electricity, Transmit Electricity, Maintain Supply Chains, Transport Cargo and Passengers by Air, Transport Cargo and Passengers by Road,* and *Transport Cargo and Passengers by Rail. Distribute Electricity* and *Transmit Electricity* are required to power operations and installations. Although many installations have backup power capabilities, such as diesel generators, they are limited in how long they can support continued operations. *Maintain Supply Chains* is required to ensure that medical supplies, spare parts, and engineering equipment can continue to operate. *Transport Cargo and Passengers by Air, Transport Cargo and Passengers by Road, and Transport Cargo and Passengers by Rail* are required to support deployment and movement of materiel and equipment between installations.

Strength of Evidence

Evidence for climate change's effect on defense installations is strong based on recent work conducted by DoD and other research organizations (DoD, undated-a; DoD, undated-b; Pinson et al., 2021). Additionally, climate drivers' effects that have occurred are documented in contemporaneous news reporting and studies (Garcia, 2022; Losey, 2020). Extreme cold has been addressed in fewer studies, likely because of the projection that extreme cold will not be a significant driver of climate-induced effects on NCFs, including *Provide Materiel and Operational Support to Defense.*

Table B.59. Strength of Evidence for *Provide Materiel and Operational Support to Defense*

Strength of Evidence	
High	Drought; extreme heat; flooding; sea-level rise; severe storm systems; tropical cyclones and hurricanes; wildfire
Medium	Extreme cold

Risk Rating Interpretations

Table B.60 provides examples of how to interpret the risk ratings for this NCF.

Table B.60. Example Interpretations of Risk Ratings for *Provide Materiel and Operational Support to Defense*

Risk Rating	NCF Example
1: No disruption or normal operations	Military units can prepare for and deploy from home base within required timelines.
2: Minimal disruption	Some military units are delayed from deploying with minimal effect on operational requirements.
3: Moderate disruption	Some military units are unable to deploy to respond to contingencies, creating mission risk.
4: Major disruption	Many military units are unable to deploy to respond to contingencies and lack operational support, leading to mission failure.
5: Critical disruption	Many military units are unable to deploy to respond to contingencies, lack critical operational support, and are unable to achieve operational objectives as directed by the commander in chief.

Perform Cyber Incident Management Capabilities

Table B.61. Scorecard for *Perform Cyber Incident Management Capabilities*

	National Risk Assessment				
	Baseline	Current Emissions		High Emissions	
Perform Cyber Incident Management Capabilities	Present	2050	2100	2050	2100
	1	2	2	2	2

Description: "Provide security systems and services that protect critical business assets and functions, including preventive guidance, simulation, testing, and warning capabilities; operate operations response centers and teams; integrate and share information; coordinate and provide response, recovery, and reconstitution services" (CISA, 2020, p. 5).

Highlights

- The risk to *Perform Cyber Incident Management Capabilities* was assessed to be minimal in 2050 and 2100 in both emissions scenarios because the diffuse nature of the function and the ability to accomplish most subfunctions remotely.
- Extreme heat, flooding, severe storm systems, tropical cyclones and hurricanes, and wildfire were assessed to contribute to minimal elevated risk because they can impede timely detection and response to emerging cyber incidents through physical damage to supporting infrastructure and effects on the workforce.
- The function's vulnerabilities stem from enduring challenges with fielding and operating inherently insecure systems and from a chronic workforce shortage.
- The time-critical subfunctions *Provide Detection and Analytics* and *Manage Containment, Eradication, and Recovery* are more at risk in circumstances in which incident responders cannot access affected systems or facilities because of physical damage.
- There are not particular regions of the country that are at greater risk than others for this function.

Synopsis

Climate drivers were assessed to have a minimal risk to the *Perform Cyber Incident Management Capabilities* NCF. The most time-critical subfunctions of the function—*Provide Detection and Analytics* and *Manage Containment, Eradication, and Recovery*—were assessed to be the most affected because of the need for quick detection and response to an emerging cyber incident. Climate drivers—particularly, flooding, hurricanes, severe storm systems, and wildfire—can cause displacement of critical workers and damage the physical infrastructure on which cyber incident responders rely.

Climate change drivers that impede responders' ability to gain access to affected systems and networks drive risk to the function, although many subfunctions for this function can be executed remotely and do not require physical access to affected systems. The exception to this is for systems and networks that are not accessible remotely, such as closed networks or operational technology (OT) that can be monitored remotely but do not allow remote remediation of cyber incidents. An interruption to the function would allow cyber incidents to spread and inflict more damage, such as disrupting the operation of critical networks and systems, whether in an enterprise IT or OT environment.

The climate drivers were assessed to present no to minimal risk of disruption compared with the current risk, but scenarios are not anticipated to change the risk to the NCF between 2050 and 2100 and between current and high emissions scenarios.

Subfunctions

The subfunctions *Provide Detection and Analytics* and *Manage Containment, Eradication, and Recovery* are the most subject to risk because of their time criticality and reliance on functioning communication to facilitate collection and analysis of data (e.g., automated alerts, threat information–sharing) and remote remediation of incidents. Some forms of incident management, such as in OT environments, might also require the physical deployment of personnel to conduct forensics, isolate systems, and place clean systems back online. *Prepare for Cyber Incidents* would not be affected because those activities (e.g., planning) can be performed prior to a climate driver occurrence. Postincident activities, such as disseminating lessons learned, are also less time critical.

Table B.62. Subfunction Risk for *Perform Cyber Incident Management Capabilities*

Subfunction	Baseline Present	Current Emissions 2050	Current Emissions 2100	High Emissions 2050	High Emissions 2100
Provide Detection and Analytics	1	2	2	2	2
Manage Containment, Eradication, and Recovery	1	2	2	2	2

NOTE: 1 = no disruption or normal operations. 2 = minimal disruption. See Chapter 2 for further explanation of ratings.

Climate Drivers

Extreme heat, flooding, severe storm systems, tropical cyclones and hurricanes, and wildfire were assessed to pose a minimal risk of disruption to the NCF. These climate drivers can impede timely detection and response to emerging cyber incidents through physical damage to supporting infrastructure and effects on the workforce (i.e., inability to access resources and systems needed to detect and respond to an incident). Demand for the function can increase as malicious actors seek to exploit an event, such as posing in emails and other digital communications as federal emergency management entities purporting to provide support to organizations and individuals affected by the event.

Table B.63. Risk from Climate Drivers for *Perform Cyber Incident Management Capabilities*

Climate Driver	Baseline	Current Emissions		High Emissions	
	Present	2050	2100	2050	2100
Drought	1	1	1	1	1
Extreme cold	1	1	1	1	1
Extreme heat	1	2	2	2	2
Flooding	1	2	2	2	2
Sea-level rise	1	1	1	1	1
Severe storm systems	1	2	2	2	2
Tropical cyclones and hurricanes	1	2	2	2	2
Wildfire	1	2	2	2	2

NOTE: 1 = no disruption or normal operations. 2 = minimal disruption. See Chapter 2 for further explanation of ratings.

Impact Mechanisms

The impact mechanisms that contribute to risk to the function are physical damage to infrastructure and facilities used to support cyber incident management. All catastrophic events and natural disasters have been used as opportunities for phishing and social engineering activities (CISA, 2022; Federal Communications Commission [FCC], 2022). Any climate event can be expected to be accompanied by cyber activity from malicious actors seeking to exploit the incident. The effects of climate change would increase demand for this NCF; however, the delays would be occasional, and routine operational needs would likely still be met.

Detection and analysis of emerging cyber incidents varies based on the security model employed: Larger organizations will typically employ internal security operations personnel, including staffing security operations centers to monitor and respond to events on corporate systems and networks. These security operations rely on active monitoring of alerts from tools on network end points and entry points. Personnel can operate in a distributed or remote fashion, potentially mitigating the effects of severe weather events that prevent access to an operations center. Smaller organizations often use a managed security services model or implement less sophisticated cybersecurity programs.

Cascading Risk

The *Perform Cyber Incident Management Capabilities* function is dependent on *Operate Core Network*; *Provide Internet Routing, Access, and Connection Services*; *Distribute Electricity*; and *Transmit Electricity* (Miro et al., 2022). Core networks and internet services are needed to receive data and alerts, conduct coordinated response, and implement mitigation procedures.

The function is interdependent with *Protect Sensitive Information* because cyber protections are required to protect such information, and *Perform Cyber Incident Management Capabilities* relies on the ability to encrypt and transmit securely sensitive information.

Strength of Evidence

The evidence for climate change's effects on the provision of cyber incident management functions is sparse. Prior natural disasters and the COVID-19 pandemic provide evidence for the increase in malicious cyber activity, which has also shown the resilience of the function in the face of chronic workforce shortages. The magnitude of the effect for the function and its subfunctions are not well understood for future climate scenarios.

Table B.64. Strength of Evidence for *Perform Cyber Incident Management Capabilities*

	Strength of Evidence
Low	Drought; extreme cold; extreme heat; flooding; sea-level rise; severe storm systems; tropical cyclones and hurricanes; wildfire

Risk Rating Interpretations

Table B.65 provides examples of how to interpret the risk ratings for this NCF.

Table B.65. Example Interpretations of Risk Ratings for *Perform Cyber Incident Management Capabilities*

Risk Rating	NCF Example
1: No disruption or normal operations	Cyber incident management continues to function as it does today, with occasional delays due to weather and operational problems.
2: Minimal disruption	There is an episodic increase of cybersecurity reporting triggered by hurricanes at local and regional levels. The time to respond to incidents increases, but incidents are still being processed at a reasonable rate.
3: Moderate disruption	Some cybersecurity incidents cannot be processed temporarily or permanently.
4: Major disruption	Most cybersecurity incident response is not viable in a timely manner, leading to cascading incidents across sectors and functions.
5: Critical disruption	Cyber incident capabilities are completely disrupted.

Protect Sensitive Information

Table B.66. Scorecard for *Protect Sensitive Information*

	National Risk Assessment				
	Baseline	Current Emissions		High Emissions	
	Present	2050	2100	2050	2100
Protect Sensitive Information	1	2	2	2	2

Description: "Safeguard and ensure the integrity of information whose mishandling, spillage, corruption, or loss would harm its owner, compromise national security, or impair competitive or economic advantage" (CISA, 2020, p. 5).

Highlights

- *Protect Sensitive Information* was assessed to be at risk of minimal disruption in 2050 and 2100 across both emissions scenarios based on risk to data centers that provide cloud services and local storage and physical records' exposure to physical damage from climate drivers. Redundancies and backups in digital records will mitigate some of the risk.
- Climate drivers that can cause significant damage to sensitive data storage facilities include flooding, severe storm systems, tropical cyclones and hurricanes, and wildfire. Drought and extreme heat put stress on cooling systems and can lead to data center or other servers to fail.
- Lack of backups, poor cybersecurity practices, operational requirements that lead to circumventing cyber and regulatory safeguards, and increases in cyber threats to sensitive data during and after climate events increase risk.
- Physical safeguards are most at risk from damage to facilities, although cyber and regulatory protections can be compromised by opportunistic malicious activity or data stewards choosing to compromise those protections to ensure timely operational response.
- Risk of disruption to the function is distributed throughout the country.

Synopsis

Climate change drivers can lead to loss, corruption, or temporary unavailability of sensitive data. Flooding, severe storm systems, hurricanes, and wildfire have the potential to damage or destroy physical records and to hinder the access to electronic records and data by causing damage to data centers and other facilities for electronic storage. Historical evidence indicates that use of digital records, such as EHRs, that are available through cloud or other distributed access means better enables the continued functioning of other NCFs (e.g., *Provide Medical Care*), as well as minimal disruption in access to such records (Abir et al., 2012; Shepard, 2017). On-premises data storage, however, is equally vulnerable to disruption from climate drivers that cause physical damage to digital media, and very large events can disrupt access even to distributed records through damage to multiple sites (Ido, Nakamura, and Nakayama, 2019). Extreme heat stresses the ability to provide adequate cooling for data centers and other forms of digital storage of sensitive data (Claburn, 2022; Quach, 2022). The use of special facilities, particularly for classified information, could also be limited by heat stresses on climate control systems. Sea-level rise is not anticipated to have a marked effect.

Extreme cold events are not projected to increase and affect the function.

Climate drivers can cause sensitive data to be unavailable, corrupted, or destroyed because of direct or indirect effects. Direct effects include flooding and high winds associated with tropical cyclones or

131

hurricanes that damage physical data stores or to digital media. Indirect effects (which are not reflected in the rating) can stem from power outages to data centers and other forms of digital storage and the loss of ability to transmit data (see "Cascading Risk"). Individuals or organizations might also consciously choose to reveal sensitive information to facilitate operational response to a disaster or stemming from malicious actors exploiting people when they are highly stressed and vulnerable to deception (Sanfilippo et al., 2020). Regulations require some sectors to develop business continuity plans that include ensuring protection of sensitive data, such as in health care as required by the Health Insurance Portability and Accountability Act of 1996 (HIPAA) (Pub. L. 104-191) and in the financial service industry (Code of Federal Regulations, Title 45, Section 164.308; Financial Industry Regulatory Authority, 2015).

Protect Sensitive Information is expected to be at risk of minimal disruption by 2050 from climate change in both the current and high emissions scenarios. These ratings are due chiefly to an expected increase in the frequency, intensity, or duration of drought, flooding, severe storm systems, tropical cyclones and hurricanes, and wildfire that could affect a larger number of sites with sensitive data. Although we assessed an increased risk of disruption between now and 2050, the risk of disruption will be minimal in 2050 and 2100 for both current and high emissions scenarios because of the geographic distribution of data centers and storage facilities for sensitive data. Cyber protections for sensitive data will experience some effect, particularly if users and other stewards of sensitive data choose to compromise the implementation of these protections for operational reasons (e.g., deciding to share sensitive data in a less secure manner to facilitate operations that require timely response). Regulatory protections for sensitive information, such as financial data, critical infrastructure information, or national security information, would be at risk of minimal disruption.

Subfunctions

Provide Physical Protections for Sensitive Information and *Provide Cyber Protections for Sensitive Information* are at risk due to damage to facilities and loss of access to data in both physical and digital forms. Improved cyber protections, including the likely deployment of postquantum cryptography, by 2050 (or 2100 at the latest) will face similar challenges due to climate drivers' likely effect on the physical infrastructure for storing and transmitting sensitive data. *Provide Regulatory Protections for Sensitive Information* was assessed to also be at risk due to failure of protections or malicious actors taking advantage of a disaster, but the climate drivers themselves were assessed to pose only a minimal risk of disruption to responsible parties' ability to develop and maintain regulatory protections, such as educating the workforce on legal, regulatory, and policy requirements for *Protect Sensitive Information*.

Table B.67. Subfunction Risk for *Protect Sensitive Information*

	Baseline	Current Emissions		High Emissions	
Subfunction	Present	2050	2100	2050	2100
Provide Physical Protections for Sensitive Information	1	2	2	2	2
Provide Cyber Protections for Sensitive Information	1	2	2	2	2
Provide Regulatory Protections for Sensitive Information	1	2	2	2	2

NOTE: 1 = no disruption or normal operations. 2 = minimal disruption. See Chapter 2 for further explanation of ratings.

Climate Drivers

Extreme heat and drought were assessed to bring risk for the storage and processing of sensitive data in digital formats in data centers which can fail due to overheating. Flooding, severe storm systems, tropical cyclones and hurricanes, and wildfire can cause physical damage to data centers, other forms of digital storage, and physical records, rendering them unavailable or damaging records. The availability of backups of different types (full, incremental, or differential) or data replication will depend on the form and location of the backups. Cloud customers, for example, may not specify that data replication must occur across data center regions instead of within a region or the same data center as the primary storage, which means a climate driver can lead to loss of primary and backup files at the same time.

Table B.68. Risk from Climate Drivers for *Protect Sensitive Information*

	Baseline	Current Emissions		High Emissions	
Climate Driver	Present	2050	2100	2050	2100
Drought	1	2	2	2	2
Extreme cold	1	1	1	1	1
Extreme heat	1	2	2	2	2
Flooding	1	2	2	2	2
Sea-level rise	1	1	1	1	1
Severe storm systems	1	2	2	2	2
Tropical cyclones and hurricanes	1	2	2	2	2
Wildfire	1	2	2	2	2

NOTE: 1 = no disruption or normal operations. 2 = minimal disruption. See Chapter 2 for further explanation of ratings.

Impact Mechanisms

The most important mechanism in affecting the ability to carry out the *Protect Sensitive Information* function is physical damage to supporting infrastructure and data itself, whether in physical or digital form.

Cascading Risk

Distribute Electricity and *Transmit Electricity* are upstream functions on which *Protect Sensitive Information* relies. For example, in 2021, power outages caused 43 percent of data center downtime (Bizo et al., 2021; Rich Miller, 2022). Data centers also rely on *Supply Water* to ensure adequate cooling. Lack of adequate cooling caused 14 percent of data center downtime in 2021 (Bizo et al., 2021). Drought will affect data centers in the western United States, where data centers are in already water-stressed areas (Siddik, Shehabi, and Marston, 2021). Cloud-based storage and processing of sensitive data also rely on *Provide Internet Routing, Access, and Connection Services.*

Downstream functions that rely on *Protect Sensitive Information* include *Maintain Access to Medical Records*, *Enforce Law*, and *Operate Government* because of the sensitive data–protection requirements for each of those functions, whether due to regulatory requirements (e.g., HIPAA for medical records) or to protect sources, methods, and investigations (e.g., case files for *Enforce Law* or intelligence for *Operate Government*).

Perform Cyber Incident Management Capabilities is interdependent with *Protect Sensitive Information* because the former encompasses detection and remediation of incidents affecting *Protect Sensitive Information* and the latter employs data-protection standards and threat information.

Strength of Evidence

The evidence for climate drivers' effect on *Protect Sensitive Information* was assessed as medium for drought, extreme heat, severe storm systems, and tropical cyclones and hurricanes and low for the remaining climate drivers. The available evidence is largely on the potential effects on hardware and physical records, while less evidence is available to address cyber and regulatory subfunctions. There is a reasonable body of literature particular to the role and challenges of accessing health and medical data in natural disasters, but other forms of sensitive data (e.g., classified information, financial data) have been studied less. Industry surveys and contemporaneous reporting have addressed climate drivers' effects, particularly drought and extreme heat, on data centers generally—logically, this can be extrapolated to assess potential effects on data centers used for storing and processing sensitive data. Reliable and evidence-based projections for the form and function of *Protect Sensitive Information* in 2050 and 2100 are lacking.

Table B.69. Strength of Evidence for *Protect Sensitive Information*

Strength of Evidence	
Medium	Drought; extreme heat; severe storm systems; tropical cyclones and hurricanes
Low	Extreme cold; flooding; sea-level rise; wildfire

Risk Rating Interpretations

Table B.70 provides examples of how to interpret the risk ratings for this NCF.

134

Table B.70. Example Interpretations of Risk Ratings for *Protect Sensitive Information*

Risk Rating	NCF Example
1: No disruption or normal operations	Protection of sensitive information functions as it does today, which can include accidental or malicious breach of sensitive data, but sensitive data are largely available for use and not affected by climatic events.
2: Minimal disruption	Sensitive information is compromised in some fashion that leads to minimal disruptions in provision of services that rely on sensitive information (e.g., health care) or financial loss to people who are subject to fraud, such as identity theft.
3: Moderate disruption	Sensitive information is compromised such that a large component of a class or classes of sensitive information are not available or trustworthy for more-extended time periods.
4: Major disruption	Sensitive information is compromised across an entire class or classes of information (e.g., health information, credit information).
5: Critical disruption	Sensitive information is compromised permanently and irreparably, rendering systems that rely on processing or making such information available useless.

Conduct Elections

Table B.71. Scorecard for *Conduct Elections*

	National Risk Assessment				
	Baseline	Current Emissions		High Emissions	
	Present	2050	2100	2050	2100
Conduct Elections	1	2	2	2	2

Description: "Conduct elections, including managing voter registration and rolls, voting infrastructure, polling places, vote counting, and certifying and publishing election results" (CISA, 2020, p. 4).

Highlights

- *Conduct Elections* was assessed to be at risk of minimal disruption from climate drivers in 2050 and 2100. The risk of disruption arises from the constrained time periods for conducting aspects of the function but is mitigated by the disaggregated oversight and execution of the function across roughly 8,800 jurisdictions in the United States.
- Extreme heat, flooding, tropical cyclones and hurricanes, and wildfire are the most-concerning climate drivers that can affect the function, particularly no-notice or short-warning events that are difficult to recover from quickly within legal timelines for casting votes and for tabulation and certification of results.
- The factors contributing to *Conduct Elections* vulnerabilities are physical damage to critical equipment and loss of access to facilities (e.g., registrars' offices, polling precincts).
- The *Conduct Election Day Activities* and *Post-Election Activities* subfunctions are potentially more vulnerable because of the time periods in which they must occur.
- Risk is distributed broadly across the United States, with some regions more susceptible to certain climate drivers than others (e.g., wildfires in the West, hurricanes in the Southeast and on the Atlantic Coast).

Synopsis

Climate change is anticipated to lead to an increase in natural hazards that have the potential to affect state and local election officials' ability to conduct elections. Climate drivers, depending on timing, can lead to disruption of preelection activities (e.g., voter registration), election activities (during primaries and general elections), and postelection activities (e.g., vote tabulation and certification). Wildfire in western states can displace voters during critical time periods or lead to damage or unavailability of polling places (or ballot drop boxes in Washington and Oregon, which are vote-by-mail states). Hurricanes can cause similar effects in states from Texas to Florida and along the Atlantic Coast as far north as Maine. Hurricane season in the Atlantic runs from June through November, the prime time period for early voting and Election Day voting, which means that a hurricane or tropical cyclone could arrive during the voting time period and disrupt voters' ability to cast their votes, election workers to run voting centers or precincts, and the tabulation and certification of election results. Flooding and other severe storm systems, particularly short- or no-notice events (e.g., derechos) when they coincide with critical election function time periods, such as Election Day, can inflict widespread damage in short time periods with little time to recover and carry out election functions properly. Extreme heat will place additional strains on climate controls in election offices and polling places and could lead to decreased voter turnout, particularly in precincts and jurisdictions that experience long waiting times at the polls (Morris and Miller, 2023).

136

The climate drivers' effect on *Conduct Elections* will not be experienced uniformly across the United States for two reasons: The types of climate-related events will vary geographically, and the conduct of elections is not uniform across the estimated 8,800 jurisdictions across the United States. The timing of a natural hazard and its location will have a large bearing on the degree to which the event represents risk to the conduct of elections. There are some historical precedents to illustrate this. Natural hazards have led to canceled elections, such as in New Orleans, Louisiana, in 2005 after Hurricane Katrina. Elections have also been postponed as occurred in Dade County, Florida, in 1992 after Hurricane Andrew (Task Force on Emergency Preparedness for Elections, 2017). New York City's primaries were delayed in 2001 because of the terrorist attacks that occurred shortly after polls opened (Morley, 2018). Hurricane Sandy made landfall on October 29, 2012, one week prior to the 2012 general election, which led to many states implementing emergency procedures. New York and New Jersey relocated polling places. Connecticut extended its voter registration deadline, and Maryland, New Jersey, New York, and Pennsylvania extended deadlines for requesting absentee ballots (Task Force on Emergency Preparedness for Elections, 2017). Some emergency provisions, such as allowing voters to fax in requests for absentee ballots or cast ballots electronically, were subsequently challenged in court (Morley, 2018). States and jurisdictions have limited options for delaying or postponing elections, but these options vary. There are also time constraints. As of 2022, deadlines for certifying election results varied by state, with five (Delaware, Louisiana, Oklahoma, South Dakota, and Vermont) requiring certification within one week of an election. Twenty-two states require certification by the end of November, 18 states in December, and five states having no fixed date. However, for the presidential election, the Electoral College electors must cast their votes on the first Monday after December 12, which constrains how long a state can delay voting in a presidential election (Ballotpedia, undated).

Conduct Elections is expected to be at risk of no to minimal disruption from climate change in both the current and high emissions scenarios by 2050. Although most climate drivers are expected to increase in frequency, intensity, or duration, climate drivers with warning times (e.g., hurricanes) mean that election officials can implement mitigations, such as moving polling places or registration offices. Equipment such as vote-casting and vote tabulation machines can be moved to safety, and alternative voting centers can be established, although there is some evidence that such moves can negatively affect voter turnout (Delaware County, undated; Morris and Miller, 2023). The NCF was assessed as being at risk of minimal disruption through 2100.

Key Considerations for *Conduct Elections*

Election law and procedures are constantly changing. State legislatures are expected to continue to change aspects of elections, such as early-voting options, polling precinct requirements, and emergency authorities. These changes will vary by state, and the potential consequences should a climate driver occur during a critical election time period cannot be projected from current circumstances. We assessed that legal challenges to election laws and changes to election procedures will continue in the future, but it is impossible to determine the outcomes of those challenges in advance.

Subfunctions

Each of the subfunctions for this NCF—*Conduct Pre-Election Activities*, *Conduct Election Day Activities*, and *Conduct Post-Election Activities*—can experience a variety of effects, depending on the type and geographic area in which the climate drivers occur. The mitigations will vary depending on local and state laws, the adequacy of prior emergency planning, and the timing of the event in relation to the election activity. Preelection activities, such as voter registration, could experience effects of a weather event that causes the closing of election offices, such as occurred during Hurricane Sandy (Morley, 2018; Task Force on Emergency Preparedness for Elections, 2017). This could affect voters' ability to update their registration information and find out election information. Preparation of election equipment, such as programming vote-casting machines and electronic pollbooks, could be affected by physical damage to storage facilities and stored equipment or during transport. Election activities could be most affected in the states that do not allow or have brief early-voting time periods. These include the states of Alabama, Connecticut, Indiana, Mississippi, New Hampshire, and South Carolina (Kavanagh, Gibson, and Cherney, 2020; National Conference of State Legislatures [NCSL], 2020). In some cases, states can allow displaced voters to cast provisional ballots, but this practice can limit voters' choices if they cast their ballots in jurisdictions with different contests from those in which they would normally be eligible to vote (Morley, 2018). Postelection activities can equally be affected, particularly where state laws require certification of election results and conduct of postelection audits during constrained time periods, as noted above. As of 2017, only a dozen states had laws authorizing delays, suspensions, or canceling of elections because of an emergency (Task Force on Emergency Preparedness for Elections, 2017).

Table B.72. Subfunction Risk for *Conduct Elections*

Subfunction	Baseline Present	Current Emissions 2050	Current Emissions 2100	High Emissions 2050	High Emissions 2100
Conduct Pre-Election Activities	1	2	2	2	2
Conduct Election Day Activities	1	2	2	2	2
Conduct Post-Election Activities	1	2	2	2	2

NOTE: 1 = no disruption or normal operations. 2 = minimal disruption. See Chapter 2 for further explanation of ratings.

Climate Drivers

Most climate drivers were assessed to pose a risk of minimal disruption to the conduct of elections. Severe storm systems, hurricanes, wildfire, and flooding can cause physical damage to polling places, close election offices, and lead to power outages (an indirect effect), affecting online services and voter registration, polling places, and vote tabulation. These events can also displace poll workers and volunteers and potentially decrease demand for the function as prospective voters deal with climate drivers' consequences on their lives. Extreme heat can strain resources, such as cooling for voting centers. Drought is not anticipated to have an effect given that drought conditions do not affect the conduct of any election functions directly. Extreme cold is not projected to increase in the climate scenarios and is therefore not likely to affect election functions. Sea-level rise could affect some low-lying areas but with sufficient warning time to allow for mitigations, such as moving polling places to higher ground.

Table B.73. Risk from Climate Drivers for *Conduct Elections*

Climate Driver	Baseline Present	Current Emissions 2050	Current Emissions 2100	High Emissions 2050	High Emissions 2100
Drought	1	1	1	1	1
Extreme cold	1	1	1	1	1
Extreme heat	1	2	2	2	2
Flooding	1	2	2	2	2
Sea-level rise	1	1	1	1	1
Severe storm systems	1	2	2	2	2
Tropical cyclones and hurricanes	1	2	2	2	2
Wildfire	1	2	2	2	3

NOTE: 1 = no disruption or normal operations. 2 = minimal disruption. 3 = moderate disruption. See Chapter 2 for further explanation of ratings.

Impact Mechanisms

The most-relevant impact mechanisms are physical damage and disruption to infrastructure and facilities supporting elections, including polling places and government offices. A significant portion of election processes are supported by volunteer poll workers, who can be displaced or otherwise affected by climate drivers and might not be able to access facilities to conduct election activities. Climate drivers can, on the other hand, reduce the demand for the function as potential voters are focused on dealing with the effects of an event and not on registering to vote or casting their ballots (Gomez, Hansford, and Krause, 2007; Morris and Miller, 2023; Stein, 2015). In such circumstances, election officials can provide for routine operations with some changes in place or timing of election functions, but the demand for the function might decrease.

139

Cascading Risk

The most-critical upstream functions that would affect *Conduct Elections* are *Provide Internet Routing, Access, and Connection Services*, *Distribute Electricity*, and *Transmit Electricity*. *Provide Internet Routing, Access, and Connection Services* is required across the subfunctions but will vary by jurisdiction. Jurisdictions that employ electronic pollbooks that require access to a centralized database (as opposed to having a static copy stored locally would be affected if *Provide Internet Routing, Access, and Connection Services* is not available. Election night reporting can also be affected but might have access to multiple forms of connection (e.g., through other forms of telecommunication). Most jurisdictions rely on electricity to carry out election functions, including maintaining voter records, running polling places, and conducting vote tabulation.

Operate Government is a downstream function that would be affected by a failure or severe disruption to *Conduct Elections*. *Operate Government* in the United States is founded on the democratic election of public officials at every level of government. A failure of *Conduct Elections* would disrupt *Operate Government*, particularly where disputes arise about the legitimacy of elected officials or about previously elected officials remaining in office past their terms.

Strength of Evidence

The strength of evidence on climate drivers' effects on *Conduct Elections* is low because it is based primarily on a small number of historical case studies and analysis of other significant events that have affected elections (e.g., terrorist attacks). The evidence for climate drivers' effect on *Conduct Elections* is most apparent for hurricanes, which is therefore rated as a medium strength of evidence. Although it is logical that extreme weather events can disrupt Election Day as well as pre– and post–Election Day activities, effects could depend largely on timing. There is less well-established literature on climate change's long-term effects on state and local officials' ability to conduct elections.

Table B.74. Strength of Evidence for *Conduct Elections*

Strength of Evidence	
Medium	Tropical cyclones and hurricanes
Low	Drought; extreme cold; extreme heat; flooding; sea-level rise; severe storm systems; wildfire

Risk Rating Interpretations

Table B.75 provides examples of how to interpret the risk ratings for this NCF.

Table B.75. Example Interpretations of Risk Ratings for *Conduct Elections*

Risk Rating	NCF Example
1: No disruption or normal operations	Elections processes are executed according to established procedures, with occasional disruptions or errors being identified and addressed in a timely manner.
2: Minimal disruption	Election procedures, such as voter registration or vote-casting, are temporarily disrupted in precincts or jurisdictions but are recoverable within legal time limits for completing election procedures.
3: Moderate disruption	Multiple jurisdictions across multiple states are unable to accomplish election procedures within legal time limits. Potential voters are unable to register to vote within time limits or are unable to cast their ballots.
4: Major disruption	Many jurisdictions across multiple states are unable to accomplish election procedures within legal time limits, leading to inability to conduct elections and allow the public to elect candidates for office.
5: Critical disruption	Most or all jurisdictions are unable to carry out election procedures, so elections are not held within legal time constraints for multiple contests across the country. This can affect elections for federal, state, and local offices.

141

Operate Government

Table B.76. Scorecard for *Operate Government*

	National Risk Assessment				
	Baseline	**Current Emissions**		**High Emissions**	
	Present	**2050**	**2100**	**2050**	**2100**
Operate Government	1	2	2	2	2

Description: "Carry out legislative, judicial, and executive government missions, including activities related to developing and enforcing codes, ordnances, rules, regulations, and laws; collecting taxes and revenues; managing records, budgets, and finances; and providing public services" (CISA, 2020, p. 4).

Highlights

- For 2050 and 2100, *Operate Government* is expected to be at risk of minimal disruption under the current and high emissions scenarios because of the increasing risk posed by all climate drivers (other than extreme cold), each of which is projected to increase in frequency, intensity, or duration.
- As climate change increases the frequency and severity of natural hazards, these events will cause physical damage to public and private infrastructure that requires repair; contribute to increases in social problems, such as crime and homelessness; and contribute to risks to physical and mental health.
- Climate drivers' effects will likely increase demand and create physical damage disruptive to *Operate Government* functions, such as social services provided through the *Provide Executive Services* subfunction, legislative action provided through the *Legislate* subfunction, and judicial services provided through the *Adjudicate* subfunction.
- As a function distributed throughout the country, *Operate Government* faces similar risks in all parts of the United States.

Synopsis

Operate Government is vulnerable to climate change in that increases in natural hazards are projected to increase demand for *Operate Government* and its legislative, judicial, and executive branch subfunctions. The increased frequency, intensity, or duration of weather events will likely increase demand for social services provided through the *Provide Executive Services* subfunction and for legislative action provided through the *Legislate* subfunction. Climate change will also likely increase the frequency and types of certain crimes, leading to greater demand for the *Adjudicate* subfunction. The increasing frequency, intensity, or duration of weather events could also lead to physical damage of public-sector infrastructure, contributing to the risk of disruption of this NCF.

For the current time period, *Operate Government* is operating normally. For 2050 and 2100, *Operate Government* is expected to be at risk of minimal disruption under both the current and high emissions scenarios. This rating increase is due to the increasing risk posed by all of the climate drivers (other than extreme cold), which are projected to increase in frequency, intensity, or duration. Because *Operate Government* is a distributed function that is provided throughout the country—and that relies on a diverse and distributed collection of facilities and infrastructure—we project that, although the climate drivers will contribute to minimal disruption of this NCF, it will be able to absorb some risk, leading to operational challenges to states and localities but not a threat to core government services.

Subfunctions

Each of the three subfunctions for this NCF—*Legislate, Adjudicate,* and *Provide Executive Services*—is expected to be at risk of minimal disruption in 2050 and 2100 due to the increased demand associated with natural disasters and physical damage to government infrastructure. As climate change increases the frequency and severity of natural hazards, these events will cause physical damage to public and private infrastructure that requires repair; contribute to increases in social problems, such as crime and homelessness; and contribute to risks to physical and mental health. These challenges will increase demand for each subfunction. The *Legislate* subfunction will likely face increased demand as federal, state, and local legislative bodies act to address these more-frequent events. The judicial system will face more demand as crime increases, putting pressure on the *Adjudicate* subfunction. Executive branch agencies at the federal, state, and local levels will face a greater need to provide executive services that promote public well-being and address social and individual needs. Climate change will also likely put pressure on public-sector budgets. At the federal level, the U.S. Government Accountability Office (GAO) has reported that the government faces serious fiscal risk from climate change (Gómez, 2019b). We are unaware of a similar analysis of state and local budgets, but these organizations will likely face similar fiscal and budgetary challenges.

Table B.77. Subfunction Risk for *Operate Government*

	Baseline	Current Emissions		High Emissions	
Subfunction	Present	2050	2100	2050	2100
Legislate	1	2	2	2	2
Adjudicate	1	2	2	2	2
Provide Executive Services	1	2	2	2	2

NOTE: 1 = no disruption or normal operations. 2 = minimal disruption. See Chapter 2 for further explanation of ratings.

Climate Drivers

Except for extreme cold, each climate driver will likely affect the *Operate Government* NCF by contributing to the need for executive services, the judicial system, and legislatures to address the effects of extreme weather events. However, the extent to which each climate driver increases demand for the NCF or physical harm is not well documented in the literature.

Flooding, severe storm systems, and tropical cyclones and hurricanes, exacerbated by sea-level rise, have the capacity to cause widespread damage or destruction to homes, businesses, and other facilities. Similarly, wildfire can cause widespread damage, requiring communities and families to rebuild. These types of effects will likely lead to demand for government programs that support the rebuilding and repair of homes, buildings, and infrastructure (Congressional Budget Office [CBO], 2016). Extreme hot weather can cause increases in the frequency and severity heat-related illnesses, and studies have also shown that hotter weather is correlated with an increase in crime and terrorist attacks. Less clear is whether there is a causal relationship and what the magnitude of that effect is (Abbott, 2008; Goin, Rudolph, and Ahern, 2017). But an increase in crime and terrorism will likely lead to increased demand for government services focused on preventing crime and moving alleged criminals through

143

the judicial system. High levels of energy usage associated with heat waves can cause blackouts and damage to power transmission facilities, which will likely lead to calls for government intervention to reduce the effect of these events. Finally, drought can negatively affect agricultural production and ranching and lead to an increase in the frequency of dust storms, which can cause respiratory disorders (Reed and Nugent, 2018). This will likely increase demand on the *Operate Government* NCF as communities and individuals seek assistance to reduce health effects and to support the agriculture and ranch sector (for example, through the federal crop insurance program) (Diffenbaugh, Davenport, and Burke, 2021; Gómez, 2019b; Office of Management and Budget [OMB], 2022).

Although extreme cold can contribute to health problems that could increase demand for government services, that driver is not expected to increase over the current level and therefore it is not expected to affect this NCF.

Table B.78. Risk from Climate Drivers for *Operate Government*

Climate Driver	Baseline Present	Current Emissions 2050	Current Emissions 2100	High Emissions 2050	High Emissions 2100
Drought	1	2	2	2	2
Extreme cold	1	1	1	1	1
Extreme heat	1	2	2	2	2
Flooding	1	2	2	2	2
Sea-level rise	1	2	2	2	2
Severe storm systems	1	2	2	2	2
Tropical cyclones and hurricanes	1	2	2	2	2
Wildfire	1	2	2	2	2

NOTE: 1 = no disruption or normal operations. 2 = minimal disruption. See Chapter 2 for further explanation of ratings.

Impact Mechanisms

The effects of climate change will likely affect the *Operate Government* NCF through two mechanisms: increasing demand and causing physical damage. On increases in demand, a large literature identifies natural disasters' negative effects on well-being. As the intensity and frequency of extreme weather increases because of climate change, these effects will likely increase, leading to increasing demand for executive services and legislative action to address them.

At the federal level specifically, GAO has identified several areas in which demand for federal services—and federal fiscal exposure—is likely to increase because of climate change (Gómez, 2019b). For example, the rising number and intensity of natural disasters are expected to increase the need for federal disaster aid. As GAO noted, the 2017 hurricane and wildfire season created a significant increase in demand for federal disaster services. Demand for these services will grow as extreme weather events become more frequent and intense. Similarly, federal insurance programs—such as those for crops and floods—will face increasing demand due to the risk that climate change creates for insured property and crops (Gómez, 2019b). Similar trends are likely at the SLTT levels.

144

Similarly, the effects and aftereffects of natural hazards affected by climate change are likely to increase demand for legislatures to take action to address them, as suggested by recent history. The National Conference of State Legislatures (NCSL) reported that state legislatures across the United States had passed a large volume of legislation in recent years in response to natural disasters. For example, in 2019 and 2020, states enacted more than 200 bills addressing natural disasters—70 of which improved infrastructure resilience and approximately 100 of which reformed aspects of emergency management (NCSL, 2020). This trend will likely continue as the risk of natural disasters increases because of climate change.

Climate change will also likely increase the frequency and types of certain crimes, leading to greater demand for the *Adjudicate* subfunction of this NCF at all levels of government. The available research indicates increases in violent and property crimes due to climate change and its social and environmental effects (Ranson, 2014). For example, researchers in a 2020 study found that the United States should expect somewhere between 2.3 million and 3.2 million additional violent crimes between 2020 and 2099 as a result of climate change (Harp and Karnauskas, 2020).

Climate change will also likely affect this NCF by causing physical damage to government infrastructure. At the federal level, as GAO has reported, the federal government "owns and operates hundreds of thousands of facilities and manages millions of acres of land that could be affected by a changing climate" (Gómez, 2019b, p. 8). In an assessment of federal facilities, the Office of Management and Budget (OMB) and NOAA identified approximately 10,250 federal buildings and structures that would be inundated or severely affected by typical high tide under an 8-foot sea-level rise scenario. In a worst-case scenario of a 10-foot sea-level rise, more than 12,195 federal buildings and structures would be inundated (OMB, 2022). Some disruption of government services is likely to result under these scenarios. Although similar analysis of SLTT-owned property has not been identified, SLTT governments likely face similar challenges that will be exacerbated by climate change.

Cascading Risk

Operate Government depends on the rule of law to function and provide services. Upstream NCFs on which *Operate Government* depends include the *Enforce Law* and *Provide Public Safety* NCFs, which support *Operate Government* by providing consistency and accountability in the application of laws. The *Conduct Elections* NCF is also critical because the failure of elections leads to illegitimacy, uncertainty, or the inability to *Operate Government* because public officials will not have been elected. The *Operate Government* NCF also relies on *Distribute Electricity* and *Provide Internet Based Content, Information, and Communication Services* to manage operations and provide services. NCFs providing these functions are necessary for the *Operate Government* NCF.

Strength of Evidence

Although the need for executive actions (e.g., services that immediately address public well-being and social needs), legislative activity (e.g., legislation to address future climate risk), or judicial adjudication (e.g., criminal trials) has been shown to increase following natural disasters, the magnitude by which specific climate drivers will increase demand for the NCF is not well documented. Similarly, there is limited evidence of climate change effects due to physical damage or disruption of government functions. Although GAO analyzed climate change's effects on the demand for federal services and federal fiscal exposure, it did not comprehensively document the effects of individual hazards and was limited to the federal level of government (Gómez, 2019b). Evidence of climate change effects on state and local government is less well documented, but these organizations will likely face similar fiscal and budgetary challenges.

Table B.79. Strength of Evidence for *Operate Government*

Strength of Evidence	
Medium	Drought; extreme heat; flooding; sea-level rise; severe storm systems; tropical cyclones and hurricanes; wildfire
Low	Extreme cold

Risk Rating Interpretations

Table B.80 provides examples of how to interpret the risk ratings for this NCF.

Table B.80. Example Interpretations of Risk Ratings for *Operate Government*

Risk Rating	NCF Example
1: No disruption or normal operations	Government organizations continue to function as they do today, with occasional delays due to capacity constraints, weather, and operational problems.
2: Minimal disruption	Government organizations in some states and localities experience time periods in which they face operational challenges, such as staff shortages and temporary disruption of systems and networks, but they can still carry out core government missions and provide public services, except for brief and localized disruptions.
3: Moderate disruption	Government organizations face operational challenges at the state level, such as staff shortages and temporary disruption of systems and networks. Disruptions to core government missions and provision of public services are more sustained and geographically widespread.
4: Major disruption	Government organizations throughout the country and at the federal level face serious operational challenges that significantly affect their ability to provide public services and carry out core government missions. These challenges include severe workforce shortages, widespread disruption of systems and networks, and budget constraints.
5: Critical disruption	Government organizations throughout the country and at the federal level cannot fulfill core operational responsibilities. Provision of public services and core government missions is severely degraded.

146

Preserve Constitutional Rights

Table B.81. Scorecard for *Preserve Constitutional Rights*

	National Risk Assessment				
	Baseline	Current Emissions		High Emissions	
	Present	2050	2100	2050	2100
Preserve Constitutional Rights	1	2	2	2	2

Description: "Secure the principles of freedom and independence and maintain the structures of American government through the protection of rights and processes prescribed in the U.S. Constitution" (CISA, 2020, p. 5).

Highlights

- By 2050, and continuing until the end of the century, natural hazards affected by climate change could increase demand for the legal, judicial, and law enforcement systems and procedures that preserve constitutional rights in the United States.
- Increased demand for this NCF could result from changing patterns of crime and additional constitutional litigation related to climate change.
- By 2050, damage to the physical infrastructure required for the routine operations of the legal, judicial, and law enforcement systems could also impede those systems' ability to protect constitutional rights, although disruptions are expected to remain temporary and localized.

Synopsis

Preserve Constitutional Rights includes critical legal, judicial, and law enforcement subfunctions, such as amending the Constitution, preserving civil rights and liberties, and using the court system to adjudicate disputes in a manner consistent with the Constitution. Although direct research on climate change's effect on the preservation of constitutional rights is limited, studies on the relationship between climate change and crime, court records from constitutional litigation related to climate change, and the historical record of physical damage to NCF infrastructure from natural hazards indicate that this NCF is vulnerable to six climate drivers (drought, extreme heat, flooding, sea-level rise, tropical cyclones and hurricanes, and wildfire) under current and high emissions scenarios (see, e.g., Agnew, 2012; *Cole v. Collier*, 2017; Interagency Security Committee, 2015; *Juliana v. United States*, 2019).

Climate change is likely to affect this NCF through two impact mechanisms. First, changing economic, social, or political conditions that are wholly or partly attributable to climate change could increase demand for the legal, judicial, and law enforcement systems that protect constitutional rights in the United States. Changing conditions could include rising crime linked to climate change and increasing climate change–related constitutional lawsuits. Second, damage to the physical infrastructure required for the routine operations of the judicial and law enforcement systems could impede those systems' ability to provide legal services and protect constitutional rights.

For both the current and high emissions scenarios, *Preserve Constitutional Rights* is expected to continue normal operations during the present-day baseline time period. *Preserve Constitutional Rights* was assessed to be at risk of minimal disruption from rising demand and physical damage for both the current and high emissions scenarios in the 2050 and 2100 time periods. Although climate change–affected natural hazards are expected to affect the courts and other constitutional and legal systems and institutions, the NCF's core infrastructure and assets, which are dispersed across regions, facilities, and systems, are expected to be able to cope with increasing demand and instances of physical damage without major disruptions to the protection of constitutional rights. As a result, the NCF is expected to generally meet routine operational demands, although there are likely to be temporary, localized delays.

Subfunctions

Three subfunctions of *Preserve Constitutional Rights—Protect Civil Rights and Civil Liberties, Address Issues of Constitutionality and Compliance*, and *Adjudicate Claims Involving Civil Rights and Civil Liberties*—face additional future risk of disruption due to the likely effects of climate drivers under both current and high emissions scenarios. Operations for these subfunctions rely on systems, infrastructure, and personnel that are expected to be in increasing demand and increasingly vulnerable to localized degradation and disruption from natural hazards affected by climate change over the coming century—namely, drought, extreme heat, flooding, sea-level rise, tropical cyclones and hurricanes, and wildfire. The *Protect Civil Rights and Civil Liberties* subfunction could experience increasing demand as the constitutional protections of the criminal justice system are affected by climate change–related changes in crime levels. Existing research suggests that some climate drivers, such as extreme heat, flooding, tropical cyclones and hurricanes, and wildfire will exacerbate individual and group stressors and strains that promote crime and lead to an increase in the number of crimes, including murder, assault, and burglary, that are committed (Agnew, 2012; Ranson, 2014). Increases in crime are likely to lead to increases in demand for the constitutional systems and procedures that protect the due-process rights of the accused. In addition, physical infrastructure, such as courthouses, jails, and legal offices, can be damaged by natural hazards, such as tropical cyclones and hurricanes or flooding.

Demand for the *Address Issues of Constitutionality and Compliance* and *Adjudicate Claims Involving Civil Rights and Civil Liberties* subfunctions is likely to increase as constitutional claims related to effects from climate change are presented, argued, and adjudicated in court. Data collection on current climate-related litigation indicate a wide variety of constitutional topics that could be implicated in future lawsuits, including allegations that harms from climate change violate plaintiffs' constitutional due-process rights (Sabin Center for Climate Change Law, Columbia Law School, and Arnold and Porter, undated). Both subfunctions are likely to face future physical disruption due to climate change as courthouses and other physical infrastructure experience damage from climate drivers. Past natural hazards, such as flooding and tropical cyclones and hurricanes, have caused physical damage to courthouses (Sloan, 2021; U.S. Courts, 2018). Court closures due to climate change–affected natural hazards could also delay the disposition of constitutional litigation. For example, federal courts in Louisiana and some areas of Mississippi were forced to close for several days after Hurricane Ida made landfall in 2021 (Sloan, 2021; U.S. Courts, 2021). The *Amend the Constitution* subfunction is less likely to face climate change–related effects because the constitutional amendment process is less exposed to natural hazards affected by climate change, generally being conducted infrequently and in only a few physical locations (i.e., state legislatures, the U.S. Capitol, and, potentially, constitutional conventions).

Table B.82. Subfunction Risk for *Preserve Constitutional Rights*

	Baseline	Current Emissions		High Emissions	
Subfunction	Present	2050	2100	2050	2100
Amend the Constitution	1	1	1	1	1
Protect Civil Rights and Civil Liberties	1	2	2	2	2
Address Issues of Constitutionality and Compliance	1	2	2	2	2
Adjudicate Claims Involving Civil Rights and Civil Liberties	1	2	2	2	2

NOTE: 1 = no disruption or normal operations. 2 = minimal disruption. See Chapter 2 for further explanation of ratings.

Climate Drivers

Six climate drivers—drought, extreme heat, flooding, sea-level rise, tropical cyclones and hurricanes, and wildfire—are expected to pose risk of disruption to *Preserve Constitutional Rights*.

Some climate drivers, such as flooding, sea-level rise, tropical cyclones and hurricanes, and wildfire, could cause physical damage to the judicial, legal, and legislative branch institutions through which constitutional rights are preserved. There is evidence from the historical record of natural hazards' effect on government operations generally and on the judicial infrastructure that considers and adjudicates constitutional claims specifically (Interagency Security Committee, 2015). Flooding can damage judicial facilities and cause them to delay operations (Commonwealth of Kentucky, undated). Sea-level rise can damage judicial facilities through both high-tide flooding and flooding due to storm surges (Lindsey, 2022). According to OMB and NOAA estimates, a sea-level rise of 8 feet would cause more than 10,000 federal facilities to be flooded or disrupted by high-tide flooding alone (OMB, 2022). In addition, tropical cyclones and hurricanes can damage judicial branch physical infrastructure (Wellborn, 2007). Wildfire can damage judicial branch facilities and cut connections between judicial branch workforce and work locations when combustible building materials are "ignited by embers, firebrands, flames, radiant energy, and convective energy" (Interagency Security Committee, 2015, p. 16).

Some climate drivers, such as flooding, drought, sea-level rise, wildfire, tropical cyclones and hurricanes, and extreme heat, could also increase demand for *Preserve Constitutional Rights*. Research indicates that extreme heat, flooding, tropical cyclones and hurricanes, and wildfire could exacerbate individual and group stresses and strains and create favorable conditions for criminal activity (Ranson, 2014). For example, temperature increases, such as those that occur during extreme heat events, have been projected to lead to increases in both violent and property crimes (Agnew, 2012). Rising crime is likely to increase demand for *Preserve Constitutional Rights* because people accused of crimes are afforded a variety of constitutional due-process protections. Some climate drivers, such as wildfire, flooding, sea-level rise, extreme heat, hurricanes, and drought, could also increase demand by spurring new constitutional litigation. Existing data on constitutional litigation related to climate change indicate a variety of constitutional topics that are implicated by climate change (Sabin Center for

Climate Change Law, Columbia Law School, and Arnold and Porter, undated). For example, lawsuits filed in federal court have alleged that natural hazards affected by climate change, including wildfire, flooding, sea-level rise, hurricanes, and drought, are currently causing or are likely to cause harms (including psychological, medical, and financial harms) to plaintiffs that affect their constitutional due-process rights (*Animal Legal Defense Fund v. United States*, 2018; *Juliana v. United States*, 2019). A lawsuit filed in federal court in 2014 alleged that extreme heat conditions in a Texas state prison violated the Eighth Amendment's prohibition of cruel and unusual punishment, with the court's opinion acknowledging that climate research indicated that heat waves were likely to become more frequent and severer in the future (*Cole v. Collier*, 2017). The projected increase in the frequency, duration, and severity of natural hazards affected by climate change under future current and high emissions scenarios appears likely to increase the overall volume of constitutional litigation related to climate change.

Neither nontropical severe storm systems nor extreme cold is expected to cause significant direct disruption to routine operations of the judicial and law enforcement systems protecting constitutional rights and liberties, either through demand increases or physical damage.

Table B.83. Risk from Climate Drivers for *Preserve Constitutional Rights*

Climate Driver	Baseline Present	Current Emissions 2050	Current Emissions 2100	High Emissions 2050	High Emissions 2100
Drought	1	2	2	2	2
Extreme cold	1	1	1	1	1
Extreme heat	1	2	2	2	2
Flooding	1	2	2	2	2
Sea-level rise	1	2	2	2	2
Severe storm systems	1	1	1	1	1
Tropical cyclones and hurricanes	1	2	2	2	2
Wildfire	1	2	2	2	2

NOTE: 1 = no disruption or normal operations. 2 = minimal disruption. See Chapter 2 for further explanation of ratings.

Climate Change and *Preserve Constitutional Rights*: Global Demand Drivers

Some climate drivers, such as drought and extreme heat, are expected to foster economic and social conditions outside the United States that could also increase demand for this NCF. For example, parts of Mexico and Central America are vulnerable to climate change–linked extreme heat and projections indicate that drought frequency and severity in the region will increase in the 21st century (Depsky and Pons, 2021; Kjellstrom et al., 2018). Climate change in Mexico and Central America has been linked to economic stresses and increased cross-border migration (Angelo, 2022; Restrepo, Werz, and Martinez, 2016). Increasing migration toward the United States could increase demand for legal and judicial services to protect the constitutional rights and adjudicate constitutional claims of migrants, who are generally entitled to certain constitutional protections, including due-process protections, once in U.S. territory (Benner and Savage, 2018). In addition, climate change has been studied as a causal factor in violent conflicts and instability outside the United States (Stockholm International Peace Research Institute, 2017). For example, a severe drought in the Middle East beginning in 2006 has been identified in some research as a potential contributory factor of the 2011 Syrian civil war (Kelley et al., 2015). U.S. involvement in future conflict linked to climate change could increase demand for the legal and judicial systems that protect civil liberties and adjudicate constitutional disputes. Historically, the perceived need to enhance security during conflicts has sometimes led the government to limit certain civil liberties domestically, provoking constitutional challenges in the courts (Stone, 2015).

Impact Mechanisms

The effects of climate change are likely to disrupt *Preserve Constitutional Rights* through two impact mechanisms. First, climate drivers are expected to damage the physical infrastructure required for the routine operations of judicial, legal, and law enforcement systems, slowing their ability to protect constitutional liberties and investigate and adjudicate claims of violations of constitutional rights. Second, demand for the systems and procedures that preserve constitutional rights is expected to increase across subfunctions as new patterns of crime increase the number of people entitled to constitutional protections through the criminal justice system and as additional climate change–related constitutional claims are made in court.

Cascading Risk

Routine operations of the *Preserve Constitutional Rights* NCF are dependent on the *Transport Cargo and Passengers by Road, Distribute Electricity, Enforce Law,* and *Operate Government* NCFs. Transportation disruptions can cause delays to the operation of court systems (*In re Response to Impact of Hurricane Ida,* 2021). Disruptions in *Distribute Electricity* may similarly affect the operations of judicial branch physical infrastructure. Disruptions to *Enforce Law* could hinder the investigation and prosecution of civil rights violations and affect the ability of the police and legal systems to ensure that constitutional rights are legally enforced. *Preserve Constitutional Rights* is critically dependent on *Operate Government,* as systems and procedures for maintaining and preserving constitutional rights are generally governmental in nature.

Strength of Evidence

Research on climate change's effects on the preservation of constitutional rights is limited. The strength of evidence supporting the climate effect of flooding and tropical cyclones and hurricanes was

151

assessed to be medium based on the empirical record of these natural hazards causing physical damage to judicial, legal, and law enforcement infrastructure. The strength of evidence on the effects of drought, extreme heat, sea-level rise, and wildfire was assessed as low because the historical record of physical damage to NCF infrastructure due to these climate drivers is less established and these climate drivers' potential effects on demand for this NCF are not well studied.

Table B.84. Strength of Evidence for *Preserve Constitutional Rights*

Strength of Evidence	
Medium	Flooding; tropical cyclones and hurricanes
Low	Drought; extreme heat; sea-level rise; wildfire

Risk Rating Interpretations

Table B.85 provides examples of how to interpret the risk ratings for this NCF.

Table B.85. Example Interpretations of Risk Ratings for *Preserve Constitutional Rights*

Risk Rating	NCF Example
1: No disruption or normal operations	The courts and constitutional system of government continue to function as they do today, with some level of disagreement about the proper interpretation of constitutional rights and liberties and delays due to judicial branch capacity constraints.
2: Minimal disruption	The courts and other constitutional and legal systems and institutions are affected by natural hazards linked to climate change, but delays are temporary and localized because of the resilience of the physical infrastructure and the fact that most constitutional procedures and operations occur indoors.
3: Moderate disruption	The operations of the courts and other constitutional and legal institutions are periodically disrupted by climate change–related events; delays are longer lasting and sometimes regional in nature.
4: Major disruption	Some state and federal courts and other constitutional and legal institutions are unable to operate, with halts in operations geographically widespread and temporally sustained; constitutional rights and liberties are sometimes infringed without redress through the court system.
5: Critical disruption	Courts and other constitutional and legal institutions are generally unable to continue operations, although certain high priority institutions may continue to operate with the focused support of the government; constitutional rights and liberties are frequently infringed; the government generally fails to operate according to the Constitution.

Energy and Infrastructure

Exploration and Extraction of Fuels

Table B.86. Scorecard for *Exploration and Extraction of Fuels*

	National Risk Assessment				
	Baseline	Current Emissions		High Emissions	
	Present	2050	2100	2050	2100
Exploration and Extraction of Fuels	3	3	3	3	3

Description: "Identify resources and collect energetic materials (including fossil fuels, nuclear materials, and others)" (CISA, 2020, p. 6).

Highlights

- *Exploration and Extraction of Fuels* was assessed to be at risk of moderate disruption through 2100 under both emissions scenarios from tropical cyclone and hurricane activity in the Gulf Coast, which can damage or disrupt the heavy concentration of coastal, near-shore, and offshore oil and gas facilities.
- In general, extraction-related activities are more sensitive to climate change than exploration is. Exploration activities can be planned and executed outside of episodic events with minimal to no effect on national-scale operations. Extraction of natural gas, in particular, is more sensitive to daily demand changes, so episodic extreme weather events are likelier to affect this subfunction.
- Drought, although assessed to pose no risk of disruption to this NCF at present, could rival the magnitude of disruptions that might occur from hurricanes by 2050 under a high emissions scenario because of water shortages.

Synopsis

Fuel exploration and extraction activities in the United States are regionally concentrated. As of 2021, Texas was the top producer for both dry natural gas (almost 24 percent of U.S. production) and oil (approximately 42 percent of U.S. production) (EIA, 2022d; EIA, 2022e). Other top-producing states are in the Gulf of Mexico, the Northeast, and the Great Plains, but a small amount of natural gas and oil exploration occurs in about two-thirds of states. This NCF's vulnerabilities to climate drivers stem from this geographic distribution of exploration and extraction activities, the NCF's reliance on water in exploration and extraction processes, and the NCF's slow pace in preparing for and hardening infrastructure against climate-related effects. For example, sea-level rise on the East Coast and hurricane events along the Gulf Coast could cause significant damage to exploration and extraction wellheads. Sustained drought conditions threaten reliable sources of water that are used for mining equipment. The increased intensity or frequency of natural hazards affected by climate change are also likely to stress the maintenance and response operations put in place to mitigate infrastructure vulnerabilities.

Because of these dynamics, this function is at risk of at least minimal disruption from three climate drivers at present (extreme cold, severe storm systems, and tropical cyclones and hurricanes) and from five climate drivers by 2050 (adding sea-level rise and drought) (Cruz and Krausmann, 2013; Office of Energy Policy and Systems Analysis, 2015). Because of the geographic distribution of the NCF and the regional nature of tropical cyclones and hurricanes, this NCF was assessed as being at risk of moderate disruption through 2100. Although the climate trends indicate increasing frequency and intensity of the natural hazards affected by climate change that affect this function, there are redundancies in the national and international markets that increase this function's resilience to national-scale cascading failures, which means that a risk of greater than moderate disruption is not anticipated into the future. Additionally, coal exploration and extraction are largely unaffected by the climate drivers except for flooding and drought by 2100 and potentially severe storm systems in a high emissions scenario. Uranium mining occurs in only one active mine today in Utah, but the facility is not anticipated to be affected by the climate drivers.

Subfunctions

This NCF has five subfunctions that are at risk of disruption from climate change: *Extract Coal, Extract Natural Gas, Explore for Natural Gas, Extract Petroleum,* and *Explore for Petroleum.* Only three of them—*Extract Natural Gas, Extract Petroleum,* and *Explore for Petroleum*—were assessed to face a risk of moderate disruption because of climate drivers by 2100. As discussed above, risk to this function is driven largely by the vulnerability of the *Extract Natural Gas* and *Extract Petroleum* subfunctions to tropical cyclones and hurricanes. Climate-related disruptions can lead to fuel shortages, prolonged outages, and economic loss at a regional scale, such as has been experienced historically with various tropical cyclones and hurricanes in the Gulf Coast (see, for example, EIA, 2011; Judson, 2013; Kirgiz, Burtis, and Lunin, 2009; and Painter, 2012). Extreme cold has also caused issues for natural gas extraction, such as the effects experienced in February 2021 in Texas (see Hilbert and Hallai, 2021, and Levin et al., 2022); however, an anticipated decrease in the frequency of extreme cold events tempers this risk at a national scale.

The *Explore for Nuclear Fuel* and *Extract Nuclear Fuel* subfunctions were not assessed to be at the same level of climate change–related risk as the fossil fuel–based subfunctions because the United States' primary source for nuclear fuel is foreign supply chains, which are outside the scope of this analysis. *Extract Coal* was assessed to face lower risk of disruption from climate drivers than other fossil fuel–based subfunctions because the geographic locations where *Extract Coal* occurs are not as vulnerable to the climate drivers assessed to be responsible for greater risk of disruption. This subfunction is vulnerable to flooding and severe storm systems, which can cause operational disruptions, including temporary closures from washed-out roads or issues with high water levels in mine-tailing dams. However, these effects have largely been local in scale (Delevingne et al., 2020).

Table B.87. Subfunction Risk for *Exploration and Extraction of Fuels*

Subfunction	Baseline Present	Current Emissions 2050	Current Emissions 2100	High Emissions 2050	High Emissions 2100
Extract Coal	1	1	2	2	2
Extract Natural Gas	3	3	3	3	3
Explore for Natural Gas	1	2	2	2	2
Extract Petroleum	3	3	3	3	3
Explore for Petroleum	1	2	2	2	3

NOTE: 1 = no disruption or normal operations. 2 = minimal disruption. 3 = moderate disruption. See Chapter 2 for further explanation of ratings.

Climate Drivers

Most climate drivers pose risks to fuel extraction and exploration; however, climate risk varies regionally because of regional differences in climatology (Wuebbles et al., 2017). According to historical impacts and projected exposure of critical fuel exploration and extraction assets, tropical cyclones and hurricanes, drought, severe storm systems, and sea-level rise are projected to pose the largest risk to fuel exploration and extraction by 2100. These climate effects have already affected the offshore wellheads in the Gulf of Mexico, as discussed above, and this trend is likely to continue. Extreme cold has also caused challenges for fuel extraction—a particularly salient example is the issues Texas had in February 2021 with extraction of natural gas, on which Texas relies heavily for power generation (see Levin et al., 2022)—but, because the trend in the number of projected extreme cold days is decreasing, the risk of disruption from this driver will also decrease through the end of the century.

Fuel exploration and extraction infrastructure is also vulnerable to climate drivers that can cause volatility in *Supply Water*, such as droughts. Drilling exploratory wells is highly dependent on availability of water resources (Zabbey and Olsson, 2017). Drought is a particularly interesting case because it does not pose a risk of disruption to this NCF at a national scale at present but, by 2100 under current emissions or by 2050 under a high emissions scenario, could rival the magnitude of disruptions brought on by hurricanes. However, the lasting effects on the industry could vary between the climate drivers because drought, coupled with policy changes, could cause more-permanent shifts in the fundamental practices in the industry. One historical example when water shortages resulting from drought affected extraction was in 2012, when a few companies with natural gas extraction operations in Kansas, North Dakota, Pennsylvania, and Texas faced high costs of water or could not access water sources for at least six weeks, reducing output or suspending operations in some locations (Hargreaves, 2012; Zamuda et al., 2013).

Table B.88. Risk from Climate Drivers for *Exploration and Extraction of Fuels*

Climate Driver	Baseline	Current Emissions		High Emissions	
	Present	2050	2100	2050	2100
Drought	1	2	3	3	3
Extreme cold	2	2	2	2	1
Extreme heat	1	1	2	2	2
Flooding	1	1	2	1	2
Sea-level rise	1	2	2	2	2
Severe storm systems	2	2	2	2	3
Tropical cyclones and hurricanes	3	3	3	3	3
Wildfire	1	1	1	1	1

NOTE: 1 = no disruption or normal operations. 2 = minimal disruption. 3 = moderate disruption. See Chapter 2 for further explanation of ratings.

Impact Mechanisms

Exploration and Extraction of Fuels is facing risk from climate change because of physical damage or disruption of the function from strong winds and heavy precipitation during tropical cyclones and hurricanes and severe storm systems and from inundation due to sea-level rise and flooding. Extreme heat and, to a lesser degree, extreme cold can physically disrupt operation by degrading performance of equipment or through physical damage to assets from overheating or freezing. Additionally, severe weather conditions during tropical cyclones and hurricanes, severe storm systems, and extreme heat and cold events could affect the workforce's ability to assess and fix damage or *Perform Maintenance* activities.

Cascading Risk

Exploration and Extraction of Fuels is critically dependent on the following NCFs:[34]

- *Distribute Electricity*
- *Transmit Electricity*
- *Generate Electricity*
- *Store Fuel and Maintain Reserves*
- *Fuel Refining and Processing Fuels*
- *Supply Water*
- *Transport Materials by Pipeline.*

---

[34] *Critically dependent* describes an NCF that, if disrupted or in failure, would severely disrupt another NCF and for which the other NCF could not recover until the upstream NCF was sufficiently operational. *Upstream* was defined as providing a key input to a downstream NCF.

Fuel extraction and exploration are part of four interdependent NCFs—*Exploration and Extraction of Fuels, Store Fuel and Maintain Reserves, Supply Water*, and *Fuel Refining and Processing Fuels*—that represent the production and commercialization of refined fuel products. If any of these four NCFs should fail, the others could experience significant disruption. Additionally, the generation and transmission of electricity are critical to continued operation of fuel extraction and exploration sites. Should any of the electricity NCFs fail, maintaining continuous fuel extraction operations would likely be exceptionally difficult. Moreover, pipelines that transport primary hydrocarbons (natural gas and crude oil) are a critical link between the fuel-related NCFs and the generation of electricity, which is dependent upon them.

Strength of Evidence

The body of existing literature and studies of climate risk to fuel extraction and exploration is moderate. The implications of tropical cyclones and hurricanes, flooding, extreme cold and sea-level rise are well studied. In particular, there are several historical examples of disruptions to this NCF from tropical cyclones and hurricanes, as detailed above (EIA, 2011; Judson, 2013; Kirgiz, Burtis, and Lunin, 2009; and Painter, 2012) and extreme cold (Hilbert and Hallai, 2021; Levin et al., 2022). Some studies have detailed risks to this NCF's infrastructure by U.S. region (Glick and Christiansen, 2019; Wakiyama and Zusman, 2021), and complementary studies have analyzed climate risk to supporting infrastructure and components of these NCFs in the context of other countries (S. Turner, Hejazi, et al., 2017); reports from those studies can be used to assess and substantiate risk ratings. Studies of climate risk to fuel extraction and exploration are less documented in the context of other climate drivers, such as wildfire, severe storm systems, extreme heat, and drought. Additionally, further research is needed to estimate climate's risk to fuel exploration and extraction infrastructure out to 2100.

Table B.89. Strength of Evidence for *Exploration and Extraction of Fuels*

Strength of Evidence	
High	Tropical cyclones and hurricanes; flooding; extreme cold; sea-level rise
Medium	Wildfire; severe storm systems; extreme heat; drought

Risk Rating Interpretations

Table B.90 provides examples of how to interpret the risk ratings for this NCF.

Table B.90. Example Interpretations of Risk Ratings for *Exploration and Extraction of Fuels*

Risk Rating	NCF Example
1: No disruption or normal operations	Exploration and fuel extraction continue to function as they do today, with occasional delays due to weather and operational problems.
2: Minimal disruption	A few companies in regions experiencing severe drought face high water costs or are denied access to water for several weeks, causing significant local effects but minimal effects at a national level because of diversity in the sources of supply.
3: Moderate disruption	A major hurricane hits the Gulf Coast, physically damaging exploration and extraction infrastructure, disrupting the regional output, and causing price increases.
4: Major disruption	Extreme weather events in near succession or simultaneously in a few regions across the United States cause widespread damage to critical wellheads; weather conditions prevent crews from accessing the damaged equipment, which impedes their ability to assess and fix damaged infrastructure.
5: Critical disruption	All exploration and extraction activities are stopped across the country because of severe drought and unprecedented hurricanes coupled with the effects of sea-level rise.

Fuel Refining and Processing Fuels

Table B.91. Scorecard for *Fuel Refining and Processing Fuels*

	National Risk Assessment				
	Baseline	**Current Emissions**		**High Emissions**	
	Present	**2050**	**2100**	**2050**	**2100**
Fuel Refining and Processing Fuels	3	3	3	3	3

Description: "Transform raw energetic materials into consumer fuels (e.g., crude cracking, gas separation, and uranium enrichment)" (CISA, 2020, p. 6).

Highlights

- *Fuel Refining and Processing Fuels* was assessed to face a moderate risk of disruption today through the end of the century, driven largely by vulnerabilities to tropical cyclones and hurricanes today and to sea-level rise and drought by mid- to late century.
- Refinery and processing infrastructure is regionally concentrated in areas of potentially high exposure to several of the climate drivers and acts as a critical node in stabilizing energy markets. Climate effects that are otherwise "local in context" have wider-reaching regional effects for this function.
- This NCF's vulnerabilities to climate drivers stem from the geographic distribution of fuel processing facilities, the sector's dependence on water in refinery processes, and the sector's slow pace in preparing for and hardening infrastructure against climate-related effects.

Synopsis

Petroleum and natural gas refining is a critical process in the oil and gas sector and plays a key role in maintaining stable energy markets (Jing et al., 2020). The activities supporting this function in the United States are regionally concentrated, with approximately half of the U.S. refinery capacity (EIA, 2002b) and nine out of the ten refineries with the most operable capacity being in the Gulf (the other is in Carson, California) (EIA, 2022b). Texas, Louisiana, and California dominate refining capacity in the United States, but, collectively, the Midwest states have about one-fifth of the refining capacity. The remaining natural gas and oil refineries are clustered in various locations in the eastern and northwestern United States and in Alaska. There are several natural gas and oil refineries along the coast in low-lying areas, particularly on the Gulf Coast and the East Coast, and thus, simply because of geographic location, susceptible to the effects of sea-level rise, tropical cyclones and hurricanes, and other severe storm systems (EIA, 2021a). Coal processing plants are similarly clustered in the eastern United States and thus vulnerable to the same climate drivers as oil and natural gas refineries in the same locations. Nuclear fuel processing is largely unaffected by the climate drivers except for drought under a high emissions scenario.

Today, risk of disruption is driven primarily by vulnerabilities to tropical cyclones and hurricanes, severe storm systems, and extreme cold. We assessed that, by 2050, sea-level rise, drought, and extreme heat would pose risk of minimal disruption (Carlson, Goldman, and Dahl, 2015; Cruz and Krausmann, 2013; Office of Energy Policy and Systems Analysis, 2015). This NCF's vulnerabilities to climate drivers stem from the geographic distribution of fuel processing facilities, which are concentrated in areas of potentially high exposure to several of the climate drivers; the sector's

159

dependence on water in refinery processes; and the sector's slow pace in preparing for and hardening infrastructure against climate-related effects. Hurricanes already cause significant effects on refineries. For example, Hurricane Harvey caused the shutdown of around 22 percent of the United States' refinery capacity and around 25 percent of oil production in the Gulf (Ramchand and Krishnamoorti, 2017). Power outages resulting from the impact of Hurricane Sandy caused the shutdown of several refineries, and there was damage directly from the hurricane to refinery infrastructure (Comes and Van de Walle, 2014; Office of Electricity Delivery and Energy Reliability, 2012b).

Sea-level rise on the East Coast and in the Gulf can exacerbate flooding and inundation from storms, including hurricanes (Strauss et al., 2021). For example, during Hurricane Ike in 2008, storm surge traveled 30 miles inland in parts of Texas and Louisiana (Berg, 2014). Prolonged drought threatens reliable sources of water required for refinery processes. Furthermore, maintenance and response capabilities will likely be stressed by the increase in frequency and intensity of extreme weather events.

Because of the regional hot spots in geographic distribution of infrastructure, this function is at risk of moderate disruption from climate change today and was assessed as remaining at risk of moderate disruption throughout the time frame considered in this analysis. Although the climate trends indicate increasing frequency and intensity of extreme weather events that affect this function, there are redundancies in the national and international markets that increase this function's resilience to national-scale cascading failures.

Subfunctions

As discussed above, risk to this function is driven largely by exposure that refining and processing infrastructure of natural gas, petroleum, and coal has to tropical cyclones and hurricanes. This infrastructure been assessed to face a moderate risk of disruption today, continuing through the end of the century. Climate-related disruptions can lead to fuel shortages, prolonged outages, and economic loss at a regional scale, such as what has been experienced with various hurricanes in the past in the Gulf (see, for example, EIA, 2011; Judson, 2013; Kirgiz, Burtis, and Lunin, 2009; and Painter, 2012). Extreme cold has also caused localized disruptions in the past—a particularly salient example is the issues Texas had in February 2021 with natural gas extraction and processing on which the state relies heavily for power generation (see Levin et al., 2022). But because the trend in the number of projected extreme cold days is decreasing, the risk of disruption from this driver will also decrease through the end of the century.

The *Process Nuclear Fuel* subfunction was not assessed to be at the same level of climate change–caused risk as the fossil fuel–based subfunctions because the United States' primary source for nuclear fuel is foreign supply chains. This subfunction could face a risk of minimal disruption by 2100 under a current emissions scenario or by 2050 in a high emissions scenario; it could face risk of moderate disruption by 2100 under a high emissions scenario.

Table B.92. Subfunction Risk for *Fuel Refining and Processing Fuels*

	Baseline	Current Emissions		High Emissions	
Subfunction	Present	2050	2100	2050	2100
Process Natural Gas	3	3	3	3	3
Process Nuclear Fuel (uranium)	1	1	2	2	3
Refine Petroleum (crude oil)	3	3	3	3	3
Process Coal	2	3	3	3	3

NOTE: 1 = no disruption or normal operations. 2 = minimal disruption. 3 = moderate disruption. See Chapter 2 for further explanation of ratings.

Climate Drivers

Most climate drivers pose a risk of at least minimal disruption by 2050, but the risk varies regionally because of regional differences in climatology (Wuebbles et al., 2017). Tropical cyclones and hurricanes pose the most near-term risk and, as discussed above, have already caused regional-scale disruptions to this NCF (see, for example, EIA, 2011; Judson, 2013; Kirgiz, Burtis, and Lunin, 2009; and Painter, 2012). Sea-level rise on the East Coast and in the Gulf can exacerbate flooding and inundation from storms, including hurricanes (Strauss et al., 2021). Increasing intensity and frequency of severe storm systems will continue to cause problems for refinery and processing infrastructure.

Because of the sector's reliance on water, *Fuel Refining and Processing Fuels* will also be increasingly affected by sustained drought conditions. Drought is a particularly interesting case because it does not pose a risk at a national scale today but, by 2100 under current emissions or by 2050 under a high emissions scenario, disruptions from drought could rival the magnitude of disruptions from hurricanes, although the lasting effects on the industry could vary between the climate drivers because drought, coupled with policy changes, could cause more-permanent shifts in the fundamental practices in the industry. One historical example of drought-induced water shortage affected extraction was in 2012, when a few companies with natural gas extraction operations in Kansas, Texas, Pennsylvania, and North Dakota faced high costs of water or could not access water sources for at least six weeks (Zamuda et al., 2013). It is a logical extension that, if extraction operations could be disrupted by drought conditions, so too could refinery operations.

Table B.93. Risk from Climate Drivers for *Fuel Refining and Processing Fuels*

Climate Driver	Baseline	Current Emissions		High Emissions	
	Present	2050	2100	2050	2100
Drought	1	2	3	3	3
Extreme cold	2	2	2	2	1
Extreme heat	1	2	2	2	3
Flooding	1	1	2	1	2
Sea-level rise	1	2	3	3	3
Severe storm systems	2	2	2	2	2
Tropical cyclones and hurricanes	3	3	3	3	3
Wildfire	1	1	1	1	1

NOTE: 1 = no disruption or normal operations. 2 = minimal disruption. 3 = moderate disruption. See Chapter 2 for further explanation of ratings.

Impact Mechanisms

Fuel Refining and Processing Fuels is facing risk from climate change because of physical damage or disruption of the function from strong winds and heavy precipitation during tropical cyclones and hurricanes and severe storm systems and from inundation due to sea-level rise and flooding. Extreme heat and, to a lesser degree, extreme cold can physically disrupt operation by degrading performance of equipment or through physical damage to assets from overheating or freezing. Additionally, severe weather conditions during tropical cyclones and hurricanes, severe storm systems, and extreme heat and cold events could affect the workforce's ability to assess and fix damage or perform maintenance activities.

Cascading Risk

Fuel Refining and Processing Fuels is critically dependent on the following NCFs:

- *Distribute Electricity*
- *Transmit Electricity*
- *Generate Electricity*
- *Store Fuel and Maintain Reserves*
- *Exploration and Extraction of Fuels*
- *Supply Water*
- *Transport Materials by Pipeline.*

Fuel Refining and Processing Fuels is one of four interdependent NCFs—*Exploration and Extraction of Fuels, Store Fuel and Maintain Reserves, Supply Water,* and *Fuel Refining and Processing Fuels*—that represent the production and commercialization of refined fuel products. If any of these four NCFs should fail, the others could experience significant disruption. Additionally, the generation and transmission of electricity are critical to continued operation of *Fuel Refining and Processing Fuels.*

162

Should any of the electricity NCFs fail, maintaining continuous fuel refining and processing operations would likely be exceptionally difficult. Moreover, pipelines that transport primary hydrocarbons (natural gas and crude oil) are a critical link between the fuel resource–related NCFs and the generation of electricity, which is dependent on these pipelines.

Strength of Evidence

The body of existing literature and studies of climate risk to refinery infrastructure is robust. The implications of sea-level rise, hurricanes, and severe storm systems are well studied through historical examples and literature. Some studies have detailed risks to this NCF's infrastructure by U.S. region (Glick and Christiansen, 2019; Wakiyama and Zusman, 2021), and complementary studies have analyzed climate risk to supporting infrastructure and components of these NCFs in the context of other countries (S. Turner, Hejazi, et al., 2017), which can be used to assess and substantiate risk ratings. However, additional research is needed to estimate the climate driver–induced risk to *Fuel Refining and Processing Fuels* infrastructure out to 2100.

Table B.94. Strength of Evidence for *Fuel Refining and Processing Fuels*

	Strength of Evidence
High	Tropical cyclones and hurricanes; severe storm systems; extreme cold; extreme heat; drought; flooding; sea-level rise
Medium	Wildfire

Risk Rating Interpretations

Table B.95 provides examples of how to interpret the risk ratings for this NCF.

Table B.95. Example Interpretations of Risk Ratings for *Fuel Refining and Processing Fuels*

Risk Rating	NCF Example
1: No disruption or normal operations	Fuel refining and processing continues to function as it does today, with occasional issues due to weather and operational problems.
2: Minimal disruption	A few companies in regions experiencing severe drought face high water costs or are denied access to water for several weeks, causing significant local effects but minimal effects at a national level because of diversity in the sources of supply.
3: Moderate disruption	Refining and processing infrastructure is physically damaged across several states that act as a key regional node in this NCF.
4: Major disruption	The increase in frequency of extreme weather events causes widespread damage to critical refinery and processing infrastructure in more than one region, which strains the system beyond existing adaptive capacity and leads to sustained price increases.
5: Critical disruption	All refinery and processing activities are stopped across the country due to severe drought and unprecedented hurricanes coupled with the effects of sea-level rise.

Store Fuel and Maintain Reserves

Table B.96. Scorecard for *Store Fuel and Maintain Reserves*

	National Risk Assessment				
	Baseline	Current Emissions		High Emissions	
	Present	2050	2100	2050	2100
Store Fuel and Maintain Reserves	2	2	2	2	2

Description: "Store energetic materials (including fossil and nuclear fuels) to reliably meet operational and strategic demands" (CISA, 2020, p. 6).

Highlights

- *Store Fuel and Maintain Reserves* was assessed to be at risk of minimal disruption at a national scale today and through the end of the century under both emissions scenarios.
- Natural gas is stored largely underground in depleted oil and gas fields and thus is not directly exposed to the climate drivers. However, the function is at risk of disruption from physical damage from flooding, tropical cyclones and hurricanes, and severe storm systems.
- The Strategic Petroleum Reserve's (SPR's) ability to mitigate effects of domestic supply disruptions is limited because it contains mostly oil, not refined products, and it is based entirely in the Gulf Coast and therefore might not be able to respond to disruptions in other regions.

Synopsis

This NCF encompasses the storage of liquid fossil fuels in both large amounts (e.g., in the SPR) and small amounts (e.g., aboveground storage tanks at refineries) and nuclear fuels (e.g., spent nuclear fuel rods stored at power plants). Flooding, severe storm systems, and tropical cyclones and hurricanes can cause physical damage to fuel storage facilities and subsequently affect service reliability (American Petroleum Institute [API], 2016). However, we have few historical examples of climate driver–induced disruption to fuel storage facilities having affected the function, even on a local scale. This is primarily because fuels are often stored underground and thus are less susceptible to the effects of various climate drivers. For example, natural gas is commonly stored in underground facilities in depleted oil and gas fields because they are widely available and already have infrastructure, such as wellheads, gathering systems, and pipelines (EPA, 2015b). Storage also occurs in aquifers and salt caverns, the latter of which supports high withdrawal and injection rates (EPA, 2015b). Fuels can also be stored in aboveground tanks—this is common at refining and processing facilities where fuels are stored temporarily. GAO has reported that storm surge flooding can affect aboveground tanks by lifting unfilled or partially filled tanks off platforms or causing corrosion (Rusco, 2014), and Godoy (2007) details flooding and wind damage from Hurricanes Katrina and Rita. However, the consequences of these disruptions have been minimal and hyperlocalized in nature.

The United States maintains the SPR, which consists of four underground salt caverns at four major oil facilities on the Gulf Coast. Although there is high storage capacity in the SPR, the SPR's ability to mitigate effects of domestic supply disruptions is limited for two reasons. First, it contains mostly oil, not refined products that are needed by users. Second, the SPR is based entirely on the Gulf Coast and might not be able to address disruptions in other regions (Rusco, 2017).

Spent nuclear fuels (rods) are stored in wet and dry casks, often on-site at the reactor. Wet casks or pools submerge the spent fuel in water continuously to keep the rods at a safe temperature and thus rely on water as an input. Storage sites are dispersed across several states but largely concentrated in the eastern half of the United States and therefore are at less risk of disruption from drought, but they still face some risk of disruption from flooding, tropical cyclones and hurricanes, severe storm systems, and, to a lesser degree, sea-level rise. Dry storage is potentially less vulnerable to certain climate drivers, such as tropical cyclones and hurricanes or severe storm systems, because dry storage facilities are built from steel and concrete (Planning and Economic Studies Section, 2018; U.S. Nuclear Regulatory Commission, 2021). About 80 percent of storage facilities at reactors are reported to be under dry conditions across the world (Planning and Economic Studies Section, 2018). Additionally, because the volume of spent fuel generated annually in the United States is quite small—on the order of the size of an Olympic pool—and is dispersed across several storage sites (Office of Nuclear Energy, 2022), the overall risk of disruption from climate drivers is less than the risk for natural gas or other liquid fuel storage.

Store Fuel and Maintain Reserves was assessed to be at minimal risk of disruption at a national scale through the end of the century for both emissions scenarios. This rating is chiefly because fuels are stored primarily underground and thus less susceptible to climate drivers. Aboveground fuel storage is distributed across many states, but historical disruptions from physical damage have largely been localized and not caused a problem for the function to meet routine operations at a national scale.

Subfunctions

All the subfunctions for this NCF—*Store Natural Gas*, *Store Liquid Fuels*, and *Store Nuclear Fuel (UO$_2$* [uranium dioxide])—were assessed to be at minimal risk by 2050, primarily from changes in anticipated frequency and intensity of flooding, tropical cyclones and hurricanes, and severe storm systems in areas where there are concentrated storage facilities, such as on the Gulf Coast and in the eastern United States. *Store Natural Gas* and *Store Liquid Fuels* have experienced hyperlocalized disruptions from tropical cyclones and hurricanes and flooding (API, 2016; Godoy, 2007; Rusco, 2017). Although the frequency and severity of these climate drivers are projected to increase through the end of the century, because the natural gas and liquid fuels are stored primarily underground, we found the risk of disruption to be only minimal at a national scale.

Store Nuclear Fuel (UO$_2$) was assessed as being at lower risk today primarily because the volume of spent fuel generated in the United States is quite small and is dispersed across several storage sites tempering the risk of disruption nationally.

Table B.97. Subfunction Risk for *Store Fuel and Maintain Reserves*

	Baseline	Current Emissions		High Emissions	
Subfunction	Present	2050	2100	2050	2100
Store Natural Gas	2	2	2	2	2
Store Liquid Fuels	2	2	2	2	2
Store Nuclear Fuel (UO₂)	1	2	2	2	2

NOTE: 1 = no disruption or normal operations. 2 = minimal disruption. See Chapter 2 for further explanation of ratings.

Climate Drivers

Flooding, tropical cyclones and hurricanes, and severe storm systems are the key drivers of risk for this function because storage facilities are concentrated on the Gulf Coast and in the eastern United States, where these climate drivers occur today (API, 2016; Rusco, 2017; Sarvestani, Ahmadi, and Alenjareghi, 2021). Although the frequency and severity of these climate drivers are projected to increase through the end of the century, because natural gas and liquid fuels are stored primarily underground, we found the risk of disruption to be only minimal at a national scale. Similarly, although drought is projected to increase dramatically and cause issues for other interdependent NCFs, such as *Exploration and Extraction of Fuels* and *Fuel Refining and Processing Fuels*, only one subfunction—*Store Nuclear Fuel (UO₂)*—relies on water as an input for wet storage and thus was not assessed to pose risk to the NCF at a national scale.

Table B.98. Risk from Climate Drivers for *Store Fuel and Maintain Reserves*

	Baseline	Current Emissions		High Emissions	
Climate Driver	Present	2050	2100	2050	2100
Drought	1	1	1	1	1
Extreme cold	1	1	1	1	1
Extreme heat	1	1	1	1	1
Flooding	2	2	2	2	2
Sea level rise	1	1	1	1	1
Severe storm systems	1	2	2	2	2
Tropical cyclones and hurricanes	2	2	2	2	2
Wildfire	1	1	1	1	1

NOTE: 1 = no disruption or normal operations. 2 = minimal disruption. See Chapter 2 for further explanation of ratings.

Impact Mechanisms

Store Fuel and Maintain Reserves is facing risk from climate drivers primarily because of the potential for physical damage or disruption of the function from flooding, tropical cyclones and hurricanes, and severe storm systems. For example, storm surge flooding can affect aboveground tanks by lifting unfilled or partially filled tanks off platforms or causing corrosion (Rusco, 2017), and strong winds can cause buckling of shells and damage to insulation (Godoy, 2007).

Cascading Risk

Store Fuel and Maintain Reserves is critically dependent on the following NCFs:

- *Distribute Electricity*
- *Transmit Electricity*
- *Generate Electricity*
- *Fuel Refining and Processing Fuels*
- *Exploration and Extraction of Fuels*
- *Supply Water*
- *Transport Materials by Pipeline.*

Store Fuel and Maintain Reserves is one of four interdependent NCFs—*Exploration and Extraction of Fuels, Store Fuel and Maintain Reserves, Supply Water,* and *Fuel Refining and Processing Fuels*—that represent the production and commercialization of refined fuel products. If any of these four NCFs should fail, the others could experience significant disruption. Additionally, the generation and transmission of electricity are critical to continued operation of *Fuel Refining and Processing Fuels.* Should any of the electricity NCFs fail, maintaining continuous fuel extraction operations would likely be exceptionally difficult. Moreover, pipelines that transport primary hydrocarbons (natural gas and crude oil) are a critical link between the fuel resource–related NCFs and the generation of electricity, which is dependent on them.

Strength of Evidence

There are gaps in the literature and studies that assess the risks of climate directly on storage infrastructure and facilities. Much of the literature is focused on exploration and production and on transport or delivery (i.e., pipeline or other means) of the commodity of interest, as described in other sections. However, we were able to draw on historical effects on the refining, processing, and transportation sectors to assess the potential risk to storage of fuels and some studies on the causes of storage tank accidents (J. Chang and Lin, 2006; Sarvestani, Ahmadi, and Alenjareghi, 2021).

Table B.99. Strength of Evidence for *Store Fuel and Maintain Reserves*

	Strength of Evidence
Medium	Flooding; tropical cyclones and hurricanes; severe storm systems
Low	Sea-level rise; extreme cold; extreme heat; wildfire; drought

Risk Rating Interpretations

Table B.100 provides examples of how to interpret the risk ratings for this NCF.

Table B.100. Example Interpretations of Risk Ratings for *Store Fuel and Maintain Reserves*

Risk Rating	NCF Example
1: No disruption or normal operations	The ability to store fuels and maintain reserves operates as it does today with normal variability in storage capacity and demand.
2: Minimal disruption	A few storage sites in flood-prone regions experience damage to facilities, but demand is met from imports from other regions domestically.
3: Moderate disruption	Some storage sites are permanently damaged by an increasing frequency of severe storms, including hurricanes, which causes temporary but prolonged fuel supply shortages in the region.
4: Major disruption	Extreme weather events in near succession or simultaneously in a few regions across the United States cause widespread damage to critical refineries and pipelines and storage facilities. Weather conditions prevent crews from accessing the damaged equipment, impeding their ability to assess and fix damaged infrastructure; as a result, the market experiences sustained price increases.
5: Critical disruption	Storage across the country is affected by climate, causing complete disruption of the function.

Generate Electricity

Table B.101. Scorecard for *Generate Electricity*

	National Risk Assessment				
	Baseline	**Current Emissions**		**High Emissions**	
	Present	**2050**	**2100**	**2050**	**2100**
Generate Electricity	2	2	3	3	3

Description: "Produce electricity from a variety of primary energy sources (including fossil fuels, nuclear materials, and renewables) to reliably meet demand" (CISA, 2020, p. 6).

Highlights

- *Generate Electricity* infrastructure is already vulnerable to climate drivers, but the disruptions have largely been localized because of built-in redundancies in energy systems with multiple generation sources, thus posing risk of only minimal disruption at a national scale through 2050. However, we assessed that, under a high emissions scenario, this NCF faces risk of moderate disruption by 2050.
- Assets in coastal zones are increasingly at risk of damage due to saltwater inundation from sea-level rise and storm surge associated with tropical cyclones and hurricanes. They might need to be moved or reestablished outside hazard zones to ensure reliable operation and to meet expected increases in demand due to rising temperatures.
- Drought coupled with increased demand because of extreme heat will pose risk to hydropower generation assets in the Northwest and Southwest, as well as thermogeneration in most regions.

Synopsis

The electric grid is a complex system of interconnected generation assets, transmission and distribution lines, substations, and transformers that ultimately delivers power to end users. Thus, some redundancies built into the system lessen the assessed risk of disruption across the relevant NCFs at a national scale and increase resilience of the grid to national cascading failures (Abi-Samra, 2017). Exceptions to this include the area covered by Electric Reliability Council of Texas (the boundaries of which closely align with Texas's state boundaries but do not line up exactly), Alaska, Hawaii, and Puerto Rico because either their grids are isolated or their connections to the national grid are limited. However, the system has known vulnerabilities to climate drivers that will continue to pose risk in the future because climate change is expected to increase the frequency and intensity of climate drivers.

The vulnerability that the country's complex network of electricity generation assets has to climate drivers varies by region, but the function as a whole was assessed to be at risk of at least minimal disruption by 2050 from all climate drivers except wildfire, which poses greater risk to transmission and distribution infrastructure. *Generate Electricity's* vulnerabilities to climate drivers stem from geographic distribution: There are thousands of power plants and distributed energy resources across the United States, the industry's reliance on water directly for *Generate Electricity*[35] or in cooling

---

[35] Hydropower and thermogeneration rely directly on *Supply Water*. Thermogeneration power plants use steam turbines to generate electricity and use a variety of fuel sources, including fossil fuel–based sources, nuclear, and renewables, such as biomass, geothermal, and concentrated solar power.

processes, and legacy designs based only on historical climate conditions and maintenance standards that do not account for increasing frequency or intensity of climate-related effects (Henry and Pratson, 2019; S. Turner, Nelson, et al., 2021). Decreasing water availability caused by sustained drought (western United States), sea-level rise (East and Gulf Coasts), and tropical cyclones and hurricanes (East and Gulf Coasts) pose the most near-term risk of disruption to generation because of the permanence of the changing condition (drought) or the potential for extensive physical damage and lengthy restoration times, as in the case of tropical cyclones and hurricanes (Office of Energy Policy and Systems Analysis, 2015; Shield et al., 2021). However, extreme heat, flooding, severe storm systems, and, to a lesser degree, extreme cold can also affect generation by reducing generation capacity through performance degradation or through physical damage to assets (Allen-Dumas, KC, and Cunliff, 2019; Office of Energy Policy and Systems Analysis, 2015). Flooding events from severe storm systems (heavy precipitation events) on the East and West Coasts and regions with dense riverine systems, sea-level rise on the East Coast, and hurricane events along the Gulf Coast are already having unprecedented on electricity generation infrastructure (Dullo et al., 2021; Gangrade et al., 2019; Qiang, 2019).

Furthermore, the increasing severity of climate drivers was found to be the primary contributor to an increase in power outage durations between 2000 and 2012 (Allen-Dumas, KC, and Cunliff, 2019; P. Larsen et al., 2015). Between 2011 and 2021, weather-related power outages increased more than 80 percent over the preceding decade (2000 to 2010) (Climate Central, 2022). Despite investments in recent years in resilience measures, including hardening, and improvements in planning and response capabilities, climate drivers remain the primary cause of power outages in the United States (Allen-Dumas, KC, and Cunliff, 2019; Reidmiller et al., 2018; Shield et al., 2021). Additionally, the projected changes in climate drivers will likely strain maintenance and response capabilities, potentially increasing restoration times and posing a risk of moderate disruption to the NCF nationally under both emissions scenarios—by 2100 under the current emissions scenario and by 2050 in a high emissions scenario.

Subfunctions

All the subfunctions under *Generate Electricity* except *Comply with Regulations* face some risk of climate-related disruption.[36] Physical damage from natural hazards affected by climate change reduces the *Operate Infrastructure* and *Enable Infrastructure* subfunctions' ability to perform routine operations, the latter encompasses the establishment, maintenance, and protection of infrastructure. These two subfunctions could also be affected by an increasing demand for maintenance because of more-frequent and -intense climate drivers, which, in turn, affects the ability to provide power access to all customers (Nateghi, 2018) and to secure operation of the system (Office of Energy Policy and Systems Analysis, 2015). Sustained drought has implications for the ability to secure primary and

---

[36] A power plant's ability to comply with regulations should not be directly affected by climate drivers to the level that the disruption poses risk as a national scale. Although there are historical examples in which power plants requested exceptions or waivers to exceed water temperature discharge levels because of extreme heat, these have been hyperlocalized and the waivers and exceptions have been granted. Moreover, during major disasters, power plant operators have been granted permission to exceed air quality and emissions standards. These historical precedents suggest that these waivers or exceptions will likely continue to be granted during disasters to minimize impact on affected communities.

auxiliary fuel sources, part of the *Obtain Resources* subfunction, because water is a key input to fossil fuel–based exploration and production and to grow biomass crops. Note that this subfunction overlaps in part with the *Provide Industrial Water* subfunction for *Supply Water*. Additionally, sea-level rise and flooding can disrupt fuel delivery to generation infrastructure. These disruptions can have implications for fuel and electricity pricing. Power outages often lead to regional economic loss, although prolonged outages can also propagate economic losses nationally or globally (Shuai et al., 2018; Yi Zhang and Lam, 2016). Although the *Secure Operation* subfunction, which relates to preventive hardening against specific threats, was not assessed to be at risk today, by 2050, sea-level rise poses a risk of minimal disruption to the subfunction and risk of moderate disruption by 2100. Unless infrastructure along coastlines that is exposed to sea-level rise is moved, hardening against this chronic, pervasive climate driver will become increasingly difficult.

Table B.102. Subfunction Risk for *Generate Electricity*

	Baseline	Current Emissions		High Emissions	
Subfunction	Present	2050	2100	2050	2100
Enable Infrastructure	2	2	3	3	3
Obtain Resources	2	2	3	3	3
Operate Infrastructure	2	2	3	3	3
Secure Operation	1	2	3	3	3

NOTE: 1 = no disruption or normal operations. 2 = minimal disruption. 3 = moderate disruption. See Chapter 2 for further explanation of ratings.

Climate Drivers

As discussed in the above sections, all of the climate drivers pose risks to *Generate Electricity* to varying degrees, depending on the region. Electricity generation infrastructure is susceptible to drought (Zohrabian and Sanders, 2018), flooding (Boggess, Becker, and Mitchell, 2014), and extreme heat and cold (Daher et al., 2018; Loew et al., 2020). According to historical effects and projected exposure of critical power generation assets, tropical cyclones and hurricanes, sea-level rise, and extreme heat are anticipated to pose the largest risk to power generation infrastructure by 2100. Hurricanes and flooding already cause major disruptions regionally (Nateghi, 2018), but these effects rarely, if ever, disrupt routine operations of the NCF nationally. Extreme heat and sea-level rise pose more chronic risk to power generation infrastructure (Burillo, Chester, Pincetl, et al., 2019; G. Griggs and Patsch, 2019; Wuebbles et al., 2017), but the effects are largely limited to the infrastructure within a region. Most power plants in the United States, regardless of fuel source (for example, coal, natural gas, nuclear, concentrated solar, or geothermal), rely on a steady supply of water for cooling (because most thermal power plants use once-through or recirculating water cooling), and operations are projected to be threatened when water availability decreases or water temperatures increase. These dynamics are difficult to mitigate. Preemptively decreasing a plant's output is a protection method, but drought effects exacerbate the magnitude and frequency of these deratings.

Table B.103. Risk from Climate Drivers for *Generate Electricity*

Climate Driver	Baseline Present	Current Emissions 2050	Current Emissions 2100	High Emissions 2050	High Emissions 2100
Drought	2	2	3	3	3
Extreme cold	2	2	1	2	1
Extreme heat	1	2	3	3	3
Flooding	1	2	2	2	3
Sea-level rise	1	2	3	3	3
Severe storm systems	1	2	2	2	2
Tropical cyclones and hurricanes	2	2	3	2	3
Wildfire	1	1	2	1	2

NOTE: 1 = no disruption or normal operations. 2 = minimal disruption. 3 = moderate disruption. See Chapter 2 for further explanation of ratings.

Impact Mechanisms

Generate Electricity could be disrupted when physical damage to power plants and other generation infrastructure occurs from strong winds and heavy precipitation during hurricanes and severe storm systems, from saltwater inundation due to sea-level rise, and, to a lesser extent, from wildfire. Extreme heat and, to a lesser degree, extreme cold can lead to performance degradation from reducing generation capacity or through physical damage to assets. There might also be input or resource constraints related to prolonged drought–induced water shortages for hydropower generation. Extreme cold or heat events can increase demand for heating and cooling and thus increase demand for this function. Additionally, severe weather conditions during hurricanes, severe storm systems, wildfire, and extreme heat and cold events could affect the workforce's ability to assess and fix damage or perform maintenance activities outside.

Cascading Risk

Generate Electricity is critically dependent on the following NCFs:

- *Distribute Electricity*
- *Transmit Electricity*
- *Exploration and Extraction of Fuels*
- *Store Fuel and Maintain Reserves*
- *Fuel Refining and Processing Fuels*
- *Transport Materials by Pipeline*
- *Supply Water.*

In the NCF framework, *Generate Electricity* is one part of three interconnected NCFs that compose bulk power delivery. If *Generate Electricity*, *Transmit Electricity*, or *Distribute Electricity* should fail, all three fail to deliver power to end users (Vartanian et al., 2018). Additionally, some NCFs—

172

Exploration and Extraction of Fuels, Store Fuel and Maintain Reserves, Supply Water, and *Fuel Refining and Processing Fuels*—maintain the development of fuel resources that are critical inputs to *Generate Electricity.* Should any of the fuel NCFs fail, it would likely be exceptionally difficult to maintain continuous power delivery through electricity generation (C. Murphy et al., 2020). The power plants in the United States with the greatest capacity rely on a continuous supply of fossil fuels to produce power. In addition, supply chain issues can also affect the on-time delivery of fuels and components meant to support and expand power production across the country. Moreover, pipelines that transport primary hydrocarbons (natural gas and crude oil) are a critical link between the fuel resource–related NCFs on which *Generate Electricity* depends.

Strength of Evidence

The body of existing literature and studies of climate risk to electricity generation infrastructure is expansive and well-studied. However, because of the variance in power plant types, some gaps are not as well studied for certain generation technologies and climate change–related vulnerabilities. Some studies have detailed risks to the NCF-related infrastructure by U.S. region (Glick and Christiansen, 2019; Wakiyama and Zusman, 2021), and complementary studies have analyzed climate risk to supporting infrastructure and components from other countries that can be used to assess and substantiate risk ratings for infrastructure in the United States (Rübbelke and Vögele, 2011; Tobin et al., 2018; S. Turner, Hejazi, et al., 2017; Zamuda et al., 2013). Because most modern systems rely on uninterrupted power supply (FEMA, 2007), there is a general interest in studying how to maintain continuous power through all types and scales of climate disasters (Boggess, Becker, and Mitchell, 2014; Huang, Swain, and Hall, 2020; Saint, 2009).

Table B.104. Strength of Evidence for *Generate Electricity*

Strength of Evidence	
High	Drought; sea-level rise
Medium	Tropical cyclones and hurricanes; severe storm systems; extreme cold; extreme heat; wildfire

Risk Rating Interpretations

Table B.105 provides examples of how to interpret the risk ratings for this NCF.

Table B.105. Example Interpretations of Risk Ratings for *Generate Electricity*

Risk Rating	NCF Example
1: No disruption or normal operations	*Generate Electricity* continues to function as it does today, with occasional short-term disruptions due to weather and operational problems.
2: Minimal disruption	A few power plants in drought-prone regions experience a reduction in generation capacity, but unmet electricity demand is met from alternative generation sources, such as a peaking power plants or through the import of electricity from other regions.
3: Moderate disruption	Several power plants close temporarily or permanently in a given region because of sea-level rise; extreme heat causes increased demand for cooling, which is not reliability met, straining the grid and causing voltage instability. Brownouts become more frequent.
4: Major disruption	Multiple regions experience regional-scale outages as a result of extreme weather events. Some grid interconnections are disrupted, which prevents the import and export of power from nonaffected regions to those without local generation capacity.
5: Critical disruption	Several power plants must close permanently because of sea-level rise and drought. New generation capacity is not being added at a rate that keeps pace with increased demand. Certain time periods are particularly challenging, such as summer and fall, when increased demand for cooling is expected to raise significantly and exposure to severe storm systems and hurricanes is expected to increase in frequency and intensity.

Transmit Electricity

Table B.106. Scorecard for *Transmit Electricity*

	National Risk Assessment				
	Baseline	Current Emissions		High Emissions	
	Present	2050	2100	2050	2100
Transmit Electricity	2	3	3	3	3

Description: "Maintain and operate high-voltage (>100kV [kilovolts]) bulk electric system to reliably supply distribution network demand for electricity from generation resources" (CISA, 2020, p. 3).

Highlights

- *Transmit Electricity* was assessed as being at risk of moderate disruption by the end of the century due to physical damage and degradation of infrastructure assets serving *Transmit Electricity*.
- Increased intensity or frequency of climate drivers, including extreme heat, tropical cyclones and hurricanes, wildfire, and sea-level rise, can damage and reduce capacity of transmission lines.
- Factors that include the condition of transmission infrastructure, implementation of infrastructure-hardening strategies, and redundancies in the transmission system affect this NCF's vulnerability to climate drivers.
- *Operate Infrastructure* and *Maintain System* are the most-vulnerable subfunctions of this NCF because of their reliance on reliable and functional equipment and assets, which can be affected by physical damage, changes in demand, and workforce shortages.
- Risk of disruption is distributed based on certain characteristics of the electric transmission provider or owner, such as a utility's ability to invest in resilience or infrastructure protection measures.

Synopsis

The electric transmission system plays a vital role in the operation of the electrical grid by delivering electricity over long distances from power generation sources to electric distribution systems. Electric transmission infrastructure is made up of many components, which are primarily transmission towers, conductors (or transmission lines), transformers, and substations. The scale of electric transmission infrastructure in the United States is considerable. For instance, there are 600,000 miles of transmission lines of which 240,000 miles are high voltage (American Society of Civil Engineers [ASCE], 2021).[37] We assessed that, by the middle and end of the century, the *Transmit Electricity* NCF will be at risk of moderate disruption from climate drivers under current and high emissions scenarios. Potential effects on this NCF from climate drivers include immediate or long-term physical damage to transmission infrastructure, increased load on electric grid system, and inability to perform maintenance activities due to unsafe working conditions. For example, high temperatures can lead to reductions in ampacity of transmission lines (the maximum current that the line can handle in the operating conditions while remaining within its temperature rating) and reduce the life span of a power transformer (Fant et al., 2020). At the same time, high temperatures increase the risk of heat-related illness, which can affect workers' ability to safely perform outdoor maintenance activities (Roh and Alamo, 2022). Although other subfunctions associated with this NCF, such as the

---

[37] Transmission lines with a voltage greater than or equal to 230 kV are categorized as high voltage.

ability to implement infrastructure-hardening measures or effectively operate the market, might be affected by climate drivers, it is less likely that these effects will lead to any disruptions.

Faults or damage to transmission lines or any other components of the transmission system can affect the ability to provide safe and reliable electricity to communities and customers. Hurricane Ida, which made landfall in Louisiana in August 2021, serves as a recent example; high wind speeds and flooding led to damaged transmission poles, transmission lines, and transformers causing electricity disruptions for more than 1 million customers (EIA, 2021b). In addition to hurricanes, other climate drivers, such as extreme heat and wildfire, can affect the ability to effectively transmit electricity because they reduce transmission capacity and physically damage poles supporting transmission lines. One estimate suggests that climate change effects resulting in reductions to transmission-line capacity and wildfire damage to electrical transmission lines will range between $0.4 billion and $1.1 billion[38] per year between 2080 and 2099 (Fant et al., 2020).

This NCF was assessed as being at risk of moderate disruption. Although the electric transmission network is distributed across multiple regions within the United States, consequences from climate drivers, such as physical damage, reduction in capacity, and increase in deferred maintenance, will occur at a local or regional scale. The electrical power grid does not operate as a single entity and instead is divided into multiple interconnected grids (e.g., the Eastern Interconnection and the Western Interconnection) that span different geographic regions. Anecdotal evidence and historical incidents illustrate how electric transmission operations and maintenance have previously experienced disruptions in select regions from exposure to climate drivers, such as flooding and tropical cyclones and hurricanes in the mid-Atlantic (Zamuda et al., 2013) or wildfires in the Southwest (Alexander-Kearns, 2015). As climate drivers increase in frequency and intensity by the middle and end of the century, the risk of disruptions to this NCF was assessed to increase from minimal to moderate because of more electric transmission components being exposed to damage or degradation leading to service disruptions across more regions.

Key Considerations for *Transmit Electricity*

Regional transmission organizations (RTOs) and independent system operators (ISOs) are entities responsible for the operation, coordination, and monitoring of electric power transmission systems in the United States. RTOs and ISOs use different strategies and approaches to manage electricity transmission and might therefore undertake different strategies and approaches toward preparing for disruptions from climate drivers. For example, RTOs and ISOs that conduct vulnerability assessments and develop plans to combat the effects of climate drivers might experience fewer disruptions in the future.

Subfunctions

Four of the six subfunctions of this NCF—*Provide Access, Operate Infrastructure, Maintain System, and Provide Electrical Protection*—were assessed as being at risk of moderate disruptions by at least one of the climate drivers by the middle and end of the century under both current and high emissions scenarios. One of the primary factors in the ability to operate, maintain, and provide access to the electric transmission system is reliable infrastructure. Damage to transmission lines, transformers, or

---

[38] Adjusted for inflation from 2017 dollars to 2022 dollars.

structures supporting transmission equipment can lead to faults within the transmission system, as well as cascading failures in other parts of the electricity system downstream (e.g., electrical distribution). Examples of vulnerabilities within the transmission system include structural failure of poles supporting transmission lines, reduction in ampacity of transmission cables, and erosion or inundation of transformers or substations (Rezaei et al., 2016). Additionally, components of the transmission system might not be designed for changing climate conditions in the future or might be overdue for replacement or maintenance, which further increases the risk of damage or degradation from climate drivers.

Other subfunctions, such as *Secure System*, were assessed as being at risk of minimal disruption for at least one of the climate drivers by the end of the century under the current and high emissions scenarios. The *Secure System* subfunction focuses primarily on activities pertaining to infrastructure hardening, such as replacing aboveground transmission lines with underground transmission lines to prevent damage from extreme wind or wildfire exposure or elevating transformers outside of flood hazard areas. Although climate drivers are expected to increase the need and demand for these activities, it is less likely that climate drivers will impede the ability to implement infrastructure-hardening or threat mitigation strategies. Economic factors pertaining to this NCF—specifically, the *Operate Market* subfunction—were assessed as being at risk of no disruption for any climate driver by the end of the century for both current and high emissions scenarios. Although the electricity transmission market might be affected by climate drivers, such as through revisions in pricing structures or mechanisms, activities pertaining to operating the electricity market are expected to continue without interruptions.

Table B.107. Subfunction Risk for *Transmit Electricity*

	Baseline	Current Emissions		High Emissions	
Subfunction	Present	2050	2100	2050	2100
Provide Access	2	3	3	3	3
Operate Infrastructure	2	3	3	3	3
Maintain System	2	3	3	3	3
Provide Electrical Protection	2	3	3	3	3
Secure System	1	1	2	2	2

NOTE: 1 = no disruption or normal operations. 2 = minimal disruption. 3 = moderate disruption. See Chapter 2 for further explanation of ratings.

Climate Drivers

All eight climate drivers are expected to cause minimal to moderate disruptions to this NCF by the middle to the end of the century. However, five climate drivers (specifically, extreme heat, flooding, sea-level rise, tropical cyclones and hurricanes, and wildfire) pose the greatest threat to electric transmission infrastructure. For instance, increased ambient air temperatures can reduce transformers' life spans by causing damage to oil-based convective heat sinks, which are used as cooling systems for transformer equipment (Fant et al., 2020). Extreme heat conditions can also reduce ampacity or capacity in transmission lines, and that reduction is exacerbated by increased demand on the electrical grid from high temperatures (Rusco, 2014). Similarly, wildfire and tropical cyclones and hurricanes threaten electric transmission lines and equipment. For example, wood poles or steel support towers carrying transmission lines are susceptible to structural failure from high wind speeds or wildfire-induced heat. One researcher estimated that extreme wind speeds can result in reductions in electric transmission system reliability by more than 30 percent (Rezaei et al., 2016). Some of the effects that some climate drivers, such as extreme heat, wildfire, and tropical cyclones and hurricanes, have on the electric transmission system are evident through recent power outage events. In October 2017, for instance, wildfires in Northern California spread to areas in transmission-line corridors, resulting in service disruptions due to damage to transmission lines (Dale et al., 2018). It is important to note that disruptions from certain climate drivers might be regional. For instance, transmission substations are typically designed to prevent flooding using the 100-year flood elevation plus an additional 0.5 meters; in some areas, such as New York City, transmission substations at ground level might experience flood damage with more-frequent and -intense flood conditions, as well as potential erosion from saltwater exposure due to rising sea levels (de Almeida and Mostafavi, 2016; Souto et al., 2022). For other climate drivers, including drought, extreme cold, and severe storm systems, *Transmit Electricity* was assessed as being at risk of minimal disruptions by the end of the century under current and high emissions scenarios. During drought conditions, for example, dry soil movement and moisture migration can reduce capacity in underground transmission cables by up to 25 percent (P. Johnston, Gomez, and Laplante, 2012). However, more than 90 percent of the in-service transmission lines in the United States are aboveground (DHS, 2022), which can result in minimal or localized disruptions for this NCF from the drought climate driver.

Table B.108. Risk from Climate Drivers for *Transmit Electricity*

Climate Driver	Baseline Present	Current Emissions 2050	Current Emissions 2100	High Emissions 2050	High Emissions 2100
Drought	1	2	2	2	2
Extreme cold	1	1	2	1	2
Extreme heat	2	3	3	3	3
Flooding	2	2	3	3	3
Sea-level rise	2	3	3	3	3
Severe storm systems	2	2	2	2	2
Tropical cyclones and hurricanes	2	3	3	3	3
Wildfire	2	3	3	3	3

NOTE: 1 = no disruption or normal operations. 2 = minimal disruption. 3 = moderate disruption. See Chapter 2 for further explanation of ratings.

Impact Mechanisms

Physical damage or disruption to transmission infrastructure is the most significant impact mechanism affecting this NCF. Physical damage can take the form of short-term effects, such as structural failure of transmission poles from exposure to fire, or long-term effects, such as saltwater exposure to transmission equipment due to sea-level rise. Additionally, workforce shortages and demand changes are impact mechanisms particular to the extreme heat climate driver for this NCF. Increased ambient air temperature results in an increased load on the electrical grid, as well as unsafe working conditions for staff performing maintenance on transmission lines or equipment due to risk of heat-related illness (Roh and Alamo, 2022).

Cascading Risk

Transmit Electricity is dependent on the following NCFs:

- *Generate Electricity*
- *Maintain Supply Chains.*

The *Transmit Electricity* NCF is closely linked to NCFs pertaining to the generation of electricity, as well as reliable operation of supply chains. *Transmit Electricity* is dependent on reliable electric power generation; disruptions or failures to electric power generation inhibits the ability to effectively transmit the electric power downstream. Additionally, the transmission of electricity requires access to equipment, materials, and supplies for operation, maintenance, and repair. Disruptions to supply chains can increase the risk of disruptions to the *Transmit Electricity* NCF.

Strength of Evidence

There is considerable evidence to indicate potential risks and disruptions from climate drivers for the *Transmit Electricity* NCF. However, the strength of evidence varies by the subfunction and climate driver. For instance, literature on the *Provide Access*, *Operate Infrastructure*, and *Maintain System* subfunctions is well documented through academic papers, case studies of electric transmission system disruptions, and reports from regulatory agencies (e.g., the Federal Energy Regulatory Commission). There is less evidence of potential effects for other activities pertaining to this NCF, such as the *Operate Market* subfunction. However, this could be due to climate drivers not having significant effect on those subfunctions. The strength of evidence also varies by climate driver. For example, evidence on the effects that extreme heat has on electrical transmission systems and equipment is well documented, whereas effects from other climate drivers, such as drought, exist but are limited.

Table B.109. Strength of Evidence for *Transmit Electricity*

Strength of Evidence	
High	Extreme heat; tropical cyclones and hurricanes
Medium	Flooding; sea-level rise; severe storm systems; wildfire
Low	Drought; extreme cold

Risk Rating Interpretations

Table B.110 provides examples of how to interpret the risk ratings for this NCF.

Table B.110. Example Interpretations of Risk Ratings for *Transmit Electricity*

Risk Rating	NCF Example
1: No disruption or normal operations	Transmission lines and equipment can operate without interruptions.
2: Minimal disruption	Transmission lines and equipment experience physical damage, resulting in limited or manageable disruptions of electric power service (i.e., disruptions lasting a few hours or a few days).
3: Moderate disruption	Transmission lines and equipment experience physical damage, resulting in electric power service disruptions in many parts of the country.
4: Major disruption	Transmission lines and equipment experience severe physical damage, leading to electrical power service disruptions throughout the country.
5: Critical disruption	The electrical power grid experiences widespread disruptions across the country because of an inability to transmit electricity from generation sources to distribution systems.

180

Distribute Electricity

Table B.111. Scorecard for *Distribute Electricity*

	National Risk Assessment				
	Baseline	Current Emissions		High Emissions	
	Present	2050	2100	2050	2100
Distribute Electricity	2	3	3	3	3

Description: "Maintain and operate medium- to low-voltage system to reliably supply consumer demand for electricity from the bulk electric power network" (CISA, 2020, p. 3).

Highlights

- *Distribute Electricity* was assessed to be at risk of moderate disruption by 2050 because of the potential damage that flooding, tropical cyclones and hurricanes, and extreme heat are anticipated to have on distribution substations, utility poles, and power conductors across the country. Wildfire and sea-level rise are secondary risk factors today but will increase the risk of disruption by 2100.
- Because *Distribute Electricity* infrastructure is, by design, diffuse across the country, the sheer number of assets, in combination with poor maintenance practices and outdated design standards, contributes to the NCF's vulnerability.
- The subfunctions of *Operate Distribution System*, *Maintain System*, and *Provide Electrical Protection for System* were assessed to be at greatest risk of disruption because these subfunctions affect the real-time direct operation and reliability of the NCF.

Synopsis

Distribute Electricity was assessed to be at risk of moderate disruption by 2050 in both emissions scenarios primarily because of the effects that flooding, tropical cyclones and hurricanes, and extreme heat are anticipated to have on distribution substations, utility poles, and power conductors. We also assessed that, by 2100, sea-level rise and wildfire will pose a risk of moderate disruption under both emissions scenarios. Sea-level rise will likely exacerbate flooding and water damage from hurricanes and severe storm systems to the point at which affected components are not able to recover sufficiently between events or could become permanently inundated. A moderate disruption to the *Distribute Electricity* function could look like the effects of severe weather in 2012, when a series of thunderstorms and strong winds caused 3 million people and businesses to lose power across the Midwest to the mid-Atlantic (Zamuda et al., 2013).

Distribute Electricity assets are increasingly at risk of disruption from climate drivers because of the sheer number of infrastructure assets across the country. This, in combination with poor maintenance practices and outdated design standards, contributes to the NCF's vulnerability. There are many documented examples of climate drivers' effects on existing electrical distribution systems, including physical damage to distribution substations and transformers, utility poles, and power conductors as a direct result of flooding, tropical cyclones and hurricanes, severe storm systems, and wildfire (Office of Energy Policy and Systems Analysis, 2015; Shield et al., 2021). For example, extreme precipitation on the West and East Coasts and regions with dense riverine systems has led to unprecedented flooding, with effects on *Distribute Electricity* infrastructure throughout the United States. Sea-level rise on the West and East Coasts, which exacerbates damage from severe storm systems, including hurricanes,

181

along the Gulf Coasts can also cause physical damage to *Distribute Electricity* assets. Notable examples include repeated flooding events in the Virginia coastal region (Virginia Secretary of Natural and Historic Resources, undated) and Hurricane Harvey in Houston, Texas, in 2017, when flooding damaged substations and other equipment and prevented access to repair and restore power (Dullo et al., 2021; Gangrade et al., 2019; Qiang, 2019; North American Electric Reliability Corporation, 2018). Extreme heat can also degrade the performance of equipment and, coupled with an increase in demand, can strain a distribution system beyond its rated capacity. Many distribution substation components are not rated for the extreme heat projections that much of the western and southeastern United States is likely to experience in the coming century (Burillo, Chester, Pincetl, et al., 2019). Moreover, legacy designs and outdated maintenance standards that consider only historical climatology exacerbate climate-related risks because many practices do not yet account for the more-frequent or -intense climate-related natural hazard events.[39]

Severe storm systems and extreme cold can also disrupt *Distribute Electricity* but often at more-local scales than other climate drivers because the infrastructure components are not as badly damaged or disrupted by these climate drivers and power is often restored more quickly. Drought does not affect *Distribute Electricity* at a national scale because water is not a critical input to the function.

Distribute Electricity NCF assets, such as substations, which serve as nodal points for the broader distribution network, are vital points susceptible to cascading failure. A substation failure can cause power outages throughout the entire power distribution network, resulting in power outages for all interconnected critical infrastructure assets. However, given the distributed nature of much of the electrical grid, there is some built-in adaptive capacity or slack in the system that increases the system's resilience to national cascading failures of the *Distribute Electricity* NCF and those NCFs with downstream dependencies on this function (Abi-Samra, 2017). However, the increased intensity and frequency of climate drivers are likely to stress maintenance and response operations put in place to mitigate vulnerabilities, leading to more-frequent and longer power outages. Furthermore, substations have components that were not designed for the climate risks of the next century (Anshari Munir, Al Mustafa, and Siagian, 2019).

Subfunctions

Of the seven *Distribute Electricity* subfunctions, all but one, *Operate in Market*, face a risk of minimal disruption or greater because of increased exposure to climate drivers. For other subfunctions, such as *Operate Distribution System, Provide Access, Maintain System, Secure System,* and *Provide Electrical Protection for System,* natural hazards affected by climate change that degrade or damage distribution lines and substation infrastructure can lead to significant interruptions in the operation of power distribution, resulting in the inability to provide power access to all customers. Physical damage can also increase demand for system maintenance and repair and for the *Secure System* subfunction. Subfunctions that are provided primarily by personnel, such as *Provide Customer Service,* are less likely than asset-heavy subfunctions to be affected by exposure to climate drivers,

---

[39] Recall that we assumed static, business-as-usual operations, with no attempt to adapt for or mitigate climate change. Routine maintenance and equipment replacement cycles might ensure that the NCF is protected in the future, but this was not considered in our analysis.

although increasing frequency and severity of climate drivers could cause issues with prompt power restoration and reduce the ability to provide consistent and high-quality customer service, posing a risk of moderate disruption by 2100.

Although the *Secure System* subfunction, which relates to preventive hardening against specific threats, was not assessed to be at risk today, by 2050, sea-level rise and wildfire pose minimal risk to the subfunction and moderate risk by 2100. Unless infrastructure along coastlines exposed to sea-level rise is moved, the ability to harden against this chronic, pervasive climate driver will become increasingly difficult. Wildfire poses a similar challenge because mitigating damage relies on improved standards and frequent use of vegetation management practices. In addition, as wildfires become more intense because of climate change, they might spread more quickly and across longer distances, negating the effect of wildfire management practices.

Table B.112. Subfunction Risk for *Distribute Electricity*

Subfunction	Baseline Present	Current Emissions 2050	Current Emissions 2100	High Emissions 2050	High Emissions 2100
Operate Distribution System	2	3	3	3	3
Provide Access	2	2	3	3	3
Maintain System	2	3	3	3	3
Provide Electrical Protection for System	2	3	3	3	3
Secure System	1	2	3	2	3
Provide Customer Service	2	2	3	3	3

NOTE: 1 = no disruption or normal operations. 2 = minimal disruption. 3 = moderate disruption. See Chapter 2 for further explanation of ratings.

Climate Drivers

All climate drivers except for drought pose risks to *Distribute Electricity* (Zuloaga and Vittal, 2021). Distribution infrastructure is susceptible to flooding, severe storm systems, and tropical cyclones that can lead to wind or water damage to structures (Boggess, Becker, and Mitchell, 2014) and extreme heat and extreme cold that can affect the performance of *Distribute Electricity* components (Anshari Munir, Al Mustafa, and Siagian, 2019; Keane, Schwarz, and Thernherr, 2013). Judging from historical effects and projected exposure of critical power distribution assets to climate drivers, we anticipate that hurricanes, sea-level rise, extreme heat, and wildfire will pose the largest risk to power distribution by 2100. Hurricanes and flooding already cause major disruptions regionally (Nateghi, 2018), but these effects rarely, if ever, disrupt routine operations to the NCF nationally. Extreme heat and sea-level rise pose more-chronic risk to power distribution infrastructure (Burillo, Chester, Pincetl, et al., 2019; G. Griggs and Patsch, 2019; Wuebbles et al., 2017), but these effects are also largely limited to the infrastructure within a region. Wildfire risk will increasingly become a concern in the western United States, but historically, wildfire's effects on *Distribute Electricity* components have been felt on only a local scale (Dale et al., 2018).

Table B.113. Risk from Climate Drivers for *Distribute Electricity*

Climate Driver	Baseline	Current Emissions		High Emissions	
	Present	2050	2100	2050	2100
Drought	1	1	1	1	1
Extreme cold	2	2	2	2	1
Extreme heat	2	3	3	3	3
Flooding	2	3	3	3	3
Sea-level rise	1	2	3	3	3
Severe storm systems	2	2	2	2	2
Tropical cyclones and hurricanes	2	3	3	3	3
Wildfire	2	2	3	2	3

NOTE: 1 = no disruption or normal operations. 2 = minimal disruption. 3 = moderate disruption. See Chapter 2 for further explanation of ratings.

Impact Mechanisms

Distribution infrastructure could be physically disrupted by damage to substations, utility poles, and power conductors from winds and heavy precipitation during hurricanes and severe storms, from inundation due to sea-level rise, and from wildfire. Extreme cold or heat events can decrease the efficiency and performance of the infrastructure and increase demand for heating or cooling, which places additional strain on the system. Additionally, severe weather conditions during hurricanes, severe storm systems, wildfire, and extreme heat and cold events could affect the workforce's ability to assess and fix damage or perform maintenance activities.

Cascading Risk

Distribute Electricity is critically dependent on the following NCFs:

- *Generate Electricity*
- *Transmit Electricity*
- *Exploration and Extraction of Fuels*
- *Store Fuel and Maintain Reserves*
- *Fuel Refining and Processing Fuels*
- *Transport Materials by Pipeline.*

In the NCF framework, *Distribute Electricity* is one part of three interconnected NCFs that compose bulk power delivery: *Generate Electricity*, *Transmit Electricity*, and *Distribute Electricity*. If any one of these NCFs fails, they all fail to deliver power to end users (Vartanian et al., 2018). Disruptions to *Distribute Electricity* infrastructure could cause cascading failures to other NCFs within a region. Cascading risk between regions is much rarer because climate drivers manifest regionally, and the redundancies present in distribution infrastructure systems across regions help mitigate this risk. We identified *Distribute Electricity* as the NCF with the most downstream NCFs depending on it to ensure normal operations. Additionally, other NCFs—*Exploration and Extraction of Fuels, Store Fuel and Maintain Reserves,* and *Fuel Refining and Processing Fuels*—maintain the development of fuel resources that are critical inputs to *Generate Electricity.* Should any of the fuel NCFs fail, it would likely be exceptionally difficult to maintain continuous power delivery through *Distribute Electricity* (C. Murphy et al., 2020). Moreover, pipelines that transport primary hydrocarbons (natural gas and crude oil) are a critical link between the fuel resource–related NCFs on which *Generate Electricity* depends.

Strength of Evidence

The body of existing literature and studies of climate risk to *Distribute Electricity* infrastructure is expansive and well studied. Previous studies have detailed risks to the NCF-related infrastructure by U.S. region (Glick and Christiansen, 2019; Office of Energy Policy and Systems Analysis, 2015; Wakiyama and Zusman, 2021), and complementary studies have analyzed climate risk to supporting infrastructure and components from other countries that can be used to assess and substantiate risk ratings for U.S. infrastructure (Rübbelke and Vögele, 2011; Tobin et al., 2018; S. Turner, Hejazi, et al., 2017; Zamuda et al., 2013). Because most modern systems rely on uninterrupted power supply (FEMA, 2007), there is a general interest in studying how to maintain continuous power through all types and scales of climate drivers (Boggess, Becker, and Mitchell, 2014; Huang, Swain, and Hall, 2020; Saint, 2009).

Table B.114. Strength of Evidence for *Distribute Electricity*

Strength of Evidence	
High	Flooding; sea-level rise; tropical cyclones and hurricanes; extreme cold; extreme heat; wildfire; drought
Medium	Severe storm systems

Risk Rating Interpretations

Table B.115 provides examples of how to interpret the risk ratings for this NCF.

Table B.115. Example Interpretations of Risk Ratings for *Distribute Electricity*

Risk Rating	NCF Example
1: No disruption or normal operations	Distribution of electricity continues to function as it does today, with occasional minimal disruptions due to weather and operational problems.
2: Minimal disruption	Distribution lines are knocked down by wind and trees, and a few substations are flooded by a severe storm in a few neighboring cities, causing power outages. These outages do not affect a substantial number of customers, and power is restored quickly.
3: Moderate disruption	A severe hurricane causes damage to distribution lines, transformers, and substations in two large metropolitan areas. Some customers in those areas are without power for several days.
4: Major disruption	Extreme weather events in near succession or simultaneously in a few regions across the United States cause widespread, prolonged power outages and significant damage to critical substations. Weather conditions prevent crews from accessing faulted and damaged equipment, impeding their ability to assess and fix damaged infrastructure.
5: Critical disruption	Distribution infrastructure is nonoperational across the majority of the country as a result of more-frequent and -intense extreme weather events, often hitting multiple regions at once and stressing the national grid and the workforce that supports maintenance and repair of damaged components. Prolonged outages occur in multiple regions.

Manage Hazardous Materials

Table B.116. Scorecard for *Manage Hazardous Materials*

	National Risk Assessment				
	Baseline	Current Emissions		High Emissions	
	Present	2050	2100	2050	2100
Manage Hazardous Materials	2	3	3	3	3

Description: "Safely identify, monitor, handle, store, transport, use, and dispose of hazardous materials (including chemical, biological, radioactive, nuclear, and explosive substances) under normal operations and in response to emergencies" (CISA, 2020, p. 4).

Highlights

- The *Manage Hazardous Materials* NCF was assessed as a moderate risk of disruption given the likelihood of various climate drivers having regional effects on this NCF.
- Flooding, sea-level rise, and tropical cyclones and hurricanes pose the greatest risk of disruption for this NCF, with a risk of moderate disruption assessed by 2050.
- More than 30 percent of facilities that make, use, or store hazardous chemicals are in areas prone to natural hazards, such as flooding, storm surge, and wildfire (Gómez, 2022).
- Extreme heat, exacerbated by the need for personal protective equipment in the use and handling of hazardous materials and waste, can cause workforce shortage issues (Favata, Buckler, and Gochfeld, 1990; Paull and Rosenthal, 1987; Spector and Sheffield, 2014).
- Two of the three subfunctions—*Manage Use of Hazardous Materials* and *Manage Hazardous Wastes*— are particularly vulnerable to climate effects because of their potential for physical disruption.
- Approximately 2,000 current or potential Superfund sites are located along the East and Gulf Coasts (J. Carter and Kalman, 2020), making this region particularly vulnerable to climate change for the *Manage Hazardous Materials* NCF.

Synopsis

Climate change is anticipated to lead to an increase in the frequency and severity of the climate drivers, which, has the potential to cause physical damage and disruption for hazardous material use and waste management (Reidmiller et al., 2018). Facilities of concern include those managed under the Resource Conservation and Recovery Act of 1976 (Pub. L. 94-580); the Comprehensive Environmental Response, Compensation, and Liability Act of 1980 (Pub. L. 96-510) (CERCLA, the bill that created the Superfund); and the Emergency Planning and Community Right-to-Know Act of 1986 (Pub. L. 99-499, Title III). The latter covers industrial and federal facilities required to monitor chemical releases under EPA's Toxics Release Inventory Program. Tropical cyclones and hurricanes, severe storm systems, flooding, and sea-level rise have the capacity to cause widespread damage or destruction to hazardous waste treatment and storage facilities and to increase the likelihood of unintended environmental releases of hazardous waste materials. Flooding, severe storm systems, wildfire, and sea-level rise are expected to affect approximately 60 percent of nonfederal Superfund National Priorities List sites (Gómez, 2019a). Additionally, more than 30 percent of facilities that make, use, or store hazardous chemicals are located in areas prone to natural hazards, such as flooding, severe storm systems, and wildfire (Gómez, 2022).

187

Extreme heat can make transport of hazardous materials difficult, affecting supply chains and hazardous waste removal and disposal for industrial manufacturing; cause increased chemical reactions of hazardous chemicals, resulting in unintended air emissions or other chemical degradation; and affect the hazardous material and waste management workforces. Extreme heat, exacerbated by the need for personal protective equipment in the use and handling of hazardous materials and waste, can cause workforce shortage issues (Favata, Buckler, and Gochfeld, 1990; Paull and Rosenthal, 1987; Spector and Sheffield, 2014). Wildfire can threaten infrastructure and increase the risk of unintentional hazardous waste releases to the environment. Drought can lead to an increase in the use of fertilizers, pesticides, insecticides, and other hazardous contaminants, increasing demand for hazardous materials (Gross, 2021; Tudi et al., 2021).

Together, the effects of climate change are expected to increase the risk of disruption to hazardous waste management treatment and disposal sites, resulting in potentially hazardous environmental releases. During disasters, emergency management personnel will need to coordinate with public health officials to contain potential releases, limit exposure to the public, and manage risk for hazardous material workers. A moderate disruption to *Manage Hazardous Materials* could result in significant environmental contamination. In 2017, for example, Hurricane Harvey devastated Houston, Texas, a major industrial city, resulting in at least 100 hazardous waste spills (T. Griggs et al., 2017).

Manage Hazardous Materials is expected to be at risk of moderate disruption from climate change in both the current and high emissions scenarios by 2050. These ratings are due chiefly to an expected increase in the frequency, intensity, and duration of flooding, tropical cyclones and hurricanes, and sea-level rise. Coastal communities have already experienced hazardous leaks, fires, or explosions close to populated areas, other critical infrastructure, and sensitive natural ecosystems (Gómez, 2019a; Summers, Lamper, and Buck, 2021), resulting in local to small-scale regional disruption. Increased exposure to the climate drivers could result in more-catastrophic regional disruption to the NCF.

Subfunctions

Two subfunctions for this NCF—*Manage Use of Hazardous Materials* and *Manage Hazardous Wastes*—are expected to be at risk of physical damage or disruption associated with natural disasters. Especially of concern is the ability to contain environmental releases of hazardous materials or wastes due to physical disruption to the system caused by flooding, tropical cyclones and hurricanes, severe storm systems, and sea level rise. Effects on the third subfunction—*Manage Transport of Hazardous Materials*—could affect hazardous waste removal and disposal from industrial facilities. Disruption to the transport of hazardous materials has the potential to affect resources for other NCFs as hazardous materials are often an important raw material in industrial manufacturing and agricultural production. However, this subfunction is not expected to be as significantly affected as the other subfunctions, primarily because of the expected physical effects on hazardous waste infrastructure and disposal sites. Specifically, stationary infrastructure is more vulnerable to physical disruption, whereas transportation of hazardous materials can be rerouted, relocated, or otherwise managed in the event of a climate driver.

Table B.117. Subfunction Risk for *Manage Hazardous Materials*

Subfunction	Baseline Present	Current Emissions 2050	Current Emissions 2100	High Emissions 2050	High Emissions 2100
Manage Use of Hazardous Materials	2	3	3	3	3
Manage Transport of Hazardous Materials	1	2	2	2	3
Manage Hazardous Wastes	2	3	3	3	3

NOTE: 1 = no disruption or normal operations. 2 = minimal disruption. 3 = moderate disruption. See Chapter 2 for further explanation of ratings.

Climate Drivers

Nearly all climate drivers will have at least a minimal effect on the management of hazardous materials. This is because flooding and tropical cyclones and hurricanes have the potential to threaten or cause physical damage and injuries at hazardous waste facilities. Additionally, extreme heat, sea-level rise, severe storm systems, and wildfire can cause physical disruption to hazardous waste infrastructure and thereby affect facility operation—which could result in accidental chemical releases into the environment. Drought can increase the need for hazardous chemicals, particularly in the agricultural sector (Gross, 2021; Tudi et al., 2021), potentially increasing the need for hazardous material transport and waste management. Extreme cold is not expected to significantly affect this NCF because hazardous material use, transport, and disposal will not be significantly affected by this climate driver.

Table B.118. Risk from Climate Drivers for *Manage Hazardous Materials*

Climate Driver	Baseline Present	Current Emissions 2050	Current Emissions 2100	High Emissions 2050	High Emissions 2100
Drought	1	2	2	2	3
Extreme cold	1	1	1	1	1
Extreme heat	1	2	2	2	3
Flooding	2	3	3	3	3
Sea-level rise	1	2	3	3	3
Severe storm systems	1	2	2	2	3
Tropical cyclones and hurricanes	2	3	3	3	3
Wildfire	1	2	2	2	3

NOTE: 1 = no disruption or normal operations. 2 = minimal disruption. 3 = moderate disruption. See Chapter 2 for further explanation of ratings.

Impact Mechanisms

The primary impact mechanism of concern for the management of hazardous materials is physical damage or disruption to the NCF due to extreme weather events. These effects have been well documented following major storms—primarily, hurricanes affecting chemical and hazardous waste management facilities along the coastlines (Manuel, 2006; Santella, Steinberg, and Sengul, 2010; TCEQ, 2017). Workforce shortages are of concern during extreme heat events, particularly because hazardous material workers must wear personal protective equipment while handling hazardous materials or waste streams, resulting in higher risk when such equipment or training in its use is unavailable. Finally, changes in demand for hazardous materials can increase during times of drought because of the increased use of fertilizers and pesticides.

Cascading Risk

The disruption of other NCFs also has the potential to cause severe disruption to the *Manage Hazardous Materials* NCF. Upstream NCFs that could cause disruption or failure to *Manage Hazardous Materials* include the following NCFs:

- *Distribute Electricity*
- *Produce Chemicals*
- *Transport Cargo and Passengers by Road*
- *Develop and Maintain Public Works and Services*.

Chemical production is a key component of the *Manage Hazardous Materials* NCFs because the use and transport of hazardous materials first requires chemical processing. This process often requires specialized processing and equipment that require energy. The International Energy Agency has identified the chemical sector as the largest industrial energy consumer (International Energy Agency, undated). Following production, chemicals are often transported by road. In fact, the Bureau of Transportation Statistics has reported that about 65 percent of U.S. hazardous material shipments occurred via truck transport (Bureau of Transportation Statistics, undated-a). Thus, disruption to *Transport Cargo and Passengers by Road* and *Develop and Maintain Public Works and Services* could result in further disruption to the *Manage Hazardous Materials* NCF.

Strength of Evidence

The strength of evidence on climate change's effect on *Manage Hazardous Materials* is modest but varies considerably by climate driver. The effect of natural disasters on the use, storage, transport, and disposal of hazardous material has been well documented in academic literature, government sources, and newspapers. In particular, there is extensive literature on the effects that flooding, sea-level rise, and tropical cyclones and hurricanes have on hazardous material use and waste management. However, other climate drivers' effects on the NCF are less well documented.

Table B.119. Strength of Evidence for *Manage Hazardous Materials*

Strength of Evidence	
High	Flooding; sea-level rise
Medium	Tropical cyclones and hurricanes; wildfire
Low	Severe storm systems; extreme cold; extreme heat; drought

Risk Rating Interpretations

Table B.120 provides examples of how to interpret the risk ratings for this NCF.

Table B.120. Example Interpretations of Risk Ratings for *Manage Hazardous Materials*

Risk Rating	NCF Example
1: No disruption or normal operations	The use of hazardous materials continues under normal conditions, with no disruption due to the event.
2: Minimal disruption	Some industrial facilities utilizing hazardous chemicals experience an incidental chemical release due to the storm surge from a hurricane, but effects are contained to a small geographic area and the NCF continues to operate normally otherwise.
3: Moderate disruption	Following a catastrophic storm, several coastal facilities are inundated by storm surge, and the resulting contamination disrupts regional production.
4: Major disruption	A major disruption would occur if the NCF were disrupted on a national scale, where most hazardous material facilities are experiencing disruption.
5: Critical disruption	A critical disruption would occur if essentially all hazardous material facilities nationwide were experiencing disruption.

Manage Wastewater

Table B.121. Scorecard for *Manage Wastewater*

| | National Risk Assessment | | | | |
| | Baseline | Current Emissions | | High Emissions | |
	Present	2050	2100	2050	2100
Manage Wastewater	2	3	3	3	3

Description: "Collect and treat industrial and residential wastewater to meet applicable public health and environmental standards prior to discharge into a receiving body" (CISA, 2020, p. 4).

Highlights

- *Manage Wastewater* is at risk of moderate disruption for all future time periods and emissions scenarios because of the regional effects that could result from several climate drivers.
- Wildfire and flooding pose the most near-term risk. Additionally, flooding, sea-level rise, and drought pose risk of moderate disruption by 2050.
- Physical damage or disruption to wastewater treatment facilities can occur during natural hazard events. For example, flooding of wastewater or stormwater collection systems can result in combined sewer overflows (CSOs).
- Unintended wastewater releases to the environment can affect ecological and human health, resulting in long-term consequences beyond a storm event itself.
- *Manage Wastewater* is facing risk of disruption from climate change primarily because of physical damage and disruption.
- Wastewater infrastructure already suffers from degradation due to deferred maintenance (ASCE, 2021), increasing vulnerability to physical damage and disruption.
- The *Treat Wastewater, Provide Facilities, Collect Wastewater, Maintain Facility Infrastructure,* and *Govern Wastewater* subfunctions are at particular risk because of the likelihood of physical damage or disruption.
- Wastewater treatment plants are often in low-lying and coastal areas, increasing the likelihood of inundation.

Synopsis

Climate change is anticipated to lead to an increase in the frequency and severity of the climate drivers (Reidmiller et al., 2018), which, in turn, has the potential to cause physical damage and disruption to wastewater infrastructure and increase the demand for wastewater management services.

Tropical cyclones and hurricanes, severe storm systems, and flooding have the capacity to cause widespread damage or destruction to wastewater collection and treatment infrastructure. For example, nearly half of Houston's wastewater facilities were inundated during Hurricane Harvey in 2017, affecting a population of more than 160,000 people (TCEQ, 2018). Additionally, the storm caused more than 25 million gallons of untreated wastewater to be released to the environment, resulting in environmental and human health effects (TCEQ, 2017). Similarly, Hurricane Sandy caused more than 10 billion gallons of untreated or partially treated wastewater to be discharged to surrounding bodies of water (Schwirtz, 2013); more recently, Hurricane Ian caused a dozen wastewater treatment plants to discharge untreated waste (Choi-Schagrin, 2022). These climate drivers can also increase the demand for wastewater treatment as regions are inundated with stormwater. Given that wastewater treatment infrastructure is generally located near bodies of water, these issues can be exacerbated by sea-level rise in coastal regions. Wildfire and drought can also affect wastewater conveyance and infrastructure systems. For example, during drought events, water use can be reduced, causing low-flow issues for wastewater treatment. This, combined with reductions in stormwater runoff, wastewater treatment designed to meet a certain capacity might struggle to adapt treatment processes for reduced load (Andersen, Lewis, and Sargent, 2004). Additionally, extreme heat can influence the wastewater treatment and collection processes and change the demand for wastewater treatment (Zouboulis and Tolkou, 2014). In locations experiencing diminished water supplies, treated wastewater effluent can serve as a useful resource for augmenting water supplies (Lachman et al., 2016).

These systems are vulnerable in part because, historically, wastewater infrastructure has been designed under the assumption of climate stationarity (i.e., based on historical climate and weather data, such as precipitation). Under changing climate conditions, the placement and design of wastewater infrastructure might require methods that consider future climate patterns to ensure adequate performance (Committee on Adaptation to a Changing Climate, 2015; Milly et al., 2008). Even under current conditions, many systems are operating beyond their intended design lives, causing failures to increase over time. Furthermore, the ubiquity of combined stormwater and wastewater sewer systems in the United States, particularly in the Midwest and along the East Coast, presents additional challenges for wastewater management. Combined sewer systems are more susceptible than separate ones to climate change because of increased severe storm systems likely in the regions where combined sewer systems continue to operate (Fortier and Mailhot, 2015). Under projected climate conditions, an increased frequency of CSOs is expected, contaminating nearby water supplies and harming ecosystems. Similarly, distributed wastewater systems along the coasts (i.e., on-site wastewater treatment and septic systems) are at greater risk than inland systems for groundwater contamination during severe storm systems and sea-level rise (J. Cooper, Loomis, and Amador, 2016).

Overall, the effects of climate change are expected to significantly affect the provision of wastewater treatment, particularly in areas prone to flooding and storm surge from tropical cyclones and hurricanes. *Manage Wastewater* is expected to be at risk of moderate disruption from climate change in both the current and high emissions scenarios by 2050. These ratings are due primarily to this NCF's vulnerability and to its exposure to an expected increase in the frequency, intensity, or duration of flooding, sea-level rise, severe storm systems, and tropical cyclones and hurricanes. Wastewater management is already vulnerable in parts of the country, primarily because of the existence of combined wastewater and stormwater systems. These systems are at greater risk of disruption due to the additional exposure to flooding and storm surge from tropical cyclones and hurricanes (Fortier and Mailhot, 2015).

Subfunctions

Of the seven subfunctions under *Manage Wastewater*, five are expected to be at risk of disruption due to physical damage and increased demand resulting from climate change: *Treat Wastewater*, *Provide Facilities*, *Collect Wastewater*, *Maintain Facility Infrastructure*, and *Govern Wastewater*. These effects will occur most acutely in the *Collect Wastewater* and *Maintain Facility Infrastructure* subfunctions as physical assets are damaged or degraded by increased incidence and severity of flooding, severe storm systems, and sea-level rise, requiring more-frequent maintenance, reconstruction, and rehabilitation. Additionally, the function could be disrupted because wastewater utilities might have difficulty maintaining environmental standards following severe storm systems. Historically, the City of Pittsburgh has suffered from CSOs; in 2021, the city reached an agreement with EPA to reduce CSOs to improve compliance with environmental regulation (EPA, 2021). As regions confront the effects of climate change, wastewater governance could become more difficult.

The remaining subfunctions—*Assure Secure Operations* and *Provide HR Services*—were assessed as being at no risk of disruption due to climate change. This assessment was generally due to these subfunctions' ability to perform operations remotely, the lack of an infrastructure footprint required for operations, or the current capacity of management entities. For example, within the subfunction *Assure Secure Operations*, the management of wastewater—the plans, protocols, and processes in place for short- and long-term operations—has long been focused on protecting and managing water resources and the operations of wastewater systems in the face of weather events. Although climate change is projected to increase the frequency or magnitude of extreme heat events for most of the United States, these changes are unlikely to directly affect localities' ability to produce and manage plans, protocols, and processes of water management in the near to long term (Godden, Ison, and Wallis, 2011). One exception to this generalization is for local and, particularly, rural areas and small wastewater managers around the United States. These entities might have a harder time addressing the challenges that enhanced risk climate drivers pose and could become overwhelmed by the demands placed on them.

Table B.122. Subfunction Risk for *Manage Wastewater*

Subfunction	Baseline Present	Current Emissions 2050	Current Emissions 2100	High Emissions 2050	High Emissions 2100
Treat Wastewater	2	2	3	3	3
Provide Facilities	2	2	3	3	3
Collect Wastewater	2	3	3	3	3
Maintain Facility Infrastructure	2	3	3	3	3
Govern Wastewater	2	2	2	2	3

NOTE: 2 = minimal disruption. 3 = moderate disruption. See Chapter 2 for further explanation of ratings.

Climate Drivers

All climate drivers except extreme cold pose a risk of at least minimal disruption to *Manage Wastewater*. Several climate drivers have the potential to cause physical damage and disruption for wastewater management services: flooding, sea-level rise, severe storm systems, tropical cyclones and hurricanes, and wildfire (EPA, 2022a; National Institute of Standards of Technology, 2016). In several circumstances, these climate drivers will also affect effluent flow patterns (e.g., from combined sewer systems) and cause degradation in surface water quality (e.g., low oxygen conditions, eutrophication, and toxic algal blooms). Most wastewater treatment plants sit along waterways or coastlines, thus subject to riverine flooding and sea-level rise. These climate drivers could temporarily or permanently damage discharge points or the treatment plants themselves (Hummel, Berry, and Stacey, 2018). These effects could halt operations or result in sewage spills that have implications for the environment and human health (Olds et al., 2018). Additionally, flooding at treatment plants or outfalls can lead to sewer backups at residences and businesses within the upstream collection system (Sandink and Robinson, 2022). Drought and extreme heat are also expected to affect wastewater management. Drought conditions can reduce wastewater treatment volumes, increasing contaminant concentrations and affecting treatment efficacy (Andersen, Lewis, and Sargent, 2004). If low-flow conditions in receiving waters occur because of drought, concentrations of pollutants could increase in water bodies, resulting in the exceedance of total maximum daily loads or other water quality thresholds and limiting wastewater treatment plants' ability to discharge into receiving waters.

Table B.123. Risk from Climate Drivers for *Manage Wastewater*

Climate Driver	Baseline	Current Emissions		High Emissions	
	Present	2050	2100	2050	2100
Drought	2	3	3	2	3
Extreme cold	1	1	1	1	1
Extreme heat	2	2	2	2	2
Flooding	2	3	3	2	3
Sea-level rise	2	3	3	2	3
Severe storm systems	2	2	3	2	3
Tropical cyclones and hurricanes	2	2	3	2	3
Wildfire	2	2	3	2	3

NOTE: 1 = no disruption or normal operations. 2 = minimal disruption. 3 = moderate disruption. See Chapter 2 for further explanation of ratings.

Impact Mechanisms

Climate change's major effect on *Manage Wastewater* will be direct physical damage or disruption to the system. Physical damage resulting from severe storm systems, tropical cyclones and hurricanes, and wildfire will affect wastewater utilities' ability to convey and treat wastewater during and following climate drivers. These can be short-term effects on wastewater management services or longer-lasting effects that require capital investments. Although other impact mechanisms, such as demand changes, could also occur under certain climate drivers, the major effects on the system will be physical damage and disruption.

Cascading Risk

Manage Wastewater is dependent on the following NCFs:

- *Distribute Electricity*
- *Develop and Maintain Public Works and Services*

195

- *Manage Hazardous Materials*
- *Produce Chemicals*
- *Maintain Supply Chains*
- *Operate Government.*

Wastewater utilities are highly dependent on other NCFs to ensure normal operations (Wasley, Jacobs, and Weiss, 2020). *Distribute Electricity* is critical to the normal operations of a wastewater utility because power is needed for many collection and treatment plant functions. The development and maintenance of infrastructure (*Develop and Maintain Public Works and Services*) are necessary to ensure asset life cycles and their proper functioning over time. The management of hazardous waste produced in wastewater treatment plants is regulated, necessitating the functioning of *Manage Hazardous Materials* (Silva, Matos, and Rosa, 2016). The production of chemicals and the supply chains that deliver chemicals and other inputs are necessary for proper treatment of waste streams. Because the majority of wastewater and stormwater agencies are public, *Operate Government* is also a necessary condition for this NCF to function.

Strength of Evidence

The effect that flooding, severe storm systems, and tropical cyclones and hurricanes can have on wastewater management has been well documented through academic literature. Specifically, the evidence from academic literature describes how wastewater infrastructure can experience physical damage and operations of wastewater systems can experience disruptions. For other climate drivers—sea-level rise, drought, and extreme heat—there is moderate evidence to document climate change's effects on *Manage Wastewater*. The effects that extreme cold and wildfire can have on *Manage Wastewater* are not well documented in the literature.

Table B.124. Strength of Evidence for *Manage Wastewater*

Strength of Evidence	
High	Flooding; severe storm systems; tropical cyclones and hurricanes
Medium	Sea-level rise; drought; extreme heat
Low	Extreme cold; wildfire

Risk Rating Interpretations

Table B.125 provides examples of how to interpret the risk ratings for this NCF.

Table B.125. Example Interpretations of Risk Ratings for *Manage Wastewater*

Risk Rating	NCF Example
1: No disruption or normal operations	Wastewater systems continue to operate normally, without significant disruption to collection, treatment, or effluent discharge.
2: Minimal disruption	A city experiences sanitary sewer overflows as a result of flooding.
3: Moderate disruption	An entire region is affected by a hurricane, resulting in significant storm surge affecting several wastewater treatment facilities. There are several unintended effluent releases to surrounding water bodies, potentially affecting human health and the environment.
4: Major disruption	A major disruption would occur if the NCF is disrupted on a national scale, where most wastewater treatment facilities are experiencing disruption.
5: Critical disruption	A critical disruption would occur if essentially all wastewater treatment facilities nationwide were experiencing disruption.

Supply Water

Table B.126. Scorecard for *Supply Water*

	National Risk Assessment				
	Baseline	**Current Emissions**		**High Emissions**	
	Present	**2050**	**2100**	**2050**	**2100**
Supply Water	3	3	4	3	4

Description: "Maintain availability of water (raw and treated)" (CISA, 2020, p. 7).

Highlights

- *Supply Water* is already stressed by drought conditions in many regions in the United States. Thus, the NCF was assessed as being at risk of major disruption given the projected challenges the NCF will continue to face as available water supplies diminish.
- Prolonged droughts limit the amount of water available for consumption and are already affecting water supplies throughout the United States.
- Drought, extreme heat, flooding, sea-level rise, severe storm systems, tropical cyclones and hurricanes, and wildfire are climate drivers that can damage water infrastructure and degrade water quality.
- *Supply Water* is vulnerable to additional challenges due to lack of investment in failing water infrastructure and deferred maintenance.
- Subfunctions related to *Supply Water*—including *Provide Raw Water*, *Provide Agricultural Water*, and *Provide Potable Water*—are all particularly vulnerable.
- The western United States is particularly vulnerable to drought, forcing states to contend with diminishing water supplies and increased demand. This region is also host to a significant portion of U.S. agriculture.

Synopsis

Climate change is anticipated to lead to an increase in the frequency and severity of climate drivers (Reidmiller et al., 2018), which, in turn, has the potential to affect *Supply Water* through lack of resources and physical damage and disruptions to water supply, treatment, and distribution infrastructure.

Flooding, sea-level rise, severe storm systems, tropical cyclones and hurricanes, wildfire, drought, and extreme heat are climate drivers that present challenges for water utilities. Tropical cyclones and hurricanes, severe storm systems, and flooding have the capacity to cause widespread physical damage or destruction to water infrastructure. For example, in 2018, 40 of the 114 drinking water treatment plants controlled by Puerto Rico's water authority were damaged during Hurricane Maria from debris or inundation (Preston et al., 2020). Flooding can also cause intrusion of harmful contaminants that pose human health risk in both surface and groundwater systems. In Jackson, Mississippi, heavy rainfall caused the Pearl River to flood, allowing contaminants to breach the system thanks to significant decreases in the water pressure (Dennis and Kaplan, 2022). Sea-level rise presents challenges for coastal water systems, including physical damage to infrastructure and saltwater intrusion of source water supplies (Climate Change Adaptation Resource Center, 2022b). Researchers in Florida evaluated the effects of saltwater intrusion in drinking water wells near the coast, and some regions are already experiencing increased salinity levels in coastal water supplies (Blanco et al., 2013). Wildfire can threaten water infrastructure, contaminate drinking supplies, and increase the need for raw water to combat wildfire (Bladon et al., 2014). The Tubbs Fire in 2017 and Camp Fire in 2018 caused drinking water to be contaminated with benzene, a known carcinogen (Proctor et al., 2020).

Drought leads to a lack of resources, and extreme heat compounds these effects by increasing water demand, putting additional pressure on an already-stressed system. Extended drought is already affecting water supplies in many regions, particularly western states, and has resulted in some areas projecting dangerously low water supplies. For example, the city of Coalinga, California, is projected to run out of water before the end of 2022 (Partlow, 2022); in the past several years, the city of Las Vegas, Nevada, has experienced 120 days or less of surface water supply allocation, requiring greater draws from already-taxed groundwater aquifers (Drought Response and Recovery Project for Water Utilities, undated). More generally, communities reliant on groundwater have had to take steps to augment supply or reduce demand, and smaller systems are often particularly vulnerable. Under extreme heat, agricultural and residential demand for water increases (X.-j. Wang et al., 2016; Zubaidi et al., 2020).

Management of short-term fluctuations in water supply (e.g., precipitation, streamflow, snowmelt) and demand patterns (e.g., population change) have long been a focus for water management. However, these systems are vulnerable in part because, historically, water infrastructure has been designed under the assumption of climate stationarity—that the variability seen in the historical climate record will hold into the future or, as an example, that the 100-year storm of the past will be the 100-year storm of the future. In many locations, these assumptions no longer hold, and, under these changing climate conditions, the placement and design of water infrastructure could require methods that consider future climate patterns to ensure adequate performance (Committee on Adaptation to a Changing Climate, 2015; Milly et al., 2008). Compounding these concerns is that water infrastructure is already vulnerable to failure because of the age and condition of key components (Folkman, 2018). One illustration of the combined effect of some climate drivers, such as tropical cyclones and hurricanes, on the current condition of water infrastructure is that, in 2017, the U.S. federal government spent $300 billion to repair damaged water and wastewater infrastructure following major disaster events (Gómez, 2020).

Overall, the effects of climate change are expected to stress water supplies, increase demand, and cause physical damage to water infrastructure systems. *Supply Water* was assessed to be at risk of moderate disruption from climate drivers in both the current and high emissions scenarios by 2050. Importantly, many of these effects are already being felt across the country, reinforcing risks posed to *Supply Water* as conditions worsen under climate change. *Supply Water* is vulnerable to physical damage and disruption due to flooding, tropical cyclones and hurricanes, severe storm systems, and sea-level rise; demand changes due to extreme heat; and loss of water availability due to drought (Lall et al., 2018). These ratings are due chiefly to an expected increase in the frequency, intensity, or duration of droughts, flooding, severe storm systems, and tropical cyclones and hurricanes. The only climate driver unlikely to affect *Supply Water* is extreme cold.

Subfunctions

Six of the nine subfunctions for this NCF are at risk of disruption due to climate drivers. Five subfunctions—*Provide Agricultural Water*, *Provide Fire Protection*, *Provide Industrial Water*, *Provide Potable Water*, and *Provide Raw Water*—are affected by lack of resources for water provision due to

199

drought and increased demand due to extreme heat under all time periods and emissions scenarios. *Provide Fire Protection* is also vulnerable to workforce response capabilities due to an anticipated increase in the number and magnitude of wildfires. A sixth subfunction, *Provide Building and Facility Water Infrastructure*, will be at risk of major disruption by 2050 due to physical damage to infrastructure, primarily from flooding and storm surge following tropical cyclones and hurricanes. The physical disruption to storage infrastructure following severe storm systems and tropical cyclones and hurricanes could also lead to a lack of resources and affect the *Provide Agricultural Water*, *Provide Fire Protection*, *Provide Industrial Water*, *Provide Potable Water*, and *Provide Raw Water* subfunctions.

The remaining three subfunctions—*Assure Secure Operations*, *Govern Water*, and *Manufacture Parts and Supplies*—were assessed as being at no risk of disruption due to climate change. This assessment was generally due to these subfunctions' ability to perform operations remotely, the lack of an infrastructure footprint required for operations, or the current capacity of management entities.

Table B.127. Subfunction Risk for *Supply Water*

Subfunction	Baseline Present	Current Emissions 2050	Current Emissions 2100	High Emissions 2050	High Emissions 2100
Provide Agricultural Water	3	3	4	3	4
Provide Building and Facility Water Infrastructure	2	3	3	3	3
Provide Fire Protection	3	3	3	3	3
Provide Industrial Water	2	2	3	3	3
Provide Potable Water	3	3	4	3	4
Provide Raw Water	3	3	4	3	4

NOTE: 2 = minimal disruption. 3 = moderate disruption. 4 = major disruption. See Chapter 2 for further explanation of ratings.

Climate Drivers

All climate drivers except extreme cold were assessed to post a risk of at least minimal disruption. Five climate drivers—flooding, sea-level rise, severe storm systems, tropical cyclones and hurricanes, and wildfire—have the potential to cause physical damage or disruption (Climate Change Adaptation Resource Center, 2022a; National Institute of Standards of Technology, 2016). In addition to effects on drinking water infrastructure, these climate drivers can affect water supplies (e.g., contamination from wildfire or inundation). Drought and extreme heat are also expected to affect water supplies. As discussed above, drought is already affecting water supplies in some regions (e.g., California, Colorado, Arizona) (Reidmiller et al., 2018). A decades-long drought in the Colorado River Basin has led to declines in Lake Mead and Lake Powell, which serve as reservoirs for downstream consumers. Because of these declines, the U.S. Department of the Interior has decreased Arizona's and Nevada's water allocations by 21 percent and 8 percent, respectively (Budryk, 2022). This example of a current disruption reinforces the projection of risk that climate change poses to *Supply Water* under all future emissions scenarios, particularly in the western states.

200

Drought, flooding, and wildfire were assessed to pose a risk of moderate disruption to *Supply Water* under current conditions. The frequency, intensity, or duration of these climate drivers is projected to increase under both current and high emissions scenarios by 2050 and 2100. As a result, risk to *Supply Water* was assessed as increasing over time. By 2050 for the current and high emissions scenarios, there is moderate risk of disruption to *Supply Water* resulting from drought, flooding, and wildfire; the NCF was assessed as being at a risk of major disruption by 2100.

Table B.128. Risk from Climate Drivers for *Supply Water*

Climate Driver	Baseline	Current Emissions		High Emissions	
	Present	2050	2100	2050	2100
Drought	3	3	4	3	4
Extreme cold	1	1	1	1	1
Extreme heat	2	2	3	3	3
Flooding	3	3	3	3	3
Sea-level rise	2	2	3	3	3
Severe storm systems	2	2	2	2	2
Tropical cyclones and hurricanes	2	2	3	3	3
Wildfire	3	3	3	3	3

NOTE: 1 = no disruption or normal operations. 2 = minimal disruption. 3 = moderate disruption. 4 = major disruption. See Chapter 2 for further explanation of ratings.

Impact Mechanisms

Physical damage and disruption and lack of resources are expected to present the most-significant challenges for *Supply Water*. Physical damage resulting from severe storm systems, tropical cyclones and hurricanes, and wildfire will affect water utilities' ability to supply adequate water during and following climate-driver events. These can be short-term effects on water provision, or they could be longer-lasting effects. Additionally, lack of resources (namely, reduced raw water supplies) will be exacerbated by drought and extreme heat. Extreme heat can also lead to increased demand and further stress limited water supplies. Furthermore, wildfire can cause water contamination and disrupt water availability locally (Robinne et al., 2021). Decreased water availability can have cascading effects on other sectors (e.g., agriculture, manufacturing, energy).

Cascading Risk

Supply Water is dependent on the following NCFs:

- *Distribute Electricity*
- *Develop and Maintain Public Works and Services*
- *Manage Hazardous Materials*
- *Produce Chemicals*

- *Maintain Supply Chains*
- *Operate Government.*

Water utilities are highly dependent on other NCFs to ensure normal operations (Wasley, Jacobs, and Weiss, 2020). *Distribute Electricity* is critical to the normal operations of a water utility because power is needed for many treatment plant functions and, in some cases, for the movement of water in and around the conveyance and distribution system. The development and maintenance of infrastructure (*Develop and Maintain Public Works and Services*) is necessary to protect the life cycle of assets, as well as their proper functioning over time. The management of hazardous waste produced in water treatment plants is regulated, and the waste must be disposed of properly, necessitating the functioning of *Manage Hazardous Materials* (Krogmann et al., 1999). The production of chemicals and the supply chains that deliver chemicals and other inputs are necessary for proper treatment of drinking water to Safe Drinking Water Act (Pub. L. 93-523, 1974) standards. Because most water agencies are public, *Operate Government* is also a necessary condition for this NCF to function.

Strength of Evidence

There is considerable evidence describing climate drivers' effect on the *Supply Water* NCF, but its strength varies depending on the specific climate driver. The effect of drought, flooding, and wildfire on the provision of water supplies has been well documented. For most of the other climate drivers—extreme heat, sea-level rise, severe storm systems, and tropical cyclones and hurricanes—there is moderate evidence to document climate change's effects on *Supply Water*. There is little evidence to suggest that extreme cold will affect the NCF.

Table B.129. Strength of Evidence for *Supply Water*

Strength of Evidence	
High	Drought; flooding; wildfire
Medium	Extreme heat; sea-level rise; severe storm systems; tropical cyclones and hurricanes
Low	Extreme cold

Risk Rating Interpretations

Table B.130 provides examples of how to interpret the risk ratings for this NCF.

Table B.130. Example Interpretations of Risk Ratings for *Supply Water*

Risk Rating	NCF Example
1: No disruption or normal operations	*Supply Water* is not disrupted and continues operation as usual.
2: Minimal disruption	Some local areas are experiencing water stress due to drought, resulting in temporary restrictions on using water for lawns and pools.
3: Moderate disruption	Significant disruption to several water treatment plants results in an entire region experiencing service disruptions and boil-water advisories.
4: Major disruption	The NCF is disrupted on a national scale: Most water treatment facilities are experiencing disruption.
5: Critical disruption	A critical disruption would occur if essentially all water facilities nationwide were experiencing disruption.

Develop and Maintain Public Works and Services

Table B.131. Scorecard for *Develop and Maintain Public Works and Services*

	National Risk Assessment				
	Baseline	Current Emissions		High Emissions	
Develop and Maintain Public Works and Services	Present	2050	2100	2050	2100
	2	3	3	3	3

Description: "Design, build, and maintain infrastructure to supply government services, including systems and assets used for transportation and traffic management, water supply, waste management, recreation, and other purposes" (CISA, 2020, p. 4).

Highlights

- The *Develop and Maintain Public Works and Services* NCF was assessed as being at risk of moderate disruption due to physical damage to infrastructure assets and increased demand in maintenance activities.
- All climate drivers except extreme cold and severe storm systems were assessed as causing a risk of moderate disruption to the NCF by 2050 under current and high emissions scenarios.
- Key factors contributing to the level of disruption to the NCF include implementation of infrastructure-hardening measures and the condition of the infrastructure systems utilized for public works and services.
- Some subfunctions of this NCF—*Transportation Construction and Maintenance, Utility Construction and Maintenance,* and *Provide and Maintain Public Buildings and Services*—are particularly vulnerable to disruptions from climate drivers.
- The NCF is not concentrated in a particular region; risk of disruption will vary depending on the economic mechanisms in place for a given community (e.g., regions with more financial resources for maintenance investment might be at lower risk than regions with fewer financial resources).

Synopsis

Public works and services are vital to providing adequate quality of life and supporting the general well-being and development of a community. The infrastructure needed to support public works and services includes transportation systems, utilities (e.g., government-owned electricity and water), public buildings (e.g., police stations, hospitals, municipal offices), and recreation spaces. The *Develop and Maintain Public Works and Services* NCF was assessed as being at risk of minimal disruption from climate drivers currently, with an increase to a risk of moderate disruption by 2050 under both current and high emissions scenarios. Climate drivers with immediate consequences, such as tropical cyclones and hurricanes, and long-term consequences, such as extreme heat and sea-level rise, pose a risk of disruption to the infrastructure systems serving public works and services through damage, destruction, or increased degradation. For example, extreme heat conditions from the past two decades in the southwestern United States have led to reductions in water availability for communities served by the Colorado River, which poses challenges for water reservoir management (Runyon, 2021). Additionally, historical incidents, such as the 2005 Gulf Coast hurricanes (Katrina, Rita, and Wilma) highlight how damage and disruptions to these systems from climate drivers are already prevalent and financially burdensome. The costs to repair and rebuild highways, transportation systems, and homes from the previously mentioned hurricanes are estimated at $7.5 billion (CBO,

2016).[40] Similarly, federal appropriations to DOT as a result of Hurricane Sandy in 2012 totaled $16.3 billion (CBO, 2016).[41]

As incidents associated with the climate drivers become more frequent and intense, the need for maintenance of infrastructure serving public works and services will further increase. This is considerable for the United States because several infrastructure systems critical to public works and services are currently in need of repair or replacement. For instance, ASCE assessed that, in the United States, 43 percent of public roadways are in "poor" or "mediocre" condition, 70 percent of electrical transmission and distribution lines are in the second half of their life spans, and infrastructure maintenance costs reached approximately $50 billion above capital in 2017 (ASCE, 2021). Consequences from climate drivers will exacerbate or accelerate the degradation, leading to additional maintenance and disruptions to the infrastructure systems, which can overwhelm local governments, municipalities, and other entities managing infrastructure that serves public works and services. For instance, Hurricane Sandy in New York City resulted in damage estimated at $5 billion to the city's subway system (Rose, 2012). Effects from sea-level rise will add to these repair and maintenance costs because of the risk of saltwater exposure to transportation equipment. The drought conditions in Coalinga, a small rural town in California's Central Valley, serves as another example. Because of depletion of local water resources, the town might need to resort to purchasing water from the open market, which will require financial resources estimated at approximately 25 percent of the town's municipal budget (Partlow, 2022).

The assessment of risk of moderate disruptions to this NCF by the middle and end of this century is based on an increase in the frequency and intensity of certain climate drivers and their effect on public works and service infrastructure (as discussed in further detail in the "Subfunctions" and "Climate Drivers" sections). Examples of historical incidents provided earlier in this section illustrate how disruptions to public works and services are already prevalent in certain regions of the country because of flooding, drought, and tropical cyclones and hurricanes. However, the severity of these effects will vary regionally. The Western Electricity Coordinating Council, for instance, estimated that costs to construct and repair electric grid infrastructure in areas with high population density were 1.59 times higher than in areas with low populations (Korbatov et al., 2017). Under a high emissions scenario, climate drivers assessed as having effects on *Develop and Maintain Public Works and Services* are expected to increase in frequency and intensity, which will pose further challenges to the operation and maintenance of public works infrastructure systems.

---

[40] The $7.5 billion cost value is adjusted for inflation from 2015 dollars ($4.5 billion) to 2022 dollars.

[41] The $16.3 billion cost value is adjusted for inflation from 2015 dollars ($13 billion) to 2022 dollars.

Subfunctions

Four out of the five subfunctions for *Develop and Maintain Public Works and Services* (specifically, *Transportation Construction and Maintenance, Utility Construction and Maintenance, Provide and Maintain Public Buildings and Services,* and *Manage Recreation Spaces and Facilities*) were assessed as being at risk of moderate disruption from at least one of the climate drivers by the end of the century under the high emissions scenario. Both *Transportation Construction and Maintenance* and *Utility Construction and Maintenance* are vulnerable to drought, flooding, sea-level rise, tropical cyclones and hurricanes, and wildfire. Fixed-node transportation systems (e.g., airports and seaports) and fixed-route transportation systems (e.g., roads, buses, and subways) provide access for commuting, emergency vehicles, and other community services. Existing transportation infrastructure in the United States is designed and managed based on historical climatology. As future climate conditions change, more-frequent disruptions and damage to transportation systems are likely (National Research Council and Committee on Climate Change and U.S. Transportation, 2008). Utility systems are also susceptible to risk of moderate disruption from some of the same climate drivers and have considerable implications for public services. For example, between 1971 and 1999, approximately 60 percent of reported hospital evacuations in Florida were due to tropical cyclones and hurricanes (Sternberg, Lee, and Huard, 2004). A more recent example is Lee Health, a county-owned and -operated health system in Florida, where damage from Hurricane Ian in 2022 disrupted water and electrical power services, leading to the evacuation and transfer of 400 patients from two of its facilities (Evans, 2022).

In addition to transportation and utility systems, climate drivers can disrupt other aspects of public works and services, such as buildings dedicated to public safety (e.g., police stations). For instance, changes in soil moisture through extreme heat can affect the structural integrity of buildings and reduce the service life of a facility (Barrelas, Ren, and Pereira, 2021). Other subfunctions for *Develop and Maintain Public Works and Services*, such as *Manage Recreation Spaces and Facilities* and *Community Revitalization*, are also at risk of potential disruption from climate drivers. Both physical components of recreation spaces, such as buildings, and natural and nature-based features can experience damage through drought, flooding, tropical cyclones and hurricanes, and wildfire. Examples include the 2020 wildfires in the western United States, which spread to 6,500 acres in Rocky Mountain National Park (Repanshek, 2020), and Hurricane Irma in 2017, which led to loss of vegetation cover and erosion of land mass on islands in Everglades National Park (Wingard et al., 2020).

Table B.132. Subfunction Risk for *Develop and Maintain Public Works and Services*

Subfunction	Baseline Present	Current Emissions 2050	Current Emissions 2100	High Emissions 2050	High Emissions 2100
Transportation Construction and Maintenance	2	3	3	3	3
Utility Construction and Maintenance	2	3	3	3	3
Provide and Maintain Public Safety Buildings and Services	2	2	2	3	3
Manage Recreation Spaces and Facilities	2	2	3	2	3
Community Revitalization	1	1	2	1	2

NOTE: 1 = no disruption or normal operations. 2 = minimal disruption. 3 = moderate disruption. See Chapter 2 for further explanation of ratings.

Climate Drivers

Six out of the eight climate drivers are expected to pose a risk of moderate disruption to the development of public works and services by 2050 and 2100 under current and high emissions scenarios. However, disruptions will have varying regional effects. Public works and services on U.S. coastlines and the Gulf of Mexico are more vulnerable to flooding, sea-level rise, and tropical cyclones and hurricanes, whereas public works and services in regions that are expected to experience high heat conditions by 2100 will be more vulnerable to wildfire, drought, and extreme heat. Additionally, it is important to note that disruptions from specific climate drivers to this NCF will also vary by subfunction. For example, drought might be more impactful than tropical cyclones and severe storm systems in disruptions to national parks because of imbalances in park ecology, transformation of vegetation, and risk of forest disease (Michalak et al., 2021). On the other hand, tropical cyclones and hurricanes might pose more of a threat to transportation and utility systems as evidenced in 2022 by Hurricane Fiona in Puerto Rico, where high winds, precipitation, and flooding disrupted electric power services to more than 1 million electricity customers and damaged roads and bridges, preventing residents from traveling to certain regions of the islands (Oxford Analytica, 2022).

Table B.133. Risk from Climate Drivers for *Develop and Maintain Public Works and Services*

Climate Driver	Baseline Present	Current Emissions		High Emissions	
		2050	2100	2050	2100
Drought	2	3	3	3	3
Extreme cold	2	2	2	2	2
Extreme heat	2	3	3	3	3
Flooding	2	3	3	3	3
Sea-level rise	2	3	3	3	3
Severe storm systems	2	2	3	3	3
Tropical cyclones and hurricanes	2	3	3	3	3
Wildfire	2	3	3	3	3

NOTE: 2 = minimal disruption. 3 = moderate disruption. See Chapter 2 for further explanation of ratings.

Impact Mechanisms

Disruptions to the *Develop and Maintain Public Works and Services* NCF are due primarily to physical damage of infrastructure systems and increases in demand for maintenance activities. Transportation systems, utility systems, and buildings utilized for the function of public works and services are susceptible to damage from some climate drivers, such as flooding, wildfire, tropical cyclones and hurricanes, and extreme heat. For instance, the costs of damage to paved roads from extreme temperature and precipitation are estimated to reach \$25 billion[42] by the end of the century (Jacobs, Culp, et al., 2018). Demand changes in maintenance might also exacerbate disruptions; as infrastructure systems continue to experience damage and degradation, maintenance requirements for these systems will continue to increase. Some municipalities and entities might be unable to fulfill these increases in demand. For instance, researchers in one study estimated that Ohio will need to increase municipal spending between 26 percent to 82 percent by 2050 to adapt to effects from certain climate drivers (Ohio Environmental Council, Power a Clean Future Ohio, and Scioto Analysis, 2022). Maintenance activities can also be inhibited by unsafe working conditions for construction laborers and maintenance staff (Pamidimukkala, Kermanshachi, and Karthick, 2020). For instance, the southern United States is vulnerable to extreme heat conditions in the future, which makes outdoor occupational hazards more dangerous because of the risk of heat-related illness (Applebaum et al., 2016). Workforce shortages due to climate drivers can delay or halt construction and maintenance activities, leading to further disruptions to this NCF.

Cascading Risk

Develop and Maintain Public Works and Services is dependent on the following NCFs:

- *Distribute Electricity*

---

[42] Adjusted for inflation from 2015 dollars to 2022 dollars.

- *Transport Cargo and Passengers by Air*
- *Transport Cargo and Passengers by Rail*
- *Transport Cargo and Passengers by Road*
- *Educate and Train*
- *Supply Water*
- *Provide and Maintain Infrastructure*
- *Operate Government*
- *Maintain Supply Chains*
- *Provide Capital Markets and Investment Activities.*

The development and maintenance of public works and services is an involved process that requires continual operation of both physical infrastructure (e.g., transportation systems, electric power, water supply) and nonphysical infrastructure (e.g., community governance, education and training of workforce, financial resources). Disruptions to these physical and nonphysical assets can have implications for construction, operation, and maintenance activities for public works and services.

Strength of Evidence

There is extensive evidence of potential risks and disruptions from climate drivers to the *Develop and Maintain Public Works and Services* NCF. Academic research reports, damage assessments from historical extreme weather events, and industry white papers describe how climate drivers affect specific subfunctions of this NCF, such as *Transportation Construction and Maintenance* and *Utility Construction and Maintenance*. Additionally, evidence on effects on buildings is well documented. However, few studies have focused specifically on buildings that service public works. Some evidence documenting climate drivers' effect on other subfunctions of this NCF, such as *Manage Recreation Spaces and Facilities* and *Community Revitalization*, exist, but the climate effects are not as well documented as effects on the physical built environment.

Table B.134. Strength of Evidence for *Develop and Maintain Public Works and Services*

Strength of Evidence	
Medium	Flooding; sea-level rise; tropical cyclones and hurricanes; severe storm systems; extreme cold; extreme heat; wildfire; drought

Risk Rating Interpretations

Table B.135 provides examples of how to interpret the risk ratings for this NCF.

Table B.135. Example Interpretations of Risk Ratings for *Develop and Maintain Public Works and Services*

Risk Rating	NCF Example
1: No disruption or normal operations	Transportation systems, utility systems, buildings, and other assets serving public works operate normally.
2: Minimal disruption	Flooding could wash roads away, or wildfire could render them inaccessible, either of which could delay ambulance services in providing emergency care.
3: Moderate disruption	Utility systems experience physical damage, making them unable to provide reliable energy and water services to some regions of the country.
4: Major disruption	Most or all regions across the country experience an inability to access energy, water, health, and community protection services.
5: Critical disruption	Construction and maintenance of utility systems are halted because of frequent effects from climate drivers and the inability to perform maintenance due to unsafe working conditions.

Provide and Maintain Infrastructure

Table B.136. Scorecard for *Provide and Maintain Infrastructure*

	National Risk Assessment				
	Baseline	Current Emissions		High Emissions	
	Present	2050	2100	2050	2100
Provide and Maintain Infrastructure	2	2	3	3	3

Description: "Design, construct, operate, repair, survey and improve private and public infrastructure" (CISA, 2020, p. 5).

Highlights

- *Provide and Maintain Infrastructure* was assessed as being at risk of moderate disruption by 2050 under a high emissions scenario because of disruptions in the ability to construct, operate, and repair infrastructure.
- Some climate drivers, such as flooding, extreme heat, and tropical cyclones and hurricanes, are expected to increase in frequency, intensity, or duration, which can lead to physical damage to infrastructure and increase the demand for repair and maintenance activities.
- Disruptions to this NCF will vary based on planning factors, such as regulatory mechanisms (e.g., adopting updated codes and standards or best management practices).
- The *Build Infrastructure* and *Perform Maintenance* subfunctions are particularly vulnerable to disruption due to persistent damage to assets, increase in maintenance needs, and loss of labor productivity.
- Communities or entities in the United States experiencing economic or governance challenges are likely at greatest risk of disruption because of their inability to implement adaptive capacity measures.

Synopsis

Infrastructure that supports socioeconomic development and growth within a community includes both physical components (e.g., buildings, utility equipment, transportation systems) and institutional components (e.g., education, legal systems). The ability to effectively provide and maintain both physical and institutional infrastructure is driven by many factors, including the availability of materials and labor, sufficient financial resources, and stable governance structures (Beckers and Stegemann, 2013). Flooding, wildfire, extreme heat, and tropical cyclones and hurricanes can hinder the development of material infrastructure because of damage to physical assets, as well as equipment and workforce shortages. For example, construction activities pertaining to earthwork and site development can be disrupted from extreme precipitation conditions due to increased soil humidity and runoff, which can lead to inundation of the site making it inaccessible. Additionally, wildfire and tropical cyclones and hurricanes can damage equipment and materials, leading to delays in infrastructure development or increasing the need for repairs of existing infrastructure (Schuldt et al., 2021).

Climate drivers' effects on this NCF are expected to include damaged or inoperable physical assets, which will increase the need for infrastructure development and maintenance. For example, damage from extreme wind and precipitation during tropical cyclones or hurricanes can lead to failures in a building's exterior wall and roof system, thereby allowing water intrusion and potential growth of mold in the interior (FEMA, 2009). Additionally, flooding and sea-level rise can increase saltwater exposure and corrosion of coastal infrastructure, which increases maintenance needs (Nasr et al., 2021). The effect of increased infrastructure maintenance is exacerbated by reduced labor availability and productivity. In extreme heat conditions, electrical power grids are stressed, which can require maintenance or replacement up to three times more often in some parts of the country, such as the Southwest (Burillo, Chester, and Ruddell, 2016); at the same time, extreme heat conditions pose significant risks for maintenance workers due to risk of heat-related illnesses or death.

Provide and Maintain Infrastructure was assessed as being at risk of minimal disruption by 2050 and moderate disruption by 2100 for the current emissions scenario. These assessments are based primarily on expected effects from flooding, tropical cyclones and hurricanes, wildfire, and extreme heat. Although these four climate drivers are expected to increase in frequency and intensity by 2050, the risk of disruption was assessed to change from minimal to moderate between the current and high emissions scenario because of an increase in damage to material infrastructure across more regions. However, disruptions are not likely to be widespread across the country; instead, disruptions will vary based on other factors, such as regulatory mechanisms (e.g., codes and standards or best management practices) in place. For instance, a study in which researchers reviewed stormwater infrastructure design standards in the United States indicated that stormwater infrastructure in 43 states was underdesigned for future precipitation conditions, making it more susceptible to flooding and therefore in greater need of ongoing maintenance as this infrastructure is affected over time (Lopez-Cantu and Samaras, 2018).

Key Considerations for *Provide and Maintain Infrastructure*

This NCF includes providing and maintaining physical infrastructure (e.g., roads, water treatment facilities, houses) and institutional infrastructure (e.g., education, workforce development, governance). The assessment of the risk of disruptions from climate drivers presented for this NCF is based primarily on the effects on physical infrastructure components. Disruptions to institutional infrastructure are covered in other NCFs, such as *Educate and Train* and *Operate Government*. Additionally, there is some overlap between the assessments for this NCF and for *Develop and Maintain Public Works and Services*. *Develop and Maintain Public Works and Services*, however, is more focused on public infrastructure, whereas this NCF includes both public and private.

Subfunctions

Three of the nine subfunctions for this NCF (specifically, *Build Infrastructure*, *Perform Maintenance*, and *Obtain Finances*) were assessed as being at risk of moderate disruption from climate drivers by the end of the century. Some activities, such as constructing or maintaining roads, bridges, or buildings, can be delayed, disrupted, or halted by unsafe working conditions or reduced labor productivity. Variations in temperature are correlated with reduced worker productivity, and some construction or maintenance tasks might not be feasible under certain conditions, such as high wind speeds (Schuldt et al., 2021). Climate drivers' negative effects on workforce productivity are combined with the increase in demand for infrastructure maintenance. As of 2019, the national total for deferred-maintenance costs for public infrastructure was estimated at approximately $1 trillion (Zhao, Fonseca-Sarmiento, and Tan, 2019). If flooding, wildfire, extreme heat, and tropical cyclones and hurricanes grow in frequency, duration, and intensity as expected, the need for infrastructure repairs will increase.

With respect to *Obtain Finances*, the effects of climate drivers can also affect jurisdictions' ability to obtain financial resources for infrastructure development and maintenance. CBO, for example, has estimated that costs from damage due to tropical cyclones and hurricanes will increase at a higher rate than economic growth in the United States, which can create a financial resource issue (CBO, 2016). It is important that funding resources for disaster recovery and preparedness efforts exist through entities, such as FEMA, but financial resources for climate change planning within the context of infrastructure development and maintenance remain minimal (Kane, Tomer, and George, 2021). For instance, in a study by the National League of Cities, 91 percent of local officials surveyed from 600 cities and towns indicated that insufficient funding was a primary factor in infrastructure decisionmaking (McFarland et al., 2021).

Other subfunctions of this NCF, such as *Develop Plans* and *Provide Security*, are critical aspects of providing and maintaining infrastructure, but these subfunctions are expected to continue operations without disruptions from climate drivers through 2100 because they are less likely to be stressed or affected by physical damage or increases in demand for repairs.

Table B.137. Subfunction Risk for *Provide and Maintain Infrastructure*

Subfunction	Baseline Present	Current Emissions 2050	Current Emissions 2100	High Emissions 2050	High Emissions 2100
Develop Plans	1	1	1	1	1
Obtain Finances	2	2	3	3	3
Build Infrastructure	2	2	2	3	3
Perform Maintenance	2	2	3	3	3
Provide Security	1	1	1	1	1
Purchase Materials	1	2	2	2	2
Store Materials	1	1	2	2	2
Dispose of Materials	1	1	1	1	1
Comply with Regulations	1	1	2	2	2

NOTE: 1 = no disruption or normal operations. 2 = minimal disruption. 3 = moderate disruption. See Chapter 2 for further explanation of ratings.

Climate Drivers

All climate drivers except extreme cold and severe storm systems are expected to pose risk of moderate disruption to this NCF by the end of the century under current and high emissions scenarios. As the frequency, intensity, or duration of climate drivers increases, activities pertaining to infrastructure development and maintenance are likely to be affected. However, disruptions to this NCF from climate drivers will vary regionally across the country. Infrastructure in coastal communities will be more susceptible to erosion and degradation from rising sea levels and flooding, whereas infrastructure development in the southwestern regions of the United States is likelier to experience disruptions due to shortages in labor productivity or availability due to extreme heat conditions (Narayanan, Willis, et al., 2016). Some of these effects are prevalent currently as evidenced by heat-related deaths of construction workers. Between 1992 to 2016, construction workers accounted for 36 percent of heat-related occupational deaths, and increasing summer temperatures during this time period were associated with increases in heat-related death rates (X. Dong et al., 2019). Exposure to other climate drivers, such as wildfire, tropical cyclones and hurricanes, and drought, can lead to the delay or halting of construction projects. For instance, unsafe air quality levels and ash from wildfires in California and Oregon in 2020 led to contractors stopping construction projects due to unsafe working conditions or inability to maintain proper construction quality control (Bousquin, 2020).

Table B.138. Risk from Climate Drivers for *Provide and Maintain Infrastructure*

| | Baseline | Current Emissions | | High Emissions | |
Climate Driver	Present	2050	2100	2050	2100
Drought	2	2	3	2	3
Extreme cold	2	2	2	2	2
Extreme heat	2	2	3	3	3
Flooding	2	2	3	3	3
Sea-level rise	2	2	3	3	3
Severe storm systems	2	2	2	2	2
Tropical cyclones and hurricanes	2	2	3	3	3
Wildfire	2	2	3	3	3

NOTE: 2 = minimal disruption. 3 = moderate disruption. See Chapter 2 for further explanation of ratings.

Impact Mechanisms

The *Provide and Maintain Infrastructure* NCF was assessed as having a risk of moderate disruption due to damage of physical assets and equipment, increase in demand for infrastructure maintenance, and reduction in labor productivity or availability. Extreme heat and wildfire create unsafe outdoor working conditions for staff performing construction work, maintenance activities, or infrastructure inspections. For instance, exposure to wildfire smoke could lead to respiratory infections or worsen existing cardiovascular conditions, and extreme heat conditions can increase the risk of heat-related illnesses or death (Roh and Alamo, 2022). The consequences of workforce shortages due to climate drivers are combined with damage to infrastructure and increases in demand for maintenance activities. Extreme heat's effect on roads serves as an example; under high-temperature conditions, concrete roads can experience thermal cracking[43] and a decrease in service life by two to eight years, thereby requiring more-frequent maintenance (Sen, Li, and Khazanovich, 2022).

Cascading Risk

Provide and Maintain Infrastructure is dependent on the following NCFs:

- *Distribute Electricity*
- *Maintain Supply Chains*
- *Educate and Train*
- *Operate Government*
- *Provide Information Technology Products and Services*
- *Provide Capital Markets and Investment Activities*
- *Supply Water*
- *Manage Hazardous Materials*
- *Manage Wastewater.*

The development and maintenance of infrastructure involve numerous stakeholders, including both public entities (e.g., local, state, and federal governments) and private entities (e.g., banks; developers; architecture, engineering, and construction firms). Failures in financial markets and governance structure pose challenges in the procurement of materials and equipment needed for infrastructure projects (Ben Ammar and Eling, 2015), while failures in education systems can disrupt workforce development and have consequences for infrastructure maintenance, which typically requires highly skilled labor (Parlikad and Jafari, 2016). Additionally, infrastructure development and maintenance depend on basic services, including access to reliable energy and water supplies.

Strength of Evidence

There is a well-documented literature that examines climate drivers' effect on existing infrastructure and the need for maintenance. ASCE, for instance, publishes a report every four years on the society's examination of the condition and improvement needs for infrastructure systems in the United States (ASCE, 2021). Additionally, some studies and anecdotal evidence provide insights on how extreme weather conditions disrupt or halt construction projects (Schuldt et al., 2021). Effects on other subfunctions of this NCF, such as *Obtain Finances*, are documented through limited sources. We identified little or no evidence that documented the effects or disruptions to the planning aspects of infrastructure development and maintenance, such as the development of plans and specifications, waste management on-site, and providing security at sites.

---

[43] Thermal cracking of concrete is a condition in which concrete expands and contracts because temperature changes and causes additional strain to the material beyond allowable limits.

Table B.139. Strength of Evidence for *Provide and Maintain Infrastructure*

Strength of Evidence	
High	Flooding; tropical cyclones and hurricanes; extreme heat; wildfire; drought
Medium	Sea-level rise; severe storm systems
Low	Extreme cold

Risk Rating Interpretations

Table B.140 provides examples of how to interpret the risk ratings for this NCF.

Table B.140. Example Interpretations of Risk Ratings for *Provide and Maintain Infrastructure*

Risk Rating	NCF Example
1: No disruption or normal operations	The maintenance and development of infrastructure will continue to operate without disruption or delays.
2: Minimal disruption	Activities pertaining to the construction and maintenance of infrastructure will experience some delays or disruptions in select regions of the country.
3: Moderate disruption	Construction projects will experience disruptions or delays in some parts of the country.
4: Major disruption	Most construction and maintenance projects will be delayed or canceled throughout the country because of material availability or worker shortage.
5: Critical disruption	Activities pertaining to the construction and maintenance of infrastructure will be halted by lack of materials, damage to equipment, and unsafe working conditions.

Transportation and Supply Chain

Transport Cargo and Passengers by Air

Table B.141. Scorecard for *Transport Cargo and Passengers by Air*

	National Risk Assessment				
	Baseline	Current Emissions		High Emissions	
	Present	2050	2100	2050	2100
Transport Cargo and Passengers by Air	3	3	3	3	3

Description: "Provide and operate aviation systems, assets, and facilities to enable a system of securely and safely conveying goods and people from place to place by air" (CISA, 2020, p. 3).

Highlights

- This NCF was assessed as being at risk of moderate disruption because air transportation is vulnerable to multiple climate drivers that can affect airport and airline operations or physical infrastructure and because flight delays and cancellations at a single critical airport can affect the operational efficiency of multiple airports and have cascading effects on the entire air transportation system.
- Air transportation is vulnerable to multiple climate drivers, with the greatest risk from sea-level rise, flooding, and tropical cyclones and hurricanes that cause damage to airport infrastructure or inundate runways and underground electrical equipment.
- Large-hub airports in low-elevation coastal zones, especially on the East Coast and Gulf Coast, are particularly vulnerable because of increased risk of coastal flooding and storm surge.
- The subfunction *Manage Transport Operations* is vulnerable to disruptions from the effects that physical damage to airport infrastructure and inclement weather and poor visibility conditions can have on operations. *Provide Diverse Energy Sources* is vulnerable to physical damage from some climate drivers, such as flooding and extreme heat.

Synopsis

Transport Cargo and Passengers by Air was assessed as being at risk of moderate disruption due to natural hazards affected by climate change because flight delays or cancellations at a single critical airport can affect the operational efficiency of multiple airports and have downstream effects on the entire air transportation system (Lykou et al., 2020). Air transportation is vulnerable to multiple climate drivers that affect airport and airline operations or physical infrastructure: flooding, sea-level rise, tropical cyclones and hurricanes, severe storm systems, extreme heat, and wildfire. Inclement weather is already the main cause of flight delays and cancellations, and the increasing frequency or intensity of flooding, severe storm systems, and tropical cyclones and hurricanes is expected to elevate the risk of runway flooding and inundation of underground electrical equipment (Burbidge, 2018). Extreme heat can reduce the allowable weight that aircraft can carry, and more high-temperature days could mean that aircraft can carry fewer goods and passengers safely (Coffel and Horton, 2015). Wildfire smoke can also reduce visibility, affecting airline and airport operations (Love, Soares, and Püempel, 2010). Finally, sea-level rise is expected to affect the physical infrastructure of some coastal airports, both through inundation and through increased storm surge.

217

The aviation system is highly interconnected. The U.S. air transportation network carries nearly 1 billion passengers each year, with more than 5,000 public-use airports supporting more than 7,000 commercial aircraft and 200,000 general aviation aircraft (Federal Aviation Administration [FAA], 2022). In 2020, airline and airport operations (including air couriers) generated about $128.5 billion in direct economic activity (FAA, 2022).[44] The U.S. air transportation industry also directly supports approximately 560,000 jobs (Division of Current Employment Statistics, undated). Flight delays and cancellations are widespread, costing the U.S. economy billions of dollars each year (FAA, 2020).

Overall, this NCF was assessed as being at risk of moderate disruption due to the vulnerability of large- and medium-hub airports with the most enplanements. Although the United States has thousands of airports, more than 85 percent of enplanements take place at just 61 large- and medium-hub airports, and delays and cancellations can cascade through the aviation system even if a climate driver affects only a few airports (B. Miller et al., 2020). The New York region is overrepresented in air travel delay propagation; eight of the 13 origin–destination pairs that result in the highest amount of delay propagation begin or end at New York–area airports (B. Miller et al., 2020). This suggests that delays in the New York region, which is particularly vulnerable to sea-level rise, can cascade throughout the air transportation system (Wright and Hogan, 2008; Yesudian and Dawson, 2021). Although the risk of disruption due to individual climate drivers (e.g., sea-level rise, extreme heat) is expected to increase over time, effects are likely to vary by region and be temporary. Moreover, natural hazards affected by climate change are not expected to permanently shut down operations for individual airlines or airports, so air transportation is not assessed as being at risk of major or critical disruption. Although disruptions are likely to be temporary, the cumulative effects of these disruptions could have long-term consequences for the airline industry in terms of, for example, operations or economics.

Subfunctions

The primary subfunctions at risk of disruption due to climate change are *Manage Transport Operations* and *Provide Diverse Energy Sources*. *Manage Transport Operations* is affected through the potential effects that both physical damage to airport infrastructure and inclement weather (particularly snow, freezing temperatures, and heavy rain and wind) can have on airport and airline operations (Burbidge, 2018), as well as the consequences of extreme heat for aircrafts' ability to transport their usual cargo and passenger loads (Coffel, Thompson, and Horton, 2017). Extreme heat can damage runways and taxiways and increase cooling requirements for airport facilities (Burbidge, 2018). *Manage Transport Operations* and *Provide Diverse Energy Sources* are at risk of moderate disruption because many of these disruptions have large effects (i.e., regional or across multiple regions) and result from common climate drivers, such as hurricanes (e.g., in September 2022, Hurricane Fiona hit Puerto Rico and the U.S. Virgin Islands, and Hurricane Ian hit Florida

---

[44] The $128.5 billion value is adjusted for inflation from 2020 dollars ($123.0 billion) to 2021 dollars using the gross domestic product implicit price deflator.

[Kelleher, 2022]). In 2019, FAA estimated that the cost of delays to airlines, passengers, and lost demand was approximately $30.5 billion (FAA, 2020).[45]

To a lesser extent, the subfunction *Obtain and Maintain Knowledgeable Workforce* could be affected by extreme heat. Although climate change might not directly affect the ability to attract and retain the skilled workers required to operate the air transportation system, it could affect workers' ability to function outdoors when temperatures are too hot to work safely (Jacklitsch et al., 2016). This suggests a risk of minimal disruption by 2100 due to high temperatures limiting or reducing the productivity of outdoor work. This effect would vary among airports depending on geographic location, environmental conditions, and other factors (Gubernot, Anderson, and Hunting, 2014).

Several subfunctions are unlikely to be directly affected by climate change. For example, the subfunction *Transport Passengers by Air* involves logistics functions (e.g., screen passengers, maintain signage to direct passengers, provide in-flight food and beverages, move baggage) that are not particularly vulnerable to climate change. Subfunctions that govern regulatory requirements, such as certifications, security, and inspections, are also unlikely to be affected by climate change.

Table B.142. Subfunction Risk for *Transport Cargo and Passengers by Air*

Subfunction	Baseline Present	Current Emissions 2050	Current Emissions 2100	High Emissions 2050	High Emissions 2100
Obtain Certifications	1	1	1	1	1
Acquire Transport Equipment	1	1	1	1	1
Acquire Transport Infrastructure	1	1	1	1	1
Administer Transport Systems	1	1	1	1	1
Obtain and Maintain Knowledgeable Workforce	1	1	2	1	2
Maintain Safe/Reliable Operations	1	1	1	1	1
Manage Transport Operations	3	3	3	3	3
Transport Passengers by Air	1	1	1	1	1
Provide Diverse Energy Sources	3	3	3	3	3

NOTE: 1 = no disruption or normal operations. 2 = minimal disruption. 3 = moderate disruption. See Chapter 2 for further explanation of ratings.

Climate Drivers

Air transportation is vulnerable to sea-level rise, tropical cyclones and hurricanes, and flooding because of the potential for damage to physical infrastructure and disruption to airport and airline operations. Sea-level rise could render some airports temporarily or permanently unusable, depending on the amount of inundation. According to one study, the United States has about 200 airports located 10 meters or less above sea level, although fewer than ten of these are large- or medium-hub

---

[45] The $30.5 billion value is adjusted for inflation from 2019 dollars ($28.8 billion) to 2021 dollars using the gross domestic product implicit price deflator.

airports (Yesudian and Dawson, 2021).[46] However, two of those airports are in the New York area, and disruptions in the New York area can affect the national aviation system by propagating delays (B. Miller et al., 2020). According to another study, 13 large- and medium-hub airports had at least one runway "within the reach of moderate to high storm surge" (Thompson, 2016, p. 107). The fact that a disproportionate share of enplanements are at large-hub airports in low-elevation coastal zones could have cascading risks for the entire air transportation system (Yesudian and Dawson, 2021). Heavy precipitation and high winds can damage landside airport infrastructure, and flooding can damage both aboveground and underground infrastructure (Burbidge, 2018). For safety reasons, aircraft are required to maintain additional separation when flying in intense precipitation (Burbidge, 2018).

Extreme heat also poses several risks to airport and airline operations. Aircraft have temperature thresholds for safe takeoff that are affected by elevation, runway length, and aircraft type and weight. Studies have projected that weight-restriction days will increase up to 50 to 100 days per year at four major airports in the United States by 2070 (Coffel and Horton, 2015) and that 10 to 30 percent of flights taking off during maximum daily temperature might require weight restrictions (Coffel, Thompson, and Horton, 2017). The number of weight-restricted days is projected to increase every decade, so effects will be greater by 2100 than by 2050. Effects will be greatest at airports with short runways or at high elevations (Coffel and Horton, 2015). Extreme heat can also damage tarmac, lead to increased cooling requirements for buildings, and create fire hazards if the flash point for aviation fuel is exceeded (Burbidge, 2018; Thompson, 2016). Extreme heat can also create unsafe conditions for outdoor airport workers, putting them at increased risk of heat exhaustion, rhabdomyolysis (breakdown of skeletal muscle tissue), heat stroke, permanent disability, and death, although the amount of time considered safe for employees to work in hot environments depends on multiple factors (Gubernot, Anderson, and Hunting, 2014; Jacklitsch et al., 2016). Additionally, the ability to efficiently conduct airport operations could decrease because extreme heat can cause workers to work fewer hours (Graff Zivin and Neidell, 2014), become less productive (Behrer and Park, 2017; Burke, Hsiang, and Miguel, 2015b; Seppänen, Fisk, and Lei-Gomez, 2006), and become more prone to injury (Park, Pankratz, and Behrer, 2021).

Given that weather is the leading cause of flight delays, an increase in the number of convective storms (storms with thunder and lightning) would likely increase delays and thus disrupt airline operations. We assessed air transportation as at risk of minimal disruption due to severe storm systems because reduced visibility alone could cause delays, but those risks are likely to be more localized and shorter than those from other climate drivers. However, we also note that there has been very little study of the risk that convective storms pose to commercial aviation, so defining more-precise effects (such as increased delay or safety hazards) is difficult.[47] Similarly, wildfire smoke can reduce visibility, affecting airline and airport operations (Love, Soares, and Püempel, 2010). Wildfire can also physically destroy electrical transmission and distribution infrastructure, temporarily

---

[46] Although our analysis focused on the U.S. air transportation system, the Yesudian and Dawson study suggested that international flights could be affected by sea-level rise at major airports in other countries.

[47] An extensive 2022 review noted the type of disruption that would occur: "longer routes, steeper climbs and descents, and less efficient airspeeds—all adversely impacting emissions, flight times and costs, and forcing more aircraft into some portions of airspace, creating potential safety issues that must be mitigated" (Gratton et al., 2022, p. 213).

disrupting airport operations, but this risk is localized to the area of the wildfire (Office of Energy Policy and Systems Analysis, 2015). In addition, wildfires can create their own weather, such as fire-induced winds, clouds, or thunderstorms, which could disrupt flights (Secaira, 2022). An increase in the frequency of large wildfires across the western United States is expected to pose a risk of minimal disruption to air transportation by 2100 but is unlikely to cause nationwide effects because the risk is usually localized.

Table B.143. Risk from Climate Drivers for *Transport Cargo and Passengers by Air*

Climate Driver	Baseline Present	Current Emissions 2050	Current Emissions 2100	High Emissions 2050	High Emissions 2100
Drought	1	1	1	1	1
Extreme cold	1	1	1	1	1
Extreme heat	2	2	3	2	3
Flooding	3	3	3	3	3
Sea-level rise	1	2	3	2	3
Severe storm systems	2	2	2	2	2
Tropical cyclones and hurricanes	3	3	3	3	3
Wildfire	2	2	2	2	2

NOTE: 1 = no disruption or normal operations. 2 = minimal disruption. 3 = moderate disruption. See Chapter 2 for further explanation of ratings.

Impact Mechanisms

The primary mechanism by which climate change affects *Transport Cargo and Passengers by Air* is physical damage and disruption. Multiple climate drivers are likely to damage airport facilities, aircraft, or ground support equipment or disrupt airline and airport operations. Coastal airports, especially those on the East Coast and Gulf Coast, are vulnerable to sea-level rise, increasing the risk of coastal flooding and storm surge, which could damage landside infrastructure, runways, and underground electrical equipment. Flooding, tropical cyclones and hurricanes, severe storm systems, and wildfire are likely to cause flight delays and cancellations and can also damage critical airport infrastructure. Extreme heat can damage runways and taxiways and can reduce cargo or passenger loads if runways are not long enough for aircraft to take off safely. Extreme heat also poses a health and safety risk to outdoor workers, which could also contribute to workforce shortages.

Cascading Risk

Transport Cargo and Passengers by Air is dependent on the following NCFs:

- *Transport Cargo and Passengers by Road*
- *Distribute Electricity*
- *Provide Positioning, Navigation, and Timing Services*

- *Provide Satellite Access Network Services*
- *Fuel Refining and Processing Fuels.*

Transportation by air requires roadway transportation for access of both goods and people (staff and passengers) to airports (Committee on Climate Change and U.S. Transportation, 2008; DOT, 2022). Aircraft are powered largely by petroleum products (jet fuel and aviation gas) (Holladay, Abdullah, and Heyne, 2020), and airport operations require electricity. The air traffic control system, which ensures safe movement of aircraft in flight, relies on both PNT and satellite-based services (Wallischeck, 2016).

Strength of Evidence

A robust scientific literature examines the relationship between numerous climate drivers and air transportation. There have been multiple analyses of the effect that extreme heat, flooding, sea-level rise, severe storm systems, and tropical cyclones and hurricanes can have on airports and aircraft operations (Burbidge, 2018; Coffel and Horton, 2015; Coffel, Thompson, and Horton, 2017; Love, Soares, and Püempel, 2010; National Research Council, 2008; Thompson, 2016; Wright and Hogan, 2008; Yesudian and Dawson, 2021). However, several studies have noted that, despite the predicted increase in the number of convective storms, there is almost no analysis of their effect on the safety of air transportation.[48] The effect of drought, extreme cold, and wildfire on air transportation is less well established. However, particular risks from these climate drivers, including disruptions to air transportation, are well-known. For example, deicing aircraft during severe storm systems is common practice, and FAA establishes minimum visibility requirements for pilots that could be affected by smoke plumes due to wildfire.

Table B.144. Strength of Evidence for *Transport Cargo and Passengers by Air*

Strength of Evidence	
High	Sea-level rise; extreme heat; flooding; tropical cyclones and hurricanes
Medium	Severe storm systems
Low	Drought; extreme cold; wildfire

Risk Rating Interpretations

Table B.145 provides examples of how to interpret the risk ratings for this NCF.

---

[48] An extensive 2022 review by Gratton et al., noted,

> Whilst there is strong evidence of increasing clear air turbulence severity and contact incidence in the vicinity of the north polar jet stream, and to a lesser extent in the vicinity of other jet streams, there appears presently to be no published analysis of the safety or other operational implications of those encounters. (Gratton et al., 2022, p. 219)

Table B.145. Example Interpretations of Risk Ratings for *Transport Cargo and Passengers by Air*

Risk Rating	NCF Example
1: No disruption or normal operations	Air transportation continues to function with only occasional delays due to inclement weather and operational problems.
2: Minimal disruption	A few airports in flood-prone regions experience longer time periods in which they cannot operate safely, but cargo and passengers can be moved with some delays or through alternative airports.
3: Moderate disruption	Flight delays and cancellations due to natural hazards affected by climate change occur with some frequency; disruptions at large- and medium-hub airports affect downstream flights and regional air transportation. Some airports close temporarily or permanently due to sea-level rise.
4: Major disruption	Flight delays and cancellations due to sea-level rise and other climate drivers occur regularly; disruptions at large- and medium-hub airports affect the entire national air transportation system. Some airlines cease to operate, and a significant number of airports close permanently.
5: Critical disruption	Air transportation is essentially halted, although some options might be available at very high prices.

Transport Cargo and Passengers by Road

Table B.146. Scorecard for *Transport Cargo and Passengers by Road*

	National Risk Assessment				
	Baseline	Current Emissions		High Emissions	
	Present	2050	2100	2050	2100
Transport Cargo and Passengers by Road	2	3	3	3	3

Description: "Provide and operate roadway systems, assets, and facilities—including commercial motor carriers and associated facilities, motorcoaches, buses, and associated systems, assets, and facilities—to enable a system of securely and safely conveying goods and people from place to place by highway" (CISA, 2020, p. 3).

Highlights

- *Transport Cargo and Passengers by Road* was assessed as being at risk of moderate disruption by 2050, primarily because of the potential for damage to physical infrastructure at a regional scale.
- Physical damage and disruption could be severe in specific areas, primarily because of effects from extreme heat, sea-level rise, flooding, severe storm systems, tropical cyclones and hurricanes, and wildfire.
- Some coastal roads could become permanently inundated by sea-level rise, particularly by 2100 under a high emissions scenario.
- The *Provide Road System* subfunction is most affected because it is primarily about construction and maintenance of physical infrastructure.

Synopsis

Transport Cargo and Passengers by Road is vulnerable to disruption by multiple climate drivers—in particular, extreme heat, coastal flooding driven by sea-level rise, flooding from storms and extreme precipitation, and wildfire. All of these drivers have the potential to cause significant damage to the roadway network itself, as well as threaten the safety of the road construction workforce (Jacobs, Cattaneo, et al., 2018; T. Wang et al., 2020; Willis et al., 2016). Surface roads are vulnerable to degradation by extreme heat events, flood inundation and storm-driven landslides, while bridges and tunnels are especially vulnerable to flooding and sea-level rise. Although transport infrastructure is vulnerable to long-term incremental changes in average temperature and precipitation, it is especially sensitive to extreme events (Rowan et al., 2013). Bridges can suffer catastrophic effects, such as collapse driven by large flooding events, or they can undergo cumulative damage from bridge scour (weakening of piers and abutments caused by sediment erosion). Tunnels at low elevation can be flooded by groundwater levels raised by sea-level rise (Bloetscher, Berry, et al., 2014; Mondoro, Frangopol, and Soliman, 2017). Extreme heat accelerates deterioration in concrete roads and bridges and can reduce pavement performance through cracking and buckling. Wildfire can cause roads to crack and asphalt to burn, poses a safety hazard to roadway workers, and increases the likelihood of debris flows if a wildfire is followed by heavy precipitation (Jacobs, Cattaneo, et al., 2018). Flooding, severe storm systems, and wildfire can also damage or destroy vehicles.

The combined damage that multiple climate drivers can cause to the roadway network is expected to cause widespread travel disruptions to passengers and freight and impede access to critical services

(Jacobs, Cattaneo, et al., 2018). Coastal flooding alone has the potential to increase the number of vehicle-hours of delay on the U.S. East Coast by a factor of ten by 2060 over 2010 levels (Jacobs, Cattaneo, et al., 2018). Over time, coastal areas could become permanently inundated by sea-level rise, and roadways in these areas will need to be moved to less threatened locations. In addition, as climate effects compound and worsen over time, the cost of maintaining transport infrastructure approaching or beyond its design life will increase (Chinowsky, Price, and Neumann, 2013). For example, maintaining pavements under increases in extreme heat could be more expensive than usual because of the added cost of material upgrades to accommodate higher temperatures (Jacobs, Cattaneo, et al., 2018). Under a high emissions scenario with no adaptation, average annual costs of damage to road systems are estimated to increase by more than $400 billion by 2090 from a 1986–2005 baseline (Neumann et al., 2021).

Overall, this NCF was assessed as being at risk of moderate disruption by 2050 under both the current and high emissions scenarios because of the potential for widespread damage to infrastructure at a regional scale. Damage from a disaster with a large regional footprint, such as a hurricane, can cause road closures along many alternative routes simultaneously. This disruption could be severe enough to make transportation by road impossible except at very local scales. Closures at critical bottlenecks, such as bridges and tunnels, are especially disruptive because routes that utilize such structures often have few alternatives. However, because the U.S. roadway network typically offers multiple routes between origins and destinations, risk is not expected to exceed the level of moderate disruption (3). In contrast, extreme heat generally does not cause widespread failure of road systems but rather increased costs and construction-related delays.

Subfunctions

The subfunctions with the largest expected effects are those directly linked to the physical infrastructure of the roadway network itself—specifically, *Provide Road System, Enable Intermodal Transport of Cargo, Transport Cargo via Road,* and *Transport Passengers by Road.* This is because the main mechanism by which climate change disrupts this NCF is damage to physical infrastructure. Of these, the *Provide Road System* subfunction is most affected because its primary activities are the construction and maintenance of roads, bridges, and tunnels. *Enable Intermodal Transport of Cargo, Transport Cargo via Road,* and *Transport Passengers by Road* critically depend on physical infrastructure but are not directly involved with construction and maintenance of the roadway system. In addition to effects from reduced access to transport infrastructure, damage to vehicles also directly affects these subfunctions.

Maintain Road Transit Workforce is also significantly affected by extreme heat and drought because high temperature and lack of access to water affect workers' ability to work outdoors safely (Jacobs, Cattaneo, et al., 2018). This could increase the cost of construction and maintenance work and might require changes to construction policies, such as working at night when temperatures are lower. Such protocols could limit construction during heat waves, resulting in lost productivity (Gordon, 2014). In the medium to long term, this could also affect transport authorities' ability to attract and retain workers. Other climate drivers that create unsafe conditions for workers, such as flooding and wildfire, could place additional demands on the workforce. There is also some risk to the *Govern Road Transport Operations* subfunction from numerous climate drivers because climate uncertainty creates challenges for road system planning. For example, planners might have limited information about how the "peak season" for hurricane and wildfire events is changing over time because of climate change, which could limit their ability to schedule routine maintenance operations and implement disaster preparedness, response, and recovery measures.

Table B.147. Subfunction Risk for *Transport Cargo and Passengers by Road*

	Baseline	Current Emissions		High Emissions	
Subfunction	Present	2050	2100	2050	2100
Provide Road System	2	3	3	3	3
Enable Intermodal Transport of Cargo	2	3	3	3	3
Transport Cargo via Road	2	3	3	3	3
Transport Passengers by Road	2	3	3	3	3
Maintain Road Transit Workforce	2	3	3	3	3
Govern Road Transport Operations	1	2	2	2	2

NOTE: 1 = no disruption or normal operations. 2 = minimal disruption. 3 = moderate disruption. See Chapter 2 for further explanation of ratings.

Climate Drivers

All climate drivers will cause at least some disruption to *Transport Cargo and Passengers by Road*, and some have the potential to cause significant damage. The most-significant climate drivers are extreme heat, flooding, sea-level rise, tropical cyclones and hurricanes, and wildfire. As described above, these climate drivers cause direct damage to roadway network infrastructure. Furthermore, these drivers are all expected to increase significantly in intensity, magnitude, or duration in the future, particularly by 2100 under a high emissions scenario. Although sea-level rise is not currently a major risk factor, it has the potential to temporarily inundate coastal roadways during storm events throughout the United States, especially in the Gulf Coast region, as early as the middle of the 21st century (Hicke et al., 2022); therefore, its risk rating rises to risk of moderate disruption by 2050. Severe (nontropical) storm systems also have potential for major damage, but their intensity is expected to increase less than that of tropical cyclones and hurricanes by 2100. Extreme cold causes asphalt to contract and shrink, leading to road damage. However, this driver is not projected to increase in the future, so its effect relative to baseline is minimal.

Table B.148. Risk from Climate Drivers for *Transport Cargo and Passengers by Road*

Climate Driver	Baseline Present	Current Emissions 2050	Current Emissions 2100	High Emissions 2050	High Emissions 2100
Drought	2	2	2	2	2
Extreme cold	2	2	2	2	2
Extreme heat	2	3	3	3	3
Flooding	2	3	3	3	3
Sea-level rise	1	3	3	3	3
Severe storm systems	2	2	2	2	2
Tropical cyclones and hurricanes	2	3	3	3	3
Wildfire	2	3	3	3	3

NOTE: 1 = no disruption or normal operations. 2 = minimal disruption. 3 = moderate disruption. See Chapter 2 for further explanation of ratings.

Impact Mechanisms

By far, the most significant impact mechanism for *Transport Cargo and Passengers by Road* is physical damage and disruption. Eight climate drivers affect at least three subfunctions through this mechanism, and all but extreme cold and drought affect five subfunctions through this mechanism (*Provide Road System*; *Enable Intermodal Transport of Cargo*; *Transport Cargo via Road*; *Transport Passengers by Road*; *Govern Road Transport Operations*). Unlike other subfunctions, *Maintain Road Transit Workforce* is not affected through the physical damage or disruption mechanism but is affected through the workforce shortage mechanism because road construction and repair operations can be disrupted if workers are unable to work safely under increasingly challenging environmental conditions.

Cascading Risk

Transport Cargo and Passengers by Road is dependent on the following NCFs:

- *Fuel Refining and Processing Fuels*
- *Provide Positioning, Navigation, and Timing Services.*

Although electric vehicles make up an increasing share of vehicles in the United States, at present, most passenger vehicles require gasoline and most trucks require diesel fuel (Davis and Boundy, 2022). PNT services are widely used in the trucking industry for vehicle location and routing (Wallischeck, 2016).

Strength of Evidence

There is strong evidence that climate change's effects due to flooding, sea-level rise, tropical cyclones and hurricanes, severe storm systems, extreme heat, and wildfire will damage roadway networks (Jacobs, Cattaneo, et al., 2018). These effects will most directly affect the *Provide Road*

227

System, Enable Intermodal Transport of Cargo, Transport Cargo via Road, and *Transport Passengers by Road* subfunctions. To a lesser extent, there is also evidence that extreme cold will damage or disrupt these subfunctions. In addition, there is a well-established body of research on extreme heat effects—and, to a lesser extent, drought effects—on the safety of outdoor workers, which could affect the *Maintain Road Transit Workforce* subfunction through potential workforce shortages.

Table B.149. Strength of Evidence for *Transport Cargo and Passengers by Road*

	Strength of Evidence
High	Extreme heat; flooding; sea-level rise; tropical cyclones and hurricanes; severe storm systems; wildfire
Medium	Drought; extreme cold

Risk Rating Interpretations

Table B.150 provides examples of how to interpret the risk ratings for this NCF.

Table B.150. Example Interpretations of Risk Ratings for *Transport Cargo and Passengers by Road*

Risk Rating	NCF Example
1: No disruption or normal operations	Transportation by road continues to function as it does today, with occasional delays due to weather that cause operational problems.
2: Minimal disruption	Road networks, including bridges and tunnels, experience temporary shutdowns or limited capacity in specific areas. Some disruption can be avoided by transporting cargo and passengers via alternative routes.
3: Moderate disruption	Widespread road closures or delays occur at a regional scale. Some coastal roadways become permanently inundated. Alternative routes in affected areas could be limited.
4: Major disruption	Traffic shutdowns and severe delays occur simultaneously at a national scale. Alternative routes are severely limited. Coastal roadways are increasingly nonviable.
5: Critical disruption	Virtually all transportation by road, except in isolated localities, is halted at a national scale.

228

Transport Materials by Pipeline

Table B.151. Scorecard for *Transport Materials by Pipeline*

	National Risk Assessment				
	Baseline	Current Emissions		High Emissions	
	Present	2050	2100	2050	2100
Transport Materials by Pipeline	2	3	3	3	3

Description: "Provide and operate systems, assets, and facilities to enable a system of securely and safely conveying materials from place to place by pipelines" (CISA, 2020, p. 4).

Highlights

- *Transport Materials by Pipeline* was assessed as being at risk of moderate disruption overall by 2050, driven primarily by the potential for catastrophic damage by hurricanes in the Gulf region.
- Extreme heat (via permafrost thaw) and wildfire are secondary risk factors but are relatively limited in scale and can be mitigated through adaptation.
- Risk is highest for the *Transport Gases* subfunction because gas pipelines make up 88 percent of all pipelines in the United States (Bureau of Transportation Statistics, undated-b).

Synopsis

Transport Materials by Pipeline was assessed as being at risk of disruption from three climate drivers: tropical cyclones and hurricanes, extreme heat, and wildfire. Hurricanes have caused enormous damage to pipelines in the U.S. Gulf Coast area, leading to significant offshore oil spills (Cruz and Krausmann, 2008; Rusco, 2021). In Alaska, extreme heat (and rising temperatures in general) thaws permafrost, causing land subsidence and decreasing the bearing capacity of pipeline foundations (Hjort et al., 2022; Markon et al., 2018). Approximately 85 percent of Alaska is underlain by permafrost (Alaska Department of Fish and Game, undated), and nearly 25 percent of near-surface permafrost in Alaska is projected to have thawed by 2100 (Pastick et al., 2015). Wildfire is a threat to infrastructure, including pipelines both in Alaska and throughout the western contiguous United States (Hicke et al., 2022). Ancillary pipeline facilities, such as control centers, pumping stations, and electric power infrastructure, can also be damaged by wildfire.

Pipeline damage disrupts transport of oil and gas supplies to domestic and export markets, causing economic losses, price increases, and environmental damage. The economy of Alaska is especially dependent on the 800-mile Trans-Alaska Pipeline System, approximately 75 percent of which passes through terrain underlain by permafrost (Hasemyer, 2021). Although land subsidence from permafrost thaw can damage many types of infrastructure, pipelines are considered especially vulnerable to permafrost thaw (Hjort et al., 2022). Permafrost-driven damage increases maintenance costs of pipelines (Melvin et al., 2016) and can require expensive engineering interventions, such as drilling pile foundations deeper into permafrost (Faki, Sushama, and Doré, 2022). Pipeline damage from both permafrost thaw and wildfire can also cause oil spills and gas leaks, leading to contamination of soil, water, and wildlife (Belvederesi, Thompson, and Komers, 2018). Hurricanes have the potential to cause significant damage to the U.S. petrochemical industry, which is concentrated in the hurricane-prone Gulf region. Hurricanes Katrina and Rita destroyed a large

229

number of offshore pipelines in 2005, leading to numerous spills; in 2021, Hurricane Ida caused pipeline shutdowns throughout the region (Cruz and Krausmann, 2008; Hampton, 2021).

Tropical cyclones and hurricanes, permafrost thaw, and wildfire are all expected to increase in severity in the future, particularly under a high emissions scenario by 2100 (Hicke et al., 2022). However, overall risk to this NCF is rated as a risk of moderate disruption by 2050 because of the potential for catastrophic damage to the Gulf Coast pipeline network from hurricanes. Because of the concentration of the U.S. petrochemical industry in the Gulf region, a Hurricane Katrina–like disaster has the potential to disrupt oil and gas transport to multiple regions within the United States. In contrast, permafrost thaw–induced subsidence in the United States is limited to Alaska and is generally a hazard that can be anticipated and managed with engineering-based adaptations (Doré, Niu, and Brooks, 2016). In addition, the U.S. pipeline network is vast, with oil and gas together making up nearly 2 million miles of pipeline (Bureau of Transportation Statistics, undated-b), enabling transport of alternative supplies to ameliorate some minimal disruptions. Pipeline structures are also not especially vulnerable to wildfire apart from some sensitive parts, such as cables (Robinson et al., 2018), so a wildfire incident near a pipeline will not necessarily lead to a disruption in service.

Subfunctions

Both the *Transport Liquids* and *Transport Gases* subfunctions are at risk of effects from tropical cyclones and hurricanes, permafrost thaw (driven by extreme heat), and wildfire. Both of these subfunctions deal primarily with the maintenance, control, and day-to-day operation of vast networks of pipeline infrastructure that can be damaged or disrupted by these climate drivers. *Transport Gases* is likely to be most affected because gas pipelines make up 88 percent of all pipeline mileage in the United States (Bureau of Transportation Statistics, undated-b).

Table B.152. Subfunction Risk for *Transport Materials by Pipeline*

Subfunction	Baseline Present	Current Emissions 2050	Current Emissions 2100	High Emissions 2050	High Emissions 2100
Transport Liquids	2	3	3	3	3
Transport Gases	2	3	3	3	3

NOTE: 2 = minimal disruption. 3 = moderate disruption. See Chapter 2 for further explanation of ratings.

Climate Drivers

Tropical cyclones and hurricanes, extreme heat, and wildfire are expected to pose significant risks to pipelines. As discussed above, the concentration of the U.S. petrochemical industry in the Gulf region means that many offshore pipelines are regularly at risk of exposure to hurricanes. Although not every hurricane will cause significant damage or disruption, a large Katrina-like storm has the potential to shut down pipeline transport to large regions of the United States simultaneously. Extreme heat does not damage pipeline infrastructure directly but rather drives land subsidence by thawing permafrost, which reduces bearing capacity of pipeline foundations (Hjort et al., 2022; Markon et al., 2018). Subsidence can be mitigated by proactively drilling pipeline foundations into deeper layers of permafrost or by installing additional adaptive measures, such as passive cooling systems that mitigate permafrost thaw. In the United States, the effect from this climate driver is limited to Alaska. In contrast, wildfire has the potential to damage pipelines in Alaska and throughout the western United States. Although pipeline structures themselves are not especially vulnerable to wildfire, ancillary infrastructure, such as cables and control systems, are comparatively vulnerable (Robinson et al., 2018). This suggests that, although pipeline vulnerability to wildfire might be minimal overall, exposure of specific critical elements of pipeline systems can nonetheless pose a significant risk of disruption to this NCF.

Table B.153. Risk from Climate Drivers for *Transport Materials by Pipeline*

	Baseline	Current Emissions		High Emissions	
Climate Driver	Present	2050	2100	2050	2100
Extreme heat	2	2	2	2	2
Tropical cyclones and hurricanes	2	3	3	3	3
Wildfire	2	2	2	2	2

NOTE: 2 = minimal disruption. 3 = moderate disruption. See Chapter 2 for further explanation of ratings.

Impact Mechanisms

The only significant impact mechanism for *Transport Materials by Pipeline* is physical damage and disruption. Tropical cyclones and hurricanes, extreme heat (through permafrost thaw), and wildfire can all cause infrastructure damage that would affect both the *Transport Liquids* and *Transport Gases* subfunctions. There is little evidence that climate drivers pose a risk to this NCF through workforce shortages, lack of resources, or demand changes.

Cascading Risk

Transport Materials by Pipeline is critically dependent on the following NCFs that provide the materials to be transported:

- *Exploration and Extraction of Fuels*
- *Fuel Refining and Processing Fuels*.

Pipelines that transport primary hydrocarbons (natural gas and crude oil) are dependent on *Exploration and Extraction of Fuels*. Pipelines that transport refined petroleum products, such as gasoline, home heating oil, diesel fuel, aviation gasoline, jet fuels, and kerosene (Pipeline and Hazardous Materials Safety Administration, 2018), are dependent on *Fuel Refining and Processing Fuels*.

Strength of Evidence

Evidence of pipeline infrastructure degradation from climate-driven permafrost thaw and damage from hurricanes, has been documented extensively in scientific literature (Cruz and Krausmann, 2008;

Hjort et al., 2022; Markon et al., 2018). The threat of wildfire to pipeline infrastructure is well-known (Robinson et al., 2018), but there have been fewer studies on the potential for climate change to increase pipeline damage from wildfire in the future. There is little evidence of significant damage to pipelines from other climate drivers.

Table B.154. Strength of Evidence for *Transport Materials by Pipeline*

	Strength of Evidence
High	Extreme heat; tropical cyclones and hurricanes
Medium	Wildfire
Low	Flooding; sea-level rise; severe storm systems; extreme cold; drought

Risk Rating Interpretations

Table B.155 provides examples of how to interpret the risk ratings for this NCF.

Table B.155. Example Interpretations of Risk Ratings for *Transport Materials by Pipeline*

Risk Rating	NCF Example
1: No disruption or normal operations	Transport of oil and gas by pipeline continues to function as it does today, with occasional delays due to supply disruptions and operational problems.
2: Minimal disruption	Pipelines experience occasional shutdowns in specific areas due to local damage.
3: Moderate disruption	Numerous shutdowns of pipelines occur in one or more petrochemical hubs; oil and gas transport to areas beyond the region of impact is slowed or suspended.
4: Major disruption	Pipeline shutdowns occur in most petrochemical hubs across the country, leading to nationwide supply disruptions.
5: Critical disruption	Virtually all transport of oil and gas by pipeline is halted at a national scale.

Transport Passengers by Mass Transit

Table B.156. Scorecard for *Transport Passengers by Mass Transit*

	National Risk Assessment				
	Baseline	Current Emissions		High Emissions	
	Present	2050	2100	2050	2100
Transport Passengers by Mass Transit	2	2	3	2	3

Description: "Provide and operate systems, assets, and facilities to enable a system of securely and safely conveying people from place to place by roads or on fixed guideways within a specified geographic area—including transit buses, trolleybuses, monorails, heavy rail (subway), light rail, passenger rail, commuter rail, and vanpool/rideshare" (CISA, 2020, p. 4).

Highlights

- *Transport Passengers by Mass Transit* is at risk of minimal disruption overall by 2050, rising to moderate disruption by 2100.
- Mass transit systems are generally local and decentralized, limiting effects to urban areas and surrounding suburban areas.
- Even so, physical damage and suspensions of service could be severe in cities vulnerable to flooding, severe storm systems, sea-level rise, and extreme heat.
- The *Manage Transit System Operations* and *Conduct Transport Operations* subfunctions are most affected because these are most directly involved in the day-to-day operations of transit systems.

Synopsis

Transport Passengers by Mass Transit is vulnerable to direct disruption by climate drivers that damage physical infrastructure. These are primarily flooding, sea-level rise, severe storm systems, tropical cyclones and hurricanes, and extreme heat (Jacobs, Culp, et al., 2018; T. Wang et al., 2020; Willis et al., 2016). Wildfires directly threaten mass transit to a lesser degree because mass transit systems are typically concentrated in urban areas rather than at the wildland–urban interface. Mass transit infrastructure vulnerable to these climate drivers includes assets and facilities owned by transit operators, including tunnels, bridges, passenger rail track, bus and rail stations, storage and maintenance locations, and rolling stock (buses and train cars).[49] Extreme heat accelerates deterioration in concrete roads and bridges and can reduce pavement performance through cracking and buckling. These effects are particularly important for mass transit in cities because of the urban heat island effect, which causes longer and more-intense heat waves in urban areas than in surrounding rural and suburban areas (Maxwell et al., 2018). Bridges can be destroyed by large flood events or suffer cumulative damage from bridge scour (weakening of piers and abutments caused by sediment erosion). Sea-level rise can cause groundwater levels to rise, causing flooding in tunnels (Bloetscher, Berry, et al., 2014; Mondoro, Frangopol, and Soliman, 2017). Over time, sea-level rise can cause some mass transit infrastructure along coastlines to be permanently inundated.

---

[49] Although roads themselves are also vulnerable to climate change, they are not typically owned by transit operators; thus, roads are considered as part of the *Transport Cargo and Passengers by Road* NCF.

Infrastructure damage caused by these climate drivers will cause disruptions to mass transit through suspensions of service, higher construction and maintenance costs, and lost fare revenues. These effects will be felt primarily in urban areas, where most transit rides occur (Fan, Wen, and Wan, 2017). Suspensions of service reduce intraurban mobility and disrupt commutes into cities from suburban areas, leading to economic losses, as was observed in the Northeast after Hurricanes Irene and Sandy (Barnes, 2015).[50] The costs of maintaining urban transport infrastructure, such as maintaining pavements under extreme heat, are expected to increase as climate effects worsen over time (Chinowsky, Price, and Neumann, 2013; Jacobs, Culp, et al., 2018). Extreme heat and precipitation events can also decrease ridership, lowering fare revenues (Ngo, 2019). Drought can affect electrified transit systems by limiting generation of hydropower (van Vliet et al., 2016), a key electricity source in the Pacific Northwest region.

Overall disruption of mass transit was rated as minimal by 2050 under both the current and high emissions scenarios. This is because transit overall is generally confined to urban areas and surrounding suburban areas such that a transit disruption in one city would be unlikely to affect transit service in other cities. Furthermore, cities vary in terms of available transit modes—although most cities have bus service, many larger cities also operate at least one type of rail (light, heavy, or commuter). The provision of multiple modes creates transportation network redundancy; for example, bus service might be able to partially compensate for flooded subway tunnels during a flood event (Jacobs, Culp, et al., 2018). Overall, mass transit is a decentralized, metropolitan-level function and is occasionally regional in scale, such as when multiple cities are connected by a transit system (such as San Francisco, Oakland, and San Jose), rather than national in scale. One climate driver, drought, has the potential to cause moderate disruption by 2100 because it can limit hydropower generation at a regional scale.

Subfunctions

The subfunctions likeliest to be affected by climate drivers are those most directly involved in the day-to-day operations of mass transit systems, including repairing and maintaining infrastructure, managing staffing, and carrying out basic functions, such as collecting fares and monitoring route operations. These functions are encompassed primarily by the *Manage Transit System Operations* and *Conduct Transport Operations* subfunctions. Suspensions of service and increased costs caused by climate-driven damage and disruption would first impair these subfunctions' ability to operate normally, then affect higher-level, planning-oriented subfunctions, such as *Management of Mass Transit System* and *Governance of Mass Transit System*. These subfunctions will nonetheless also be affected because climate drivers increase the difficulty of developing, managing, and regulating transit systems resilient enough to withstand extreme events. The *Provide Diverse Energy Sources* subfunction would also be affected by loss of hydropower generation, as mentioned above.

---

[50] Such losses would be less likely in any other metropolitan area because the New York City region has a far greater number of transit riders than that of any other area in the United States (Federal Transit Administration, undated).

Table B.157. Subfunction Risk for *Transport Passengers by Mass Transit*

Subfunction	Baseline Present	Current Emissions 2050	Current Emissions 2100	High Emissions 2050	High Emissions 2100
Management of Mass Transit System	1	2	2	2	2
Governance of Mass Transit System	1	2	2	2	2
Manage Transit System Operations	2	2	2	2	2
Conduct Transport Operations	2	2	2	2	2
Provide Diverse Energy Sources	2	2	3	2	3

NOTE: 1 = no disruption or normal operations. 2 = minimal disruption. 3 = moderate disruption. See Chapter 2 for further explanation of ratings.

Climate Drivers

All climate drivers have the potential to disrupt *Transport Passengers by Mass Transit* to varying degrees. Many large U.S. cities are in areas currently at risk of seasonal riverine flooding and coastal flooding from tidal events (FEMA, undated), which can damage tunnels, rolling stock, track and stations, bus lots, and maintenance facilities. These effects are magnified by sea-level rise and severe storm systems (tropical and nontropical) and could be particularly severe where municipal sewer and storm sewer systems are not designed to withstand the capacity demand of heavy rainfall events (Maxwell et al., 2018). Flooding also necessitates power shutoff to electrified third-rail systems and can erode soil directly adjacent to paved areas. Heavy rain, even if it does not result in flooding, can cause landslides that damage roads, tracks, and facilities.

In addition to damaging pavements, extreme heat can cause tracks to buckle, leading to derailments (Chinowsky, Helman, et al., 2019), and overhead catenary wires to fail, leading to loss of power (Hodges, 2011). High temperatures can also cause critical physical infrastructure for train control, monitoring, and communication to overheat and malfunction. Wildfire can directly damage vehicles, buildings, tracks, and roads, although its impact is limited to areas that include a wildland–urban interface. As mentioned above, drought has the potential to cause regionwide disruption by limiting water available to generate hydropower, which would have severe consequences for hydropower-dependent regions, such as the Pacific Northwest. As a result, its risk rating is moderate under a high emissions scenario by 2100. Extreme cold also causes seasonal damage to roads, tracks, and other infrastructure. However, this driver is not projected to increase in the future, so its impact relative to baseline is minimal.

Table B.158. Risk from Climate Drivers for *Transport Passengers by Mass Transit*

Climate Driver	Baseline	Current Emissions		High Emissions	
	Present	2050	2100	2050	2100
Drought	2	2	3	2	3
Extreme cold	2	2	2	2	2
Extreme heat	2	2	2	2	2
Flooding	2	2	2	2	2
Sea-level rise	1	2	2	2	2
Severe storm systems	2	2	2	2	2
Tropical cyclones and hurricanes	2	2	2	2	2
Wildfire	1	2	2	2	2

NOTE: 1 = no disruption or normal operations. 2 = minimal disruption. 3 = moderate disruption. See Chapter 2 for further explanation of ratings.

Impact Mechanisms

The most common impact mechanism for *Transport Passengers by Mass Transit* is physical damage or disruption. All climate drivers except drought affect the *Manage Transit System Operations* and *Conduct Transport Operations* subfunctions through this mechanism, and five climate drivers affect the *Management of Mass Transit System* and *Governance of Mass Transit System* subfunctions through this mechanism. The *Conduct Transport Operations* subfunction is also affected by numerous climate drivers through the workforce shortage mechanism because this subfunction encompasses most of the day-to-day operations of transit systems that would be most affected in the event of a workforce shortage caused by workers being unable to work outdoors safely because of drought or extreme heat. Drought also affects the *Provide Diverse Energy Sources* subfunction through the lack-of-resources mechanism by limiting hydropower generation.

Cascading Risk

Transport Passengers by Mass Transit is critically dependent on the following NCFs:

- *Transport Cargo and Passengers by Road*
- *Distribute Electricity*
- *Fuel Refining and Processing Fuels*
- *Provide Positioning, Navigation, and Timing Services.*

Transportation by mass transit often requires road-based transportation for passengers to stations. In addition, buses, the transit mode with the highest ridership, operate on roads (Office of Budget and Policy, undated). Most transit modes operate on some kind of fossil fuel or electricity (Davis and Boundy, 2022). PNT services are widely used for vehicle location (Wallischeck, 2016).

Strength of Evidence

Evidence of physical damage and disruption to transit systems by extreme heat, flooding, sea-level rise, and heavy precipitation from storms (tropical and nontropical) is well documented (Jacobs, Culp, et al., 2018). Effects from wildfire, drought, and extreme cold are less well documented but highly plausible. Aside from research on direct effects on infrastructure, there is relatively little research on how climate change can affect transit systems' ability to maintain their workforces and manage and regulate transit operations. Overall, research on mass transit systems tends to be local rather than national, meaning that studies on one transit system might not be easily generalizable to other systems.

Table B.159. Strength of Evidence for *Transport Passengers by Mass Transit*

Strength of Evidence	
High	Extreme heat; flooding; sea-level rise; tropical cyclones and hurricanes; severe storm systems
Medium	Extreme cold; wildfire; drought

Risk Rating Interpretations

Table B.160 provides examples of how to interpret the risk ratings for this NCF.

Table B.160. Example Interpretations of Risk Ratings for *Transport Passengers by Mass Transit*

Risk Rating	NCF Example
1: No disruption or normal operations	Mass transit operations continue to function as they do today, with occasional suspensions of service due to weather and operational problems.
2: Minimal disruption	Mass transit systems experience temporary shutdowns longer than a routine suspension of service. Some disruption can be avoided by utilizing alternative transport modes.
3: Moderate disruption	Widespread shutdowns of mass transit systems, either total or partial, occur sporadically in cities across the country or are concentrated in specific regions. Some coastal systems are no longer viable because of permanent inundation. Alternative transport modes are limited.
4: Major disruption	Mass transit system shutdowns occur simultaneously in cities across the country. Alternative transport modes are severely limited. Coastal transit systems are increasingly nonviable.
5: Critical disruption	Virtually all mass transit systems across the country cease to operate. Alternative transport modes are unavailable.

Transport Cargo and Passengers by Rail

Table B.161. Scorecard for *Transport Cargo and Passengers by Rail*

	National Risk Assessment				
	Baseline	Current Emissions		High Emissions	
	Present	2050	2100	2050	2100
Transport Cargo and Passengers by Rail	2	2	3	3	3

Description: "Provide and operate freight and passenger railroad systems, including conveyances, infrastructure, and management systems to enable a system of securely and safely conveying goods and people from place to place by rail" (CISA, 2020, p. 3).

Highlights

- The U.S. rail network is exposed to several climate drivers and is at risk of moderate disruption in later time periods due to projected increases in climate-driver severity.
- Severer flooding, extreme heat, sea-level rise, tropical cyclones and hurricanes, and wildfire can destroy tracks and create bottlenecks that disrupt the ability to transport goods and people. Drought can disrupt intermodal transport of cargo.
- Rail nodes located in coastal regions are vulnerable to sea-level rise and hurricanes, while important rail nodes in the Midwest are vulnerable to flooding.
- The subfunctions related to movement of people and freight, including intermodal freight, are particularly vulnerable to climate change.
- Some rail nodes and rail passages are more important than others to the regional and national flow of cargo, making the NCF vulnerable to climate drivers that affect specific parts of the rail system.

Synopsis

Transport Cargo and Passengers by Rail was assessed as being at risk of minimal to moderate disruption from climate change for several climate drivers. For instance, flooding can inundate tracks, cause erosion, and make it difficult or impossible to move cargo and passengers through some rail segments ("Amtrak," 2019; Chinowsky, Helman, et al., 2019; Rossetti, 2003; Rossetti, 2007). For example, the riverine floods of 1993 resulted in more than 4,000 miles of railroad tracks being either flooded or idled, which led to more than an estimated $200 million in losses (Rossetti, 2007). Similar incidents have been reported by rail operator BNSF, which experienced service delays in the Northwest following flooding in the region (BNSF Railway, 2021). Hurricanes can also disrupt rail activities in coastal regions, inflicting damage to rail infrastructure and assets (Grenzeback and Lukmann, 2008; Rossetti, 2003), while sea-level rise is expected to negatively affect coastal rail and intermodal rail activities in the future (Allen, McLeod, and Hutt, 2022; Nazarnia et al., 2020). Extreme heat can lead to track buckling, forcing rail operators to reduce speeds (Rossetti, 2003). This slows deliveries and causes supply chain disruptions (Chinowsky, Helman, et al., 2019). The popular press has reported on wildfire disrupting rail operations in the western United States (Franz, 2021; Marsh, 2021) and on railway operators seeing wildfire as a meaningful threat to future operations (Chavez, 2021).

Taken together, climate drivers are expected to continue to disrupt *Transport Cargo and Passengers by Rail* at minimal to moderate levels in the next century. As climate drivers become more frequent and severer, increased degradation or damage to rail assets from some drivers, such as flooding, hurricanes, sea-level rise, wildfire, and extreme heat, are likely to cause longer delays or suspension of cargo or passenger transportation along some corridors. These disruptions could rise to a regional level in some instances: Because railways make up an interconnected network, local physical damage to assets could lead to regional delays or interruptions if critical nodes are affected. For instance, a sizable amount of intermodal freight shipped from the Ports of Los Angeles and Long Beach passes through Chicago by rail (Capra, 2006), so a disruption to rail operations in Chicago could have regional implications. However, not all nodes and links in the network have regional importance. In most cases, the impacts would be limited to local effects. Intermodal transport could be disrupted by sea-level rise, particularly if rail yards and tracks in major U.S. seaports are flooded with increased regularity.

The projected rise in flooding occurrence and severity, over both time frames and across scenarios, is expected to lead to risk of moderate disruption to *Transport Cargo and Passengers by Rail*. Hurricanes are projected to become severer by 2100, contributing to elevated risk ratings in this time period. Similarly, sea-level rise and extreme heat are expected to increase over the century for both emissions scenarios; both of these drivers are expected to increase disruptions to the NCF.

Subfunctions

Several subfunctions are at risk of moderate disruption by 2100 under current emissions and by 2050 under high emissions. Specifically, subfunctions that are highly reliant on rail infrastructure— *Provide Rail Systems, Provide Intermodal Transport of Freight, Transport Cargo via Rail,* and *Transport Passengers via Rail*—might be unable to operate normally in some regions because of disruptions from flooding, sea-level rise, hurricanes, and extreme heat. Other subfunctions are at risk of minimal disruption. For instance, *Maintain Rail Workforce* could face minimal disruptions due to extreme heat because working outdoors may become difficult. This is likely to pose a minimal disruption in later time periods because extreme heat is expected to increase over the century, but not all work in rail transportation is outdoors, so the effects are diminished.

Table B.162. Subfunction Risk for *Transport Cargo and Passengers by Rail*

	Baseline	Current Emissions		High Emissions	
Subfunction	Present	2050	2100	2050	2100
Provide Rail Systems	2	2	3	3	3
Provide Intermodal Transport of Freight	2	2	3	3	3
Transport Cargo via Rail	2	2	3	3	3
Transport Passengers via Rail	2	2	3	3	3
Maintain Rail Workforce	1	1	2	1	2

NOTE: 1 = no disruption or normal operations. 2 = minimal disruption. 3 = moderate disruption. See Chapter 2 for further explanation of ratings.

Climate Drivers

The climate drivers that pose the greatest risk to the NCF are those that damage tracks or make them impassable. These include flooding, which inundates tracks and causes track-destroying erosion; sea-level rise, which can inundate coastal tracks, cause erosion, and limit intermodal transport; and tropical cyclones and hurricanes, which can damage rail infrastructure from high winds or storm surge (Grenzeback and Lukmann, 2008; Rossetti, 2003). These climate drivers can disrupt rail operations in critical regions, creating national choke points, such as rail operations at the Port of New York and New Jersey, one of the country's largest ports for containerized trade, that delay movement of goods and people. Flooding in the Midwest could disrupt the movement of agriculture products by rail, with Chicago being a critical node for the national rail network. Wildfire can limit visibility in some areas (Rossetti, 2003), which could force operators to reduce speeds. Additionally, wildfire can destroy tracks or make corridors impassable (Chavez, 2021; Franz, 2021; Marsh, 2021). Wildfire in Southern California can disrupt rail operations on critical corridors moving goods from the Ports of Los Angeles and Long Beach into the interior of the country. Extreme heat also causes damage to rail infrastructure, causing tracks to buckle and forcing operators to reduce speeds (Chinowsky, Helman, et al., 2019). Other climate drivers are expected to pose risk of minimal disruptions. For instance, extreme cold can make tracks impassable if frozen, but extreme cold temperatures are less likely than other climate drivers to occur in the future. Similarly, severe storm systems (e.g., tornadoes, hail) could cause minimal disruptions. Drought can lead to increased demand for rail transport as water levels in critical rivers fall (Marufuzzaman and Ekşioğlu, 2017; Micik, 2013), creating a shift from barge transportation to rail and causing a minimal disruption to provision of rail services for cargo transportation.

Table B.163. Risk from Climate Drivers for *Transport Cargo and Passengers by Rail*

Climate Driver	Baseline Present	Current Emissions 2050	Current Emissions 2100	High Emissions 2050	High Emissions 2100
Drought	1	1	2	1	2
Extreme cold	2	1	1	1	1
Extreme heat	1	2	3	2	3
Flooding	2	2	3	3	3
Sea-level rise	1	2	3	2	3
Severe storm systems	1	2	2	2	2
Tropical cyclones and hurricanes	2	2	3	2	3
Wildfire	2	2	3	2	3

NOTE: 1 = no disruption or normal operations. 2 = minimal disruption. 3 = moderate disruption. See Chapter 2 for further explanation of ratings.

Impact Mechanisms

Transport Cargo and Passengers by Rail faces risks of disruption through physical damage to tracks and rail infrastructure from flooding, sea-level rise, extreme heat, hurricanes, and wildfire; workforce disruptions from extreme heat; and demand changes due to modal switching during time periods of drought when water levels fall in inland rivers.

Cascading Risk

Transport Cargo and Passengers by Rail is critically dependent on the following NCFs:

- *Transport Cargo and Passengers by Road*
- *Fuel Refining and Processing Fuels*
- *Distribute Electricity*
- *Transport Cargo and Passengers by Vessel*
- *Provide Positioning, Navigation, and Timing Services.*

Transportation by rail requires road-based transportation for both goods and people (staff and passengers) to reach rail stations; it is also heavily reliant on carrying shipments to and from seaports (Bureau of Transportation Statistics, 2020). Freight rail is powered largely by diesel fuel, while some passenger rail service requires electricity (Nunno, 2018). Rail operations rely on PNT services (DOT, undated; Wallischeck, 2016).

Strength of Evidence

There is an extensive literature on climate change's effects on *Transport Cargo and Passengers by Rail*. There is a relatively large body of research on damage and disruption of rail transportation due to flooding, sea-level rise, and tropical cyclones and hurricanes (Grenzeback and Lukmann, 2008; Rossetti, 2003). Similarly, there is a robust literature on wildfire and extreme heat, which can damage tracks and force rail operators to reduce speeds (Chavez, 2021; Chinowsky, Helman, et al., 2019; Franz, 2021; Marsh, 2021). In addition to the scientific literature, numerous reports in the popular press have addressed wildfire's effects on rail operations. Several studies have analyzed disruptions to rail transportation due to severe storm systems and extreme cold. To a lesser extent, some studies also have examined disruptions from drought.

Table B.164. Strength of Evidence for *Transport Cargo and Passengers by Rail*

Strength of Evidence	
High	Flooding; sea-level rise; tropical cyclones and hurricanes; extreme heat; wildfire
Medium	Severe storm systems; extreme cold
Low	Drought

Risk Rating Interpretations

Table B.165 provides examples of how to interpret the risk ratings for this NCF.

Table B.165. Example Interpretations of Risk Ratings for *Transport Cargo and Passengers by Rail*

Risk Rating	NCF Example
1: No disruption or normal operations	Rail transportation continues to function as it does today, with occasional delays due to weather and operational problems.
2: Minimal disruption	A few rail lines in flood-prone regions experience longer time periods in which they cannot operate safely, but cargo and passengers can be moved through alternative corridors.
3: Moderate disruption	Some rail lines close temporarily or permanently because of sea-level rise around critical seaports on the Gulf Coast; rail delays and cancellations due to extreme heat, severe storm systems, and floods (and the damage they cause to railroad infrastructure) become far more frequent.
4: Major disruption	A severe flood in Chicago halts cross-country intermodal freight transport temporarily. Heat waves across the country make most commercial rail travel and freight movement nonviable. Continued flooding, sea-level rise, and damage from hurricanes cause some railroads cease to operate for longer time periods, and a significant number of rail lines close permanently.
5: Critical disruption	Rail travel and freight movements by rail are essentially halted. Some corridors remain operational, but critical corridors are not operational.

Transport Cargo and Passengers by Vessel

Table B.166. Scorecard for *Transport Cargo and Passengers by Vessel*

	National Risk Assessment				
	Baseline	Current Emissions		High Emissions	
	Present	2050	2100	2050	2100
Transport Cargo and Passengers by Vessel	3	3	3	3	3

Description: "Provide and operate maritime systems, assets, and facilities to enable a system of securely and safely conveying goods and people from place to place by the Maritime Transportation System" (CISA, 2020, p. 4).

Highlights

- This NCF was assessed as being at risk of moderate disruption due to the extreme vulnerability of coastal ports to sea-level rise and the potential for operational delays to significantly affect regional or national supply chains.
- Ports are particularly vulnerable to flooding and tropical cyclones and hurricanes, which can render some port infrastructure at least temporarily inoperative and disrupt navigation channels. Drought can reduce the navigability of inland waterways.
- Coastal ports are particularly vulnerable to infrastructure damage because of their locations, and many ports are vulnerable to disruption because of outdated port infrastructure and insufficient channel depths.
- The *Build Marine Transportation System* and *Operate Marine Transportation System* subfunctions are particularly vulnerable because sea-level rise could render some coastal infrastructure unusable and increase the costs of operations and maintenance, repair, and replacement.

Synopsis

Transport Cargo and Passengers by Vessel is vulnerable to multiple climate drivers that can result in infrastructure damage, operational delays, and, to a lesser extent, reductions in navigability of inland waterways. Coastal ports are particularly vulnerable to sea-level rise, which is expected to increase the risk of damage due to tropical cyclones and hurricanes and storm surges (Fleming et al., 2018; Tebaldi, Strauss, and Zervas, 2012). Sea-level rise is also expected to increase coastal erosion and make ports more vulnerable to inundation, which could render some port infrastructure at least temporarily inoperative. Inland riverine transportation is vulnerable to drought, which could reduce the navigability of inland waterways and increase transport prices (Jonkeren et al., 2014). Extreme heat events are also expected to increase in intensity and duration, which could damage port infrastructure, affect labor productivity, and disrupt maritime operations.

Disruptions due to climate drivers can have cascading effects throughout the maritime transportation system, intermodal connections, and supply chains. For example, tropical cyclones and hurricanes pose a risk of moderate disruption because the closure of a single port can cause supply chain delays to an entire region, or even nationally, depending on the size and function of the port. Seaports provide critical links in global supply chains, handling more than 80 percent of world trade by volume (United Nations Conference on Trade and Development, 2021). Ocean shipping is an inherently international function, so disruptions due to natural hazards affected by climate change at major ports in other countries that serve as important origins or destinations for U.S. goods could also affect the U.S. maritime transportation system.

Tropical cyclones and hurricanes, which are expected to increase in intensity, can severely damage ports, resulting in service disruptions that can last for a week or longer (Verschuur, Koks, and Hall, 2020) and causing economic losses in hundreds of millions or billions of dollars per day (Vilchis et al., 2006). Two trends in cargo shipping—increasing specialization in ports (e.g., container, bulk cargo, petroleum) and ports with shipping lanes deep enough to accommodate increasing vessel size—mean that vessels cannot easily load or offload at other ports if operations are disrupted (Verschuur, Koks, and Hall, 2020). For this reason, disruptions at ports moving the highest volume of goods—the ports of Los Angeles and Long Beach and of New York and New Jersey (for containerized goods), South Louisiana (for bulk cargo), and Houston (for petroleum) (Bureau of Transportation Statistics, 2020)—could have a disproportionate economic effect on regional or national supply chains.

Subfunctions

The primary affected subfunctions, *Build Marine Transportation System* and *Operate Marine Transportation System*, were assessed as being at risk of moderate disruption because they are particularly vulnerable to climate drivers damaging port facilities; contributing to flooding, storm surge, and coastal erosion; and at least temporarily disrupting operations and affecting supply chains because of outdated port infrastructure and insufficient channel depths (ASCE, 2021). In some regions, the increasing frequency or intensity of tropical cyclones and hurricanes and severe storm systems could reduce the efficiency of the maritime transportation system and increase accident risk in ports and at sea (Van Houtven et al., 2022; World Shipping Council, undated).

Maintain Workforce was assessed as being at risk of minimal disruption by 2100 due to extreme heat's effects on working conditions in outdoor environments. Extreme heat could lead to more-challenging working conditions, increasing the risk of heat-related illness and injury and decreasing labor productivity among maritime workers (Behrer and Park, 2017; Park, Pankratz, and Behrer, 2021), and could significantly reduce the number of days considered safe to work outdoors (Jacklitsch et al., 2016; Licker, Dahl, and Abatzoglou, 2022). *Transport Cargo* and *Transport Passengers* generally deal with logistics functions (e.g., procedures and capabilities to load cargo and passengers, security and safety processes, locations and processes to support intermodal transportation), which are not specifically vulnerable to climate drivers in the same way that port infrastructure and operations are. Similarly, *Govern Marine Transportation*, which addresses regulatory processes, guidance, and enforcement, is unlikely to be affected by climate drivers.

Table B.167. Subfunction Risk for *Transport Cargo and Passengers by Vessel*

Subfunction	Baseline Present	Current Emissions 2050	Current Emissions 2100	High Emissions 2050	High Emissions 2100
Build Marine Transportation System	3	3	3	3	3
Operate Marine Transportation System	3	3	3	3	3
Transport Cargo	1	1	1	1	1
Transport Passengers	1	1	1	1	1
Govern Marine Transportation	1	1	1	1	1
Maintain Workforce	1	1	2	1	2

NOTE: 1 = no disruption or normal operations. 2 = minimal disruption. 3 = moderate disruption. See Chapter 2 for further explanation of ratings.

Climate Drivers

Transport Cargo and Passengers by Vessel is vulnerable to six of the eight climate drivers: drought, extreme heat, flooding, sea-level rise, severe storm systems, and tropical cyclones and hurricanes (Committee on Climate Change and U.S. Transportation, 2008; Jacobs, Culp, et al., 2018; Van Houtven et al., 2022). Coastal ports, because of their locations, are particularly vulnerable to tropical cyclones and hurricanes and sea-level rise, which are expected to increase the risk of damage to port infrastructure and vessels through storm surge and higher waves (Douglass and Webb, 2020). Sea-level rise can also cause inundation, nuisance flooding, or high-tide flooding, rendering coastal port landside infrastructure unusable either temporarily or permanently. Many ports are expected to be frequently inundated by 2100 during even moderate storm events (Becker, Acciaro, et al., 2013). Researchers performing a multihazard (including high winds, extreme heat, heavy precipitation, storm surge, flooding, and sea-level rise) climate analysis of coastal ports around the world found that, by 2100, 21 ports in the United States would be considered "very high risk" (the second-highest category) and another 127 considered "medium risk" (the fourth category) (Izaguirre et al., 2021).[51]

Tropical cyclones and hurricanes can damage landside infrastructure and vessels or blow debris into navigation channels, leading to temporary port closures (Verschuur, Koks, and Hall, 2020). A recent study of 141 port disruptions from 2011 to 2019 due to tropical cyclones and hurricanes across 74 ports in 12 countries found that roughly half of the events led to a complete shutdown of port operations, with a median duration of six days (Verschuur, Koks, and Hall, 2020). Shipments can be delayed by flooding, heavy rains, or high winds affecting crane operations and truck service (Becker, Acciaro, et al., 2013). Tropical cyclones and hurricanes are expected to increase in intensity by 2100.

Severe storm systems (such as nor'easters), intense precipitation, and flooding can also damage port infrastructure at both coastal and inland ports (Van Houtven et al., 2022). Drought could affect barge shipments that travel by river or cargo vessels that travel on the Great Lakes because vessels require a minimum water level to operate. However, climate drivers' effects on the Great Lakes are

---

[51] This analysis was conducted using the representative concentration pathway 8.5 scenario, which corresponds to the high emissions scenario in this report.

difficult to predict because the water levels generally fluctuate on an annual basis, and climate change might be increasing these fluctuations (Gronewold et al., 2013; Posey, 2012). Maritime transportation was assessed as being at risk of minimal disruption by 2100 due to drought because of the risk to navigability of inland waterways.

Extreme heat can damage port infrastructure and equipment and create unsafe working conditions for port workers and shipping crews, disrupting operations in port and at sea (Becker, Acciaro, et al., 2013; Van Houtven et al., 2022). Extreme heat can put workers at increased risk of heat exhaustion, rhabdomyolysis (breakdown of skeletal muscle tissue), heat stroke, permanent disability, and death (Gubernot, Anderson, and Hunting, 2014; Jacklitsch et al., 2016). Additionally, the ability to efficiently operate the maritime transportation system could decrease because extreme heat can cause workers to work fewer hours (Graff Zivin and Neidell, 2014), become less productive (Behrer and Park, 2017; Burke, Hsiang, and Miguel, 2015b; Seppänen, Fisk, and Lei-Gomez, 2006), and become more prone to injury (Park, Pankratz, and Behrer, 2021). Maritime transportation was assessed as being at risk of minimal disruption by 2100 due to extreme heat because the number of days above 90 degrees Fahrenheit each year is expected to increase.

Extreme cold events are not expected to increase over current levels, so their effect on this NCF is not expected to change. Similarly, a literature review of climate change's effect on maritime transportation did not identify any effects with specific regard to wildfire, nor were we able to independently identify any scenarios that we assessed as meriting a higher risk rating for this climate driver.

Table B.168. Risk from Climate Drivers for *Transport Cargo and Passengers by Vessel*

Climate Driver	Baseline Present	Current Emissions 2050	Current Emissions 2100	High Emissions 2050	High Emissions 2100
Drought	1	1	2	1	2
Extreme cold	1	1	1	1	1
Extreme heat	1	1	2	1	2
Flooding	3	3	3	3	3
Sea-level rise	1	2	3	2	3
Severe storm systems	2	2	2	2	2
Tropical cyclones and hurricanes	3	3	3	3	3
Wildfire	1	1	1	1	1

NOTE: 1 = no disruption or normal operations. 2 = minimal disruption. 3 = moderate disruption. See Chapter 2 for further explanation of ratings.

Impact Mechanisms

The primary mechanisms by which climate change affects *Transport Cargo and Passengers by Vessel* is physical damage or disruption. Multiple climate drivers are expected to damage or disrupt the operations of port structures, facilities, vessels, equipment, and aids to navigation with increasing frequency or intensity. Coastal ports are exceptionally vulnerable to tropical cyclones and hurricanes and sea-level rise causing damage or disrupting operations. Increased precipitation due to severe storm systems or drought could disrupt the navigability of inland waterways because of extremely high or low waters. Extreme heat could create unsafe working conditions and decrease labor productivity, causing operational delays and contributing to workforce shortages.

Cascading Risk

Transport Cargo and Passengers by Vessel is dependent on the following NCFs:

- *Transport Cargo and Passengers by Road*
- *Transport Cargo and Passengers by Rail*
- *Provide Positioning, Navigation, and Timing Services*
- *Provide Satellite Access Network Services*
- *Fuel Refining and Processing Fuels*.

Transportation by road and rail are considered crucial because these are the modes via which cargo and passengers move between ports and other parts of the country. Port operations would be severely disrupted if there were no modes through which to move goods via land because warehouse space is finite and would quickly become overwhelmed (Goodman, 2021). PNT and satellite services are critical to shipping operations for vessel positioning the way wire-line services are for landside operations (DOT, undated; MarineTraffic, undated; Wallischeck, 2016). Finally, fuel-related NCFs are important because almost all shipping vessels require diesel fuel to operate (EIA, 2015a). Some climate drivers, such as inland flooding and droughts, could cause indirect damage or disruption to ports and the shipping industry via the inland supply chains that are critical to the maritime transportation system (Van Houtven et al., 2022). At its peak during the COVID-19 pandemic, the supply chain bottleneck at the Ports of Los Angeles and Long Beach forced container ships to wait for two weeks or more (Murray, 2022). A recent study estimated that the closure of the Ports of Los Angeles and Long Beach for 15 days would lead to an economic loss of $7.6 billion, or about $0.5 billion per day in 2022 (Inforum, 2022).

Strength of Evidence

There is an extensive literature on the effects that sea-level rise and hurricanes have on ports, with some rigorous analysis of the different risk levels by port (Becker, Ng, et al., 2018; Izaguirre et al., 2021; Verschuur, Koks, and Hall, 2020; Wright and Hogan, 2008). In addition, climate change effects on *Transport Cargo and Passengers by Vessel* linked to flooding, severe storm systems, and extreme heat have been well documented (Committee on Climate Change and U.S. Transportation, 2008; Jacobs, Culp, et al., 2018; Van Houtven et al., 2022). There is mixed evidence on whether and how drought might affect inland cargo shipping (Gronewold et al., 2013; Posey, 2012). There is less research available on the effects that extreme cold and wildfire could have on water transportation specifically.

Table B.169. Strength of Evidence for *Transport Cargo and Passengers by Vessel*

Strength of Evidence	
High	Tropical cyclones and hurricanes; sea-level rise
Medium	Extreme heat; flooding; severe storm systems
Low	Drought; extreme cold; wildfire

Risk Rating Interpretations

Table B.170 provides examples of how to interpret the risk ratings for this NCF.

Table B.170. Example Interpretations of Risk Ratings for *Transport Cargo and Passengers by Vessel*

Risk Rating	NCF Example
1: No disruption or normal operations	Transportation by vessel continues to function with only occasional delays due to inclement weather and operational problems.
2: Minimal disruption	Coastal ports experience damage and disruption due to severe storm systems and flooding, but cargo and passengers can be moved through alternative routes or modes of transportation.
3: Moderate disruption	Port damage and disruption due to extreme heat, flooding, and hurricanes occur with some frequency, and shipping delays can last a week or more. Temporary shutdowns of major ports significantly disrupt regional supply chains. Some ports must rebuild or elevate structures or relocate because of sea-level rise.
4: Major disruption	Port closures due to flooding and storm damage occur regularly; shutdowns of major ports (e.g., coastal ports that are permanently inundated by sea-level rise) and reduced navigability of waterways disrupt supply chains in multiple regions concurrently. Some ocean and inland cargo shipping companies and ferry and cruise operators cease to operate.
5: Critical disruption	Water transportation is essentially halted because most critical ports are inoperative and major waterways are not navigable, although routes might be available to smaller vessels.

Maintain Supply Chains

Table B.171. Scorecard for *Maintain Supply Chains*

| | National Risk Assessment | | | | |
| | Baseline | Current Emissions | | High Emissions | |
	Present	2050	2100	2050	2100
Maintain Supply Chains	3	3	4	3	4

Description: "Manage and sustain the networks of assets, systems, and relationships that enable movement of goods and services from producers to consumers" (CISA, 2020, p. 3).

Highlights

- *Maintain Supply Chains* was assessed to be at risk of moderate disruption in the baseline time period and of major disruption by 2100.
- The NCF is vulnerable to several climate drivers, but hurricanes present the largest current and future risk to supply chain activities, assets, and logistics.
- Several climate drivers disrupt the flow of goods between producers and consumers, destroy warehousing facilities and other supply chain assets, and impede supply chain logistics.
- Several subfunctions are at risk of moderate disruption by 2100, but *Maintain Supply Chain Operations* and *Manage Product Distribution* are at the greatest risk of disruption.
- Critical coastal infrastructure needed for product distribution make the NCF vulnerable to hurricanes and sea-level rise, particularly along the East and Gulf Coasts.

Synopsis

A changing climate is expected to lead to an increase in the frequency and severity of extreme weather events, several of which have the potential to disrupt supply chain activities. Tropical cyclones and hurricanes and coastal flooding can damage supply chain assets, such as shipping containers, commercial trucks, and distribution centers (Scholz et al., 2021; Sturgis, Smythe, and Tucci, 2014; Xia and Lindsey, 2021); destroy stockpiles and warehouses (Smythe, 2013); and interrupt freight operations and logistics activities, causing shipping delays and price increases (DOT, 2022; Sturgis, Smythe, and Tucci, 2014; Van Houtven et al., 2022). Hurricanes that are severe enough to disrupt supply chains are projected to become two to four times more frequent by 2040 (Scholz et al., 2021). Similarly, inland flooding can disrupt product distribution by slowing or stopping the movement of products along rail, roads, and rivers (Angel et al., 2018; Pierce, 2020; Rossetti, 2003; Scholz et al., 2021), which contributes to difficulties in managing optimal inventory. In 2020, inland flooding in the Midwest caused supply chain disruptions, interrupting the flow of goods from producers to consumers (Pierce, 2020). Extreme heat can reduce worker productivity in distribution activities (Yuqiang Zhang and Shindell, 2021), which can cause shipment delays and potential shortages of some commodities. Additionally, extreme heat can reduce the speed at which products can be moved from producers to consumers (Rossetti, 2003) and slow the pace of production at different points along the supply chain (Scholz et al., 2021). Droughts disrupt product distribution as well (K. Chang, 2022; Hirtzer, Elkin, and Deaux, 2022; Scholz et al., 2021; Zamuda et al., 2013) and can make procuring important production inputs difficult in some regions (W. Phillips, 2022; Scholz et al., 2021). Sea-level rise and resulting extreme tides could result in lower freight throughput capacity at key logistics hubs in low-lying regions (AECOM, 2018; Allen, McLeod, and Hutt, 2022; Scholz et al., 2021; DOT, 2022), which might necessitate new supply chain configurations (Slay and Dooley, 2020). Transitioning to these new configurations will likely involve a time period of increased logistics disruptions and impede inventory management. Finally, wildfire poses risk to product distribution in the West (Resilinc, 2021). Given the importance of the Ports of Los Angeles and Long Beach to regional and national retailers, disruptions from wildfire that delay the distribution of shipments from these ports could cause larger-scale disruptions as wildfires intensify in the next century.

Climate change's effects on supply chain assets, logistics, and processes are likely to disrupt the flow of goods between producers and consumers. Disruptions to product distribution can limit retail inventories (Chopra and Sodhi, 2014; Helper and Soltas, 2021), cause price spikes for critical inputs (Leibovici and Dunn, 2021), or force product distributors to substitute alternative routes or shipment modes (J. Dong et al., 2015), causing disruptions to day-to-day freight operations and logistics. Additional disruptions due to physical damage to retail facilities and manufacturing plants (Cachon, Gallino, and Olivares, 2012) can also interrupt normal supply chain operations.

Maintain Supply Chains was assessed as facing a risk of moderate disruption from climate change in the baseline time period, but that risk was assessed to increase by 2100 in both current and high emissions scenarios. The baseline risk of moderate disruption is driven by hurricanes, which already affect supply chain assets, activities, and processes. For instance, Hurricanes Katrina and Sandy disrupted normal supply chain processes in multiple states and metropolitan areas (Skipper, Hanna, and Gibson, 2010; Sturgis, Smythe, and Tucci, 2014). Hurricanes are projected to become severer in the next century (Kossin et al., 2017), and the importance of the Ports of New York and New Jersey for national distribution networks (LaRocco, 2022; Shen et al., 2020; Tomer and Kane, 2015) means that a particularly severe hurricane could pose a risk of major disruption to *Maintain Supply Chains*.

Key Considerations for *Maintain Supply Chains*

The ability to transport goods is one of the most vital aspects of supply chains because producers and consumers are rarely in the same location. Ports are critical distribution nodes, and roads, railways, and river barges are essential modes of distribution, without which most supply chains would not exist. As a result, climate drivers that disrupt transportation networks also directly affect *Maintain Supply Chains*. For instance, Hurricane Sandy caused significant supply chain disruptions due to shipping delays at the Ports of New York and New Jersey (Kashiwagi, Todo, and Matous, 2021).

Subfunctions

Most subfunctions are expected to experience increased risk of disruption by the end of the century, in both current and high emissions scenarios. Some subfunctions are at risk of moderate disruption in the baseline time period, an assessment due largely to historical experiences with hurricanes. Specifically, *Maintain Supply Chain Operations*, which involves maintaining the flow of products from suppliers to customers within an acceptable time frame, and *Manage Product Distribution*, which involves managing the logistics of transporting goods and day-to-day freight operations, were both assessed to be at risk of moderate disruption from hurricanes in the baseline time period. Disruptions from hurricanes are likely to grow in the next century because hurricanes are projected to become severer (Kossin et al., 2017). As a result, the risk of disruption for these subfunctions was assessed as increasing in 2100, reflecting the potential for a particularly devastating storm hitting a nationally critical distribution hub.

Other subfunctions were assessed to be at risk of minimal disruption in the baseline but at moderate risk of disruption in later time periods. Specifically, *Manage Product Development and Manufacturing*, which involves maintaining the optimum number of products in inventory; *Manage Retailers*, which involves retail functions of selling goods to end users; and *Manage Purchasing*, which involves procurement of raw materials and goods used in production, were assessed as facing growing risk of disruption as climate drivers become severer. These subfunctions were assessed to be at lower risk of disruption than *Maintain Supply Chain Operations* and *Manage Product Distribution* because, for these two, climate drivers tend to create bottlenecks and disruptions in distribution networks (Committee on Building Adaptable and Resilient Supply Chains After Hurricanes Harvey, Irma, and Maria, 2020), which, in turn, affect procurement, inventory, and retailers. The climate drivers responsible for the growing risk of disruption to these subfunctions are flooding, sea-level rise, hurricanes, extreme heat, and drought.

Table B.172. Subfunction Risk for *Maintain Supply Chains*

Subfunction	Baseline	Current Emissions		High Emissions	
	Present	2050	2100	2050	2100
Maintain Supply Chain Operations	3	3	4	3	4
Manage Product Development and Manufacturing	2	2	3	2	3
Manage Product Distribution	3	3	4	3	4
Manage Retailers	2	2	3	2	3
Manage Purchasing	2	2	3	2	3

NOTE: 2 = minimal disruption. 3 = moderate disruption. 4 = major disruption. See Chapter 2 for further explanation of ratings.

Climate Drivers

Most climate drivers will have some effect on *Maintain Supply Chains*. Flooding, hurricanes, wildfire, and drought already pose risks of disruptions, while others do not. For example, several hurricanes have disrupted normal supply chain operations across states and major metropolitan areas.

Under current emissions, the risk of disruption from other climate drivers will increase as well. Specifically, drought, inland flooding, and wildfire pose a risk of minimal disruption in the baseline time period, but these climate drivers are expected to pose a risk of moderate disruption to *Maintain Supply Chains* as they intensify in the next century. Under high emissions, extreme heat and flooding are expected to pose a moderate risk of disruption by 2050.

Sea-level rise does not currently pose a risk of disruption, but sea levels are projected to rise by 4 to 6 feet by 2100 (Sweet, Horton, et al., 2017). This degree of sea-level rise will inundate portions of coastal seaports (AECOM, 2018; Kafalenos et al., 2008), potentially limiting capacity and the ability to distribute goods along supply chains. Severe storm systems (e.g., thunderstorms, hailstorms, tornadoes) currently do not pose a risk to supply chain activities and are generally not cited in the literature on extreme weather's effects on supply chains. However, these storms are projected to become severer in the next century (Kossin et al., 2017), which could increase the risk of disruption to supply chain operations and distribution.

Extreme cold is not expected to pose a risk of disruption to *Maintain Supply Chains*. This climate driver is not identified as a risk to supply chains in the literature currently. Most critical supply chain nodes are not in areas that experience extremely cold temperatures, although there are some exceptions (e.g., distribution centers on the Great Lakes). Extremely cold temperatures are expected to decline in the next century, limiting supply chains' exposure to this driver.

Table B.173. Risk from Climate Drivers for *Maintain Supply Chains*

Climate Driver	Baseline Present	Current Emissions 2050	Current Emissions 2100	High Emissions 2050	High Emissions 2100
Drought	2	2	3	2	3
Extreme cold	1	1	1	1	1
Extreme heat	1	2	3	3	3
Flooding	2	2	3	3	3
Sea-level rise	1	2	3	2	3
Severe storm systems	1	2	2	2	2
Tropical cyclones and hurricanes	3	3	4	3	4
Wildfire	2	2	3	2	3

NOTE: 1 = no disruption or normal operations. 2 = minimal disruption. 3 = moderate disruption. 4 = major disruption. See Chapter 2 for further explanation of ratings.

Impact Mechanisms

Physical damage or disruption to critical supply chain infrastructure and assets from hurricanes, flooding, and sea-level rise could limit the ability to move goods from producers to consumers. Extreme heat could also limit the ability to move goods around the country through its effect on worker productivity in outdoor product distribution activities. Finally, as product distributors attempt to manage disruptions from a climate driver, demand for alternative distribution modes or ports is likely to increase and could increase freight rates even on unaffected routes.

Cascading Risk

Maintain Supply Chains is critically dependent on the following NCFs:

- *Distribute Electricity*
- *Transport Cargo and Passengers by Vessel*
- *Transport Cargo and Passengers by Road*
- *Transport Cargo and Passengers by Rail.*

Maintain Supply Chains requires access to electricity to manage retail and purchasing and to power warehousing and distribution centers. Additionally, supply chain operations require maintaining distribution networks and functioning transportation networks to carry products to consumers (Scholz et al., 2021). A failure of any one of these NCFs would lead to a near-immediate failure of supply chains.

Strength of Evidence

There is a large body of research on climate change's projected effect on supply chains. Many reports and academic research articles provide evidence of historical and likely future effects that hurricanes and flooding have had and could have on supply chains (Scholz et al., 2021; Sturgis, Smythe, and Tucci, 2014; Xia and Lindsey, 2021). Similarly, there is a large body of literature on sea-level rise's projected effect on supply chains (AECOM, 2018; Kafalenos et al., 2008). There is less research on the implications that drought and extreme heat could have for supply chain activities, although some studies provide evidence of potential effects, such as effects on inland riverine transportation. Evidence of the effects that severe storm systems, extreme cold, and wildfire could have on supply chains is less well documents, and there is significantly less research on these climate drivers than on others.

Table B.174. Strength of Evidence for *Maintain Supply Chains*

Strength of Evidence	
High	Tropical cyclones and hurricanes; sea-level rise; flooding
Medium	Drought; extreme heat
Low	Severe storm systems; extreme cold; wildfire

Risk Rating Interpretations

Table B.175 provides examples of how to interpret the risk ratings for this NCF.

Table B.175. Example Interpretations of Risk Ratings for *Maintain Supply Chains*

Risk Rating	NCF Example
1: No disruption or normal operations	Supply chains continue to function as they do today, with occasional delays due to weather and operational problems.
2: Minimal disruption	Floods cause local shipment delays and supply chain disruptions; a hurricane temporarily reduces capacity at a seaport, but other ports can absorb diverted shipments, causing a relatively small price increase in some goods locally but having limited effect regionally or nationally.
3: Moderate disruption	Some small seaports close permanently because of sea-level rise. Drought reduces water levels in the Mississippi River, delaying shipments of some commodities for several months during the summer. Hurricanes damage coastal infrastructure and destroy commodity inventories, limiting product distribution for several weeks out of some ports. Extreme heat reduces port worker productivity and limits employers' ability to hire for key outdoor positions.
4: Major disruption	Some small and medium-sized seaports on the East and Gulf Coasts close or significantly reduce storage capacity because of sea-level rise. Hurricanes frequently disrupt major container terminals at key seaports, causing long-term disruptions to commodity flows across large parts of the country. Drought reduces water levels in several key rivers, making it impossible to transport commodities by barge in summer months.
5: Critical disruption	Major seaports completely shut down because of sea-level rise, severely limiting distribution capacity and reducing the flow of goods across the country. Extreme heat makes working outdoors nearly impossible several months out of the year, limiting the product distribution workforce and causing severe commodity price spikes; severe commodity shortages in retailers across the country result in consumers' inability to purchase many goods for long time periods.

Manufacture Equipment

Table B.176. Scorecard for *Manufacture Equipment*

	National Risk Assessment				
	Baseline	Current Emissions		High Emissions	
	Present	2050	2100	2050	2100
Manufacture Equipment	2	2	3	3	3

Description: "Fabricate and assemble components to produce tangible property" (CISA, 2020, p. 6).

Highlights

- *Manufacture Equipment* was assessed to be at risk of moderate disruption by the end of the century because of exposure to hurricanes, flooding, sea-level rise, and extreme heat.
- Severer hurricanes, floods, heat waves, and extreme tides and sea-level rise are expected to disrupt the ability to carry out manufacturing processes.
- The reliance on having operable physical facilities to carry out manufacturing processes make the NCF vulnerable to disruptions from climate drivers that can deteriorate facilities.
- Subfunctions dedicated to the manufacturing processes are likely the most vulnerable to climate change because of their need for operable facilities and machinery and time-sensitive nature.
- There are equipment manufacturing plants across the country, making the NCF vulnerable to numerous climate drivers but limiting the national effect of disruptions.

Synopsis

Manufacture Equipment faces risk from several climate drivers, across different subfunctions, and through several mechanisms. The risks that *Manufacture Equipment* faces from climate drivers are likely similar to those faced in other manufacturing industries because the manufacturing processes and activities that this NCF incorporates are similar to those in other manufacturing industries. For instance, assembling components, processing raw materials, and maintaining equipment quality are likely present in most manufacturing industries.[52] Many of the metropolitan areas that are highly specialized in manufacturing (including computer and electronic equipment, transportation equipment, and machinery) lie along major waterways or on coastlines (Helper, Krueger, and Wial, 2012), which could become more vulnerable to flooding, hurricanes, or sea-level rise (Kossin et al., 2017; Mallakpour and Villarini, 2015; Slater and Villarini, 2016; Sweet, Horton, et al., 2017). There is a large production technology and heavy machinery industrial cluster in the upper Midwest and the Great Lakes region (U.S. Cluster Mapping Project, undated), which is a region expected to experience severer flooding in the next century (Mallakpour and Villarini, 2015; Slater and Villarini, 2016). Evidence from the manufacturing sector indicates that floods and hurricanes, exacerbated by sea-level rise, have the capacity to cause widespread damage or destruction to facilities and disrupt production activities, regardless of industry (Cachon, Gallino, and Olivares, 2012; Scholz et al., 2021; Seetharam,

---

[52] In 2021, equipment manufacturing made up 40 percent of total manufacturing output (Bureau of Economic Analysis, 2022). Equipment manufactured includes fabricated metal products; machinery; computer and electronic products; electrical equipment, appliances, and components; motor vehicles, bodies and trailers, and parts; other transportation equipment; and medical equipment and supplies.

2018). There are production technology and heavy machinery facilities in regions prone to these hazards, such as the Gulf Coast and other parts of Florida, as well as the Northeast (U.S. Cluster Mapping Project, undated). Similarly, to the extent that equipment manufacturing facilities lie in coastal floodplains, sea-level rise could force some coastal manufacturing facilities to relocate or close permanently (Adapting to Rising Tides, 2017; Hirschfeld and Holland, 2012; Scholz et al., 2021). For instance, there are numerous aerospace equipment and transportation equipment manufacturing firms in Florida (Enterprise Florida, undated), a state particularly vulnerable to sea-level rise (Sweet, Horton, et al., 2017). Where outdoor labor is a critical input to production, extreme heat could limit worker productivity (Hsiang et al., 2017; Scholz et al., 2021). Some occupations in equipment manufacturing require workers to work outdoors or indoors in non–environmentally controlled settings (National Center for O*NET Development, 2023).[53] Some research has found evidence that extreme heat reduces worker productivity in indoor manufacturing activities as well (P. Zhang et al., 2018), and evidence from Canada indicates that extreme temperatures reduce manufacturing output by 2.2 percent (Kabore and Rivers, 2020). An analysis of U.S. automobile manufacturing plants found that six days with a temperature of 90 degrees can reduce weekly production by approximately 8 percent (Cachon, Gallino, and Olivares, 2012).

Together, the effects of climate change are expected to increase damage to manufacturing facilities (Scholz et al., 2021; Seetharam, 2018), which could disrupt the processes of manufacturing equipment, such as mold casting, joining materials, and assembling components to equipment. Most climate drivers are likely to cause temporary disruptions to these production processes, although disruptions can last more than a week in some cases (Cachon, Gallino, and Olivares, 2012). Additionally, some evidence shows that repeated disruptions can result in long-term degradation of manufacturing facilities (broadly), physical capital, and productivity required to carry out some production processes (Letta and Tol, 2019). Additionally, climate change is expected to reduce labor productivity because of extreme heat, limiting the ability to perform manufacturing processes and potentially the ability to maintain equipment quality and remain in compliance with regulations.

In the current emissions scenario, increased severity and frequency of flooding and hurricanes could disrupt manufacturing processes by 2100. There are many equipment manufacturing facilities in the upper Midwest and the Great Lakes region, where flooding is projected to become severer in the next century (Mallakpour and Villarini, 2015; Slater and Villarini, 2016). Processes that require functional physical facilities and involve repeated processes that are time sensitive are likely at most risk of direct disruption. As the disruptions from these climate drivers grow, *Manufacture Equipment* might not be able to fill all routine needs in some highly exposed regions. For instance, equipment manufacturing facilities on the Gulf and East Coasts might become more exposed to damage from hurricanes—making it more difficult to carry out production, processing, and assembly actions in a timely manner—because these storms are projected to become severer in the next century (Kossin et al., 2017). Under current emissions, sea-level rise, severe storm systems, and extreme heat are likely to pose a risk of minimal disruption by 2100 if these climate drivers affect equipment manufacturing in

---

[53] For instance, inspectors, testers, sorters, samplers, and weighers; first-line supervisors of production and operation workers; welders, cutters, solderers, and brazers involve outdoor or indoor but non–environmentally controlled settings (National Center for O*NET Development, 2023). These occupations made up 10 percent of all equipment manufacturing in 2020, according to data from the U.S. census (Office of Occupational Statistics and Employment Projections, 2022).

similar ways that they are expected to affect the manufacturing sector as a whole (Scholz et al., 2021). Sea-level rise under current emissions could affect some coastal facilities; nationally, though most manufacturing facilities will be left unaffected. Severe storms (thunderstorms, hailstorms, and tornadoes) are likely to become more frequent and severer but generally cause less damage than hurricanes and flooding. Similarly, extreme heat could reduce labor productivity in manufacturing, although most workers in the sector likely work indoors and in environmentally controlled settings.

The high emissions scenario looks similar to the current emissions scenario, apart from the risk of moderate disruption from flooding by 2050 and the risk of moderate disruption from sea-level rise by 2100. Flooding is projected to increase by a similar amount over the baseline in current emissions 2100 and high emissions 2050 (Easterling et al., 2017), making the risk of disruption in the high emissions scenario greater in earlier time periods. Sea-level rise could reach up to 6 feet in this scenario (Sweet, Horton, et al., 2017), putting more infrastructure and facilities at risk of frequent coastal flooding. For instance, 6 feet of sea-level rise would inundate large parts of Miami, south Louisiana, and several cities in the Northeast (Climate Central, undated). In this scenario, it might be difficult to operate equipment manufacturing facilities in these regions, leading to moderate disruption to the NCF.

Subfunctions

Subfunctions dedicated to the physical, repeated processes of manufacturing (*Processing Raw Materials for Manufacturing*, *Produce Components*, and *Assemble Components to Equipment*) are likely to face the greatest risk of disruption from climate change. These subfunctions are the likeliest to be directly affected by physical damage to manufacturing facilities from flooding, hurricanes, and, later in the century, sea-level rise. Without operational facilities, it will likely be difficult to carry out related activities, such as developing casts, molding materials, performing thermal and chemical joining, or coating and treating materials. In regions that are highly exposed to these climate drivers, these subfunctions could fail to meet routine operational needs for longer than a week (Cachon, Gallino, and Olivares, 2012). For instance, a severe hurricane in the Gulf of Mexico could disrupt the ability to produce components in local facilities for weeks because of facility damage. Evidence suggests that manufacturing automobiles is adversely affected by extreme heat (Cachon, Gallino, and Olivares, 2012), which suggests that subfunctions pertaining to manufacturing processes could be adversely affected by extreme heat as well.

Some subfunctions are likely less exposed to climate change. For instance, *Maintain Equipment Quality*—which involves conducting audits—and *Maintain Compliance with Manufacturing Regulations* are likely less reliant on physical structures that climate drivers can damage. Additionally, these actions might be less time sensitive than production-related processes, although the literature does not provide evidence of this. Some evidence from the literature does suggest that the quality of some manufacturing equipment is highly sensitive to extremely high temperatures (Scholz et al., 2021), which could make maintaining equipment quality more difficult as extreme temperatures become more frequent across the country (Vose et al., 2017). Extreme heat can also reduce productivity of workers conducting audits, even if they work indoors (Cachon, Gallino, and Olivares, 2012). Additionally, degradation of manufacturing facilities from floods could make it more difficult to operate automated equipment, although there is little research on this topic. The subfunction

Conduct Equipment Production Planning is likely not vulnerable to climate change. This subfunction involves establishing a bill of materials—or an inventory of items needed to build a product—which can likely be performed off-site and is not tied directly to production facilities.

Table B.177. Subfunction Risk for *Manufacture Equipment*

	Baseline	Current Emissions		High Emissions	
Subfunction	Present	2050	2100	2050	2100
Processing Raw Materials for Manufacturing	2	2	3	3	3
Produce Components	2	2	3	3	3
Assemble Components to Equipment	2	2	3	3	3
Maintain Equipment Quality	1	2	2	2	2
Maintain Compliance with Manufacturing Regulations	1	2	2	2	2
Operate Automation for Manufacturing	1	2	2	2	2

NOTE: 1 = no disruption or normal operations. 2 = minimal disruption. 3 = moderate disruption. See Chapter 2 for further explanation of ratings.

Climate Drivers

Flooding, hurricanes, extreme heat, and sea-level rise pose the greatest direct risk of disruption to the *Manufacture Equipment* NCF. However, disruptions from these climate drivers are not expected to cause national-level disruptions because manufacturing is relatively distributed across the country, although regions specialize in particular subsectors and activities (Helper, Krueger, and Wial, 2012). By later time periods, the disruption from these drivers could result in the inability to meet operational needs of the NCF in some regions, particularly those that are highly exposed to the drivers (e.g., flooding in the Midwest, hurricanes and sea-level rise on the Gulf Coast, extreme heat in the Southwest).

Flooding and hurricanes have historically disrupted broadly defined manufacturing processes by degrading facilities and limiting access to inputs from suppliers (Scholz et al., 2021). There are many equipment manufacturing facilities in the Great Lakes region, where flooding is expected to become severer in the next century (Easterling et al., 2017; Mallakpour and Villarini, 2015; Slater and Villarini, 2016). Similarly, there are equipment manufacturing facilities located along the Gulf Coast (U.S. Cluster Mapping Project, undated), a region where hurricanes are expected to become severer in the next century (Kossin et al., 2017). A particularly severe hurricane, causing widespread damage to multiple coastal regions, could inhibit equipment manufacturing facilities in the affected region from operating normally.

Nonhurricane storm systems (tornadoes, thunderstorms, and hailstorms) are not reported in the literature to be particularly damaging to manufacturing facilities but could pose some minor delays to equipment manufacturing in the Midwest, where these types of storms are projected to become severer (Kossin et al., 2017).

Extreme heat poses a risk of minimal disruption currently: Evidence suggests that high temperatures reduce labor productivity in manufacturing sectors (Scholz et al., 2021), including in automobile manufacturing (Cachon, Gallino, and Olivares, 2012). Extreme heat is projected to become severer across the country in the next century but to increase the most in the Southwest, Southeast, and Gulf area (Vose et al., 2017). Although not all labor in the sector involves outdoor or non–environmentally controlled settings, increased heat stress in these regions could pose a risk of minimal disruption to the NCF under current emissions. Heat waves are projected to become more frequent in 2100 under high emissions. The number of days above 90 degrees is expected to increase by 60 to 80 percent across the United States, with the Southwest, Gulf Coast, and Southeast experiencing the largest increases (Vose et al., 2017). According to evidence from automobile manufacturing, this increase in extreme heat could reduce productivity and output in equipment manufacturing, causing the NCF to fail to complete all normal routine operations in these regions.

Sea-level rise could cause a risk of moderate disruption as well in the high emissions scenario by 2100. In this scenario, sea levels are expected to be 6 feet higher than they are today. This level of sea-level rise would inundate large portions of Miami, south Louisiana, parts of the Northeast, and parts of the San Francisco Bay area (Climate Central, undated). In these regions, operating equipment manufacturing facilities might be difficult, and completing all routine operations might not be feasible.

There is little evidence that wildfire has affected manufacturing facilities, although it is possible that more-intense and -widespread fires in the West could directly affect the ability to perform routine equipment manufacturing functions by damaging facilities and indirectly by displacing workers. Similarly, there is little evidence that extreme cold presents a risk to manufacturing, and this climate driver is likely to be less frequent in future scenarios (Vose et al., 2017). Drought likely poses a risk to equipment manufacturing processes through its effect on *Supply Water*. This is discussed in the "Cascading Risks" section below.

Table B.178. Risk from Climate Drivers for *Manufacture Equipment*

Climate Driver	Baseline	Current Emissions		High Emissions	
	Present	2050	2100	2050	2100
Drought	1	1	1	1	1
Extreme cold	1	1	1	1	1
Extreme heat	2	2	2	2	3
Flooding	2	2	3	3	3
Sea-level rise	1	2	2	2	3
Severe storm systems	1	2	2	2	2
Tropical cyclones and hurricanes	2	2	3	2	3
Wildfire	1	1	1	1	1

NOTE: 1 = no disruption or normal operations. 2 = minimal disruption. 3 = moderate disruption. See Chapter 2 for further explanation of ratings.

Impact Mechanisms

The mechanisms likely to drive the risks to *Manufacture Equipment* are physical damage from some climate drivers, such as hurricanes, flooding, and sea-level rise. Extreme heat could also cause workforce shortages because heat reduces labor productivity and degrades equipment quality.

Cascading Risk

Manufacture Equipment is critically dependent on the following NCFs:

- *Distribute Electricity*
- *Maintain Supply Chains*
- *Provide and Maintain Infrastructure*
- *Supply Water.*

Manufacturing requires timely access to intermediate inputs and raw materials (P. Goldberg et al., 2010), infrastructure (Cohen and Paul, 2004), and electric power (Acar and Berk, 2022). Although these upstream NCFs have some resilience to temporary disruptions, any sustained time period of failure would cause significant disruption to the *Manufacture Equipment* NCF. For instance, power infrastructure is critical to manufacturing performance (Acar and Berk, 2022). Similarly, if *Maintain Supply Chains* or *Provide and Maintain Infrastructure* were to fail, most manufacturing processes would struggle to source important imported inputs, causing significant production delays (Wuebbles et al., 2017). Finally, some manufacturing processes rely intensively on water as an intermediate input (Chatain et al., 2021; Doyle et al., 2017; National Integrated Drought Information System, undated; Scholz et al., 2021), and disruptions to the *Supply Water* NCF could cause production delays for some products.

Strength of Evidence

The strength of evidence of climate-driver effects on the *Manufacture Equipment* NCF was assessed as medium. There is a considerable body of literature on climate change's effects on manufacturing, but evidence specific to equipment manufacturing is scarce. Most of the literature focuses on the effects of hurricanes, flooding, drought, and extreme heat. Evidence of other climate drivers' effects on manufacturing is limited. For instance, extreme cold's effect on manufacturing is not widely studied in the literature and not mentioned as a risk factor in analyses that examine multiple drivers. Similarly, there is little research on climate change's potential effects on the ability to operate automated manufacturing equipment. Additional studies on whether technological advances leading to more automation in the manufacturing sector can provide insight on this NCF's adaptability to climate change.

Table B.179. Strength of Evidence for *Manufacture Equipment*

	Strength of Evidence
High	Flooding; tropical cyclones and hurricanes; extreme heat
Medium	Drought
Low	Severe storm systems; extreme cold; wildfire; sea-level rise

Risk Rating Interpretations

Table B.180 provides examples of how to interpret the risk ratings for this NCF.

Table B.180. Example Interpretations of Risk Ratings for *Manufacture Equipment*

Risk Rating	NCF Example
1: No disruption or normal operations	*Manufacture Equipment* continues to function as it does today, with occasional disruptions due to weather and operational problems.
2: Minimal disruption	A few facilities in flood-prone regions, such as the Great Lakes, experience production disruptions in which they cannot operate safely, but the NCF is able to meet all routine operational needs.
3: Moderate disruption	Local disruptions become severer and more frequent, but the NCF is able to perform all routine operations in most of the country. Some coastal facilities in the Gulf Coast and Miami close temporarily or permanently because of sea-level rise. Production delays due to extreme heat, flooding, hurricanes, and drought become far more frequent.
4: Major disruption	Production is not viable in most regions. Some firms relocate, but extracting natural resources becomes costly, and many firms go out of business.
5: Critical disruption	Manufacturing is essentially halted, although some products might be available at very high prices.

Produce Chemicals

Table B.181. Scorecard for *Produce Chemicals*

| | National Risk Assessment | | | | |
| | Baseline | Current Emissions | | High Emissions | |
	Present	2050	2100	2050	2100
Produce Chemicals	2	2	3	3	3

Description: "Manufacture basic chemicals from raw organic and inorganic materials and manufacture intermediate and final products from basic chemicals" (CISA, 2020, p. 6).

Highlights

- *Produce Chemicals* was assessed as being at risk of minimal disruption and is expected to be at risk of moderate disruption by the end of the century as climate drivers intensify.
- *Produce Chemicals* is particularly vulnerable to flooding, hurricanes, and sea-level rise, all of which destroy and inundate facilities.
- Many of the country's chemical facilities are near rivers and on the Gulf Coast, making the NCF vulnerable to flooding and tropical cyclones and hurricanes.
- Subfunctions dedicated to chemical production and distribution are at the highest risk of disruption because so many chemical production and distribution facilities are on the Gulf Coast.
- The NCF is clustered in flood-prone areas around water bodies and along coasts. Facilities on the Gulf Coast are particularly vulnerable to climate change given the projected rise in hurricane intensity and sea-level rise.

Synopsis

Chemical production facilities are vulnerable to disruptions from several climate drivers. A 2022 report by GAO indicates that 31 percent of U.S. chemical plants had been exposed to disruption from flooding, storm surge, wildfire, or sea-level rise (Gómez, 2022). Historically, tropical cyclones and hurricanes and flooding have damaged facilities, causing the release of toxic chemicals and slowing production (ACC, 2021; Trager, 2017; DiChristopher, 2017). Nearly 25 percent of basic chemical manufacturing facilities and petroleum and coal product manufacturing facilities are in areas that could be inundated by storm surge from a category 4 or 5 hurricane (Gómez, 2022). Many chemical facilities are near rivers, which reduces transport costs and provides access to water. However, this proximity puts many facilities at risk of disruption due to inland flooding (Tabuchi et al., 2018). GAO also identified 350 chemical facilities, including 61 facilities in California, in areas with high or very high wildfire hazard potential (Gómez, 2022). Finally, other climate drivers, such as extreme heat, which can damage facilities when chemicals combust (World Health Organization, 2018), are likely to cause disruptions to facilities in the next century as heat waves become severer. Extreme heat can also reduce the productivity of workers, particularly those in outdoor or non–temperature-controlled work settings (L. Carter et al., 2018). For instance, chemical equipment operators and tenders—the most common occupation in the chemical production industry (Office of Occupational Statistics and Employment Projections, undated)—are frequently exposed to outdoor weather and work indoors in non–environmentally controlled facilities (National Center for O*NET Development, 2023).

Together, the effects of climate change are expected to increase physical damage to chemical production facilities and disrupt chemical distribution. The ability to produce and distribute basic and applied chemicals depends on the integrity of physical facilities and on reliable distribution channels, both of which are vulnerable to disruptions due to climate change.

Produce Chemicals is expected to be at risk of moderate disruption from climate change by 2100 in both the current and high emissions scenarios. These ratings are due chiefly to an expected increase in the frequency, intensity, or duration of flooding, sea-level rise, wildfire, and hurricanes. Chemical production was assessed to be at risk of minimal disruptions from flooding and hurricanes in the baseline time period due to the damage these climate drivers cause to facilities and disruptions they create for production operations. These risks are expected to increase in the next century as these climate drivers intensify. Sea-level rise does not currently pose a risk to the NCF but is expected to pose an increasing risk of disruption as higher tides and sea levels threaten many coastal facilities on the Gulf and East Coasts. In the high emissions scenario, flooding, hurricanes, and sea-level rise are expected to pose a risk of moderate disruption by 2050.

Subfunctions

Because of the expected increase in physical damage to facilities, *Produce Basic Chemicals* and *Produce Applied Chemicals* were assessed to be at risk of moderate disruption from flooding, tropical cyclones and hurricanes, and sea-level rise by 2050 in the high emissions scenario. The geographic concentration of these production activities on the Gulf and East Coasts (Gómez, 2022) and the projected growth in climate-driver intensity or frequency currently drive risk to these subfunctions. Facility damage and inundation, which disrupt the ability to produce basic and applied chemicals, will likely grow in the future because of climate change and could result in an inability to meet operational needs in some regions, such as the Gulf and East Coasts. Flooding, hurricanes, and sea-level rise are also expected to increase disruptions to the ability to distribute commodities in some regions (AECOM, 2018; Allen, McLeod, and Hutt, 2022; Sturgis, Smythe, and Tucci, 2014; Xia and Lindsey, 2021), and, by the end of the century, these disruptions could disrupt some regions' ability to distribute chemical products, resulting in a risk of moderate disruption to *Distribute Chemicals* as well.

Table B.182. Subfunction Risk for *Produce Chemicals*

Subfunction	Baseline	Current Emissions		High Emissions	
	Present	2050	2100	2050	2100
Produce Basic Chemicals	2	2	3	3	3
Produce Applied Chemicals	2	2	3	3	3
Distribute Chemicals	2	2	3	3	3
Regulate Chemical Production	1	1	2	1	2

NOTE: 1 = no disruption or normal operations. 2 = minimal disruption. 3 = moderate disruption. See Chapter 2 for further explanation of ratings.

Climate Drivers

Given their proximity to inland rivers and the coasts, chemical production facilities are vulnerable to flooding, sea-level rise, and hurricanes. As these climate drivers become more frequent or intense in the next century, chemical production could fail to meet routine operational needs in some regions. Additionally, the projected increase in wildfire occurrence increases the risk of disruption in the next century as well. Finally, extreme heat is expected to reduce labor productivity in many sectors (L. Carter et al., 2018) and could reduce labor productivity in chemical production, particularly in occupations that might require outdoor labor, such as chemical equipment operators (National Center for O*NET Development, 2023).

Table B.183. Risk from Climate Drivers for *Produce Chemicals*

Climate Driver	Baseline	Current Emissions		High Emissions	
	Present	2050	2100	2050	2100
Drought	1	1	1	1	1
Extreme cold	1	1	1	1	1
Extreme heat	1	2	2	2	2
Flooding	2	2	3	3	3
Sea-level rise	1	2	3	3	3
Severe storm systems	1	2	2	2	2
Tropical cyclones and hurricanes	2	2	3	3	3
Wildfire	1	2	3	2	3

NOTE: 1 = no disruption or normal operations. 2 = minimal disruption. 3 = moderate disruption. See Chapter 2 for further explanation of ratings.

Impact Mechanisms

Disruptions from sea-level rise, flooding, hurricanes, and wildfire are likely to affect *Produce Chemicals* by damaging facilities, making operating difficult or impossible in some areas. Extreme heat could reduce the productivity of workers in the sector, potentially reducing sectoral output. Finally, hurricanes, flooding, and sea-level rise are expected to increase disruptions to chemical distribution channels.

Cascading Risk

Produce Chemicals is dependent on the following NCFs:

- *Distribute Electricity*
- *Generate Electricity*
- *Transmit Electricity*
- *Maintain Supply Chains*
- *Manage Hazardous Materials*

264

- *Manage Wastewater*
- *Provide and Maintain Infrastructure*
- *Provide Capital Markets and Investment Activities*
- *Supply Water.*

The physical process of chemical production requires access to electricity and complex supply chain operations (DiChristopher, 2017). Additionally, producing chemicals involves the disposal of hazardous materials and wastewater used in some production processes (Tabuchi et al., 2018). The *Supply Water* NCF also plays a critical role in supplying chemical production facilities with water as an intermediate input or cooling agent (Centers for Disease Control and Prevention, 2016; Chatain et al., 2021). Finally, access to infrastructure (e.g., roads, ports) and capital markets is required to transport products and finance production activities (ACC, 2021; DiChristopher, 2017). Failures of these upstream NCFs would likely cause significant disruptions to chemical production. However, depending on the upstream NCF, some of the disruptions to chemical production could occur over longer time scales. For instance, an inability to provide access to capital markets could affect future chemical production more than current production. On the other hand, the inability to dispose of hazardous materials could force chemical production facilities to shut down immediately.

Strength of Evidence

There is a growing body of research on climate change's effect on chemical production. Most research in this area has focused on climate change's potential effects from flooding, but some researchers have analyzed multiple climate drivers (see, e.g., Rusco, 2021). Other information comes in the form of reporting after extreme weather events (e.g., DiChristopher, 2017, on the effects of Hurricane Harvey). Reports from industry stakeholder groups provide some evidence on the potential effects of climate change (e.g., ACC, 2021) as well. The effects that sea-level rise, hurricanes, and flooding could have on the ability to distribute products through supply chains have not been specifically on chemicals, although some studies have analyzed supply chain vulnerabilities more generally. The potential effects of hurricanes, wildfire, and flooding have the most support in the literature. Other climate drivers are less discussed as risk factors for chemical production.

Table B.184. Strength of Evidence for *Produce Chemicals*

Strength of Evidence	
High	Flooding; tropical cyclones and hurricanes; wildfire
Medium	Drought; extreme heat; sea-level rise
Low	Severe storm systems; extreme cold

Risk Rating Interpretations

Table B.185 provides examples of how to interpret the risk ratings for this NCF.

Table B.185. Example Interpretations of Risk Ratings for *Produce Chemicals*

Risk Rating	NCF Example
1: No disruption or normal operations	*Produce Chemicals* continues to function as it does today, with occasional delays due to weather and operational problems.
2: Minimal disruption	A few chemical plants in flood-prone regions experience longer time periods in which they cannot operate safely, but basic and applied chemicals are produced at a sufficient level by other facilities and can be moved through supply chains sufficiently to meet needs.
3: Moderate disruption	Some chemical plants close temporarily or permanently because of sea-level rise or continued flooding. Distributing chemicals becomes costly in some regions due to flooding, hurricanes, and sea-level rise. Extreme heat reduces worker productivity, and drought increases the cost of some chemical products.
4: Major disruption	Most chemical production is not viable, and prices for basic and applied chemicals rise across the country. Some chemical plants close permanently.
5: Critical disruption	Chemical production is essentially halted, although some basic and applied chemicals might be available at very high prices.

Provide Metals and Materials

Table B.186. Scorecard for *Provide Metals and Materials*

	National Risk Assessment				
	Baseline	Current Emissions		High Emissions	
	Present	2050	2100	2050	2100
Provide Metals and Materials	2	2	3	3	3

Description: "Manufacture iron, steel, and ferro-alloy products; alumina and aluminum products; non-ferrous metals; and materials as primary components for other industries" (CISA, 2020, p. 6).

Highlights

- *Provide Metals and Materials* is currently at risk of minimal disruption and at risk of moderate disruption in later time periods due to disruptions to mining operations and distribution channels.
- Flooding, sea-level rise, and severe storm systems pose the greatest risk to the NCF through their effects on mineral extraction and distribution.
- *Acquire Raw Materials*, *Process Raw Materials*, and *Distribute Metals and Materials* were assessed to be at the greatest risk of disruption from climate change.
- The outdoor nature of *Acquire Raw Materials* makes the NCF vulnerable to climate drivers, as does the reliance on specific coastal distribution centers in the Gulf Coast that are vulnerable to sea-level rise.
- Mining operations in the Midwest are vulnerable to increased disruption from flooding, and critical metal and material distribution nodes are vulnerable in the Gulf Coast to sea-level rise.

Synopsis

Climate change is likely to affect the stability and effectiveness of mining infrastructure and equipment, site closure, and transportation routes used to distribute and import raw minerals and other materials (Nelson and Schuchard, undated). Flooding can inundate mines, delaying the extraction of raw minerals and materials (Nelson and Schuchard, undated). There have been historical instances when flooding has delayed or disrupted mining for nonferrous metals. For instance, flooding in 2022 led some palladium and platinum mines in Montana to shut down activity for seven weeks (Njini, 2022). Steel manufacturing is also vulnerable to flooding (EPA, 2022b). Flooding has led to localized disruptions to production operations at steel mills, where even a minor amount of water can cause explosions and destruction (Metallurgprom, 2019). Heavy rain from severe storm systems can affect slope stability and cause erosion of opencast mines (Nelson and Schuchard, undated), particularly in mountainous regions where large deposits are generally found (Mohaddes et al., 2022).

The *Distribute Metals and Materials* subfunction definition has significant overlap with those of the transportation NCFs. Therefore, climate drivers' direct effects on transportation modes used to distribute metals and materials were considered in this assessment. Reports from other countries indicate that flooding can inundate rail corridors and lead to erosion around rail tracks, increasing the costs of distributing raw materials and minerals (Flint and Varriale, 2021), although it is not clear whether the same outcome would occur in the United States. Sea-level rise and hurricanes disrupt the ability to import raw materials and minerals (Becker, Acciaro, et al., 2013; L. Carter et al., 2018). Imports play a critical role in U.S. consumption of some products. For instance, 25 percent of steel consumed in the United States was imported in 2019 (International Trade Administration, 2020) and was imported primarily through seaports on the Gulf Coast (Coyne and Oktay, 2021; U.S. Census Bureau, undated), which is the most vulnerable region to sea-level rise and hurricanes (Sweet, Horton, et al., 2017). Hurricane Ida disrupted port operations at the Port of New Orleans, a major hub for steel and steelmaking raw materials (Coyne and Oktay, 2021). Extreme heat can cause adverse chemical reactions for some minerals and can reduce labor productivity in the sector (L. Carter et al., 2018; Nelson and Schuchard, undated) because many of the jobs require outdoor work. Wildfire can also delay mining activities, disrupting the ability to acquire raw materials needed in metal production. For instance, wildfires in the Pacific Northwest in 2021 suspended production in local mines for uranium (a nonferrous metal) (Focus FS, 2022). There is some evidence that companies mining nonferrous material view wildfire as a risk to mining facilities in the future (Nelson and Schuchard, undated).

The projected increase in flooding, sea-level rise, severe storm systems, wildfire, and extreme heat in the next century is expected to increase the risk of disruption to *Provide Metals and Materials*. Disruptions to the metal and material distribution system from sea-level rise and flooding pose risks for distributing and importing metals and materials, as well as acquiring raw materials needed in downstream production. Additionally, floods can disrupt acquiring raw materials by inundating mines, and extreme heat can reduce the productivity of workers in outdoor mines, making it more difficult to acquire and process raw materials.

Under the current emissions scenario, all climate drivers except drought and extreme cold are expected to pose a risk of minimal disruption by 2050. The increase in flooding, sea-level rise, and severe storm systems is expected to increase disruptions to the NCF by 2100, posing a risk of moderate disruption to *Provide Metals and Materials*. This risk rating is driven by climate change's effect on the ability to operate outdoor mines, disruptions to metal distribution networks, and damage to physical production facilities. Given the distributed nature of the NCF, climate change is likely to cause regional, but not national, disruptions to *Provide Metals and Materials*. The risk ratings are similar under the high emissions scenario, apart from the likely risk of moderate disruption from flooding by 2050 because flooding is expected to increase more rapidly in this scenario.

Subfunctions

Acquire Raw Materials is at the greatest risk of disruption from climate change because this subfunction involves operating outdoor mining, which is vulnerable to disruptions from flooding and severe storm systems. There is also some limited evidence that wildfire could pose risks to mining operations. The disruptions caused by these climate drivers could cause *Acquire Raw Materials* to not meet all routine needs in some regions. However, given the geographically distributed nature of the subfunction, the disruptions caused by these climate drivers are unlikely to lead to a national-level disruption. This subfunction also includes transporting material inputs, which is vulnerable to flooding and sea-level rise because these climate drivers will likely delay and disrupt inland and coastal distribution networks in the next century (AECOM, 2018; Allen, McLeod, and Hutt, 2022; Xia and Lindsey, 2021). Similarly, the subfunction *Distribute Metals and Materials* is expected to be at risk of moderate disruption by 2100, largely due to expected disruptions to regional metal and material distribution networks and transportation networks from floods, hurricanes, and sea-level rise.

Process Raw Materials (which includes such functions as casting, physically processing, rolling, and combining minerals) is expected to be at risk of minimal disruption from climate change in most time periods due to the need for operable physical facilities to carry out this action. Processing facilities might be vulnerable to damage from flooding, severe storm systems, and hurricanes, which are projected to be severer under the high emissions scenario. For instance, there are metal casting facilities across the United States (EPA, 2016), although many are concentrated in the upper Midwest (where flooding is projected to increase) and the Northeast (vulnerable to hurricanes and sea-level rise), and some are on the Gulf Coast (where sea-level rise and hurricanes are expected to increase). There are also nonferrous metal processing plants on the Gulf Coast (DHS, 2018). However, unlike mining, processing activities are largely indoors and potentially more protected from climate drivers. In the high emissions scenario, sea levels are expected to be approximately 6 feet higher in 2100, which would inundate large portions of south Florida and southern Louisiana, as well as parts of New Jersey. *Process Raw Materials* might fail to perform normal routine operations in these regions under this scenario, but, given the distributed nature of the subfunction, the disruption will likely not escalate to a national level.

Other subfunctions are likely to face less risk from climate change. For example, flooding might disrupt *Identify Raw Material Sources* by limiting the ability to locate viable reserves temporarily in some local areas. However, climate drivers are unlikely to disrupt other aspects of the subfunction, such as reviewing public market prices or identifying geologic properties.

Table B.187. Subfunction Risk for *Provide Metals and Materials*

Subfunction	Baseline Present	Current Emissions 2050	Current Emissions 2100	High Emissions 2050	High Emissions 2100
Identify Raw Material Sources	1	1	2	2	2
Acquire Raw Materials	2	2	3	3	3
Process Raw Materials	1	2	2	2	3
Distribute Metals and Materials	1	2	3	2	3

NOTE: 1 = no disruption or normal operations. 2 = minimal disruption. 3 = moderate disruption. See Chapter 2 for further explanation of ratings.

Climate Drivers

Flooding and severe storm systems have disrupted mining and metal production operations historically, although these disruptions have been highly localized. As a result, the NCF is at risk of minimal disruption from these climate drivers in the baseline time period. These climate drivers are projected to increase in frequency and intensity in the next century, in both current and high emissions scenarios (Easterling et al., 2017; Kossin et al., 2017). The upper Midwest and Great Lakes regions are particularly vulnerable to increased flood severity (Mallakpour and Villarini, 2015; Slater and Villarini, 2016). Mining and metal production operations in the upper Midwest and around the Mississippi River could see increased disruption from increased flood and severe storm events. As the frequency and severity of these climate drivers grow, the risk posed to the NCF will likely grow as well.

In later time periods, the disruptions caused by these climate drivers could result in the inability to perform all routine operations in some regions.

Other climate drivers do not pose a risk currently but might pose one in future time periods. For instance, sea-level rise poses a risk to ports on the Gulf Coast, which are important hubs for metal and material imports (Coyne and Oktay, 2021; U.S. Census Bureau, undated). Researchers found that a 4-foot rise in sea level would affect three-quarters of Gulf Coast port facilities (Kafalenos et al., 2008). Four feet of sea-level rise is projected to occur by 2100 in both current and high emissions scenarios (Sweet, Horton, et al., 2017). This level of disruption could result in the NCF's inability to perform all routine operations in the Gulf region. At present, tropical cyclones and hurricanes do not appear to pose a threat to the NCF. However, given the proximity of some mining and mineral processing facilities to hurricane-prone regions (U.S. Geological Survey, undated), there is a potential for local disruptions to grow modestly in the future as these storms become severer (Kossin et al., 2017). There is some limited evidence that wildfire disrupts nonferrous metal mining (Focus FS, 2022), and these disruptions could grow as wildfire becomes more frequent and severer in the West. Finally, extreme heat could limit worker productivity, but not all workers in the NCF work outdoors, so this effect might be relatively minor. Extremely cold temperatures can also reduce worker productivity, but this climate driver is expected to become less frequent and severe in the next century (Vose et al., 2017) and will likely not lead to disruptions to normal mining operations.

Table B.188. Risk from Climate Drivers for *Provide Metals and Materials*

Climate Driver	Baseline Present	Current Emissions 2050	Current Emissions 2100	High Emissions 2050	High Emissions 2100
Drought	1	1	1	1	1
Extreme cold	1	1	1	1	1
Extreme heat	1	2	2	2	2
Flooding	2	2	3	3	3
Sea-level rise	1	2	3	2	3
Severe storm systems	2	2	3	2	3
Tropical cyclones and hurricanes	1	2	3	3	3
Wildfire	1	2	3	3	3

NOTE: 1 = no disruption or normal operations. 2 = minimal disruption. 3 = moderate disruption. See Chapter 2 for further explanation of ratings.

Impact Mechanisms

Disruptions to *Provide Metals and Materials* are expected to come from physical damage or disruption to mines and facilities, workforce disruptions due to extreme heat, and input shortages due to increased disruptions to the flow of critical imports.

Cascading Risk

Provide Metals and Materials is critically dependent on the following NCFs:

- *Supply Water*
- *Distribute Electricity*
- *Maintain Supply Chains*
- *Manage Wastewater*
- *Transport Cargo and Passengers by Vessel*
- *Transport Cargo and Passengers by Rail*
- *Transport Cargo and Passengers by Road*
- *Provide and Maintain Infrastructure*
- *Provide Capital Markets and Investment Activities.*

Drought might pose a risk to the NCF in future time periods through the *Supply Water* NCF because many mining operations are in drought-prone regions (Bullock, 2020), and droughts are projected to intensify in the next century (Wehner et al., 2017). There are many mines and mineral processing plants, including copper mines, located in the Southwest (Bullock, 2020), where droughts are projected to become more intense in the next century (Wehner et al., 2017). Thus, the NCF is vulnerable to drought, which could limit the availability of this critical input (Nelson and Schuchard, undated). Because of the intensive use of water in mining activities, managing the runoff is important as well. Thus, the NCF is critically dependent on *Manage Wastewater*. Running metal refineries and production facilities requires electricity, creating a dependence on *Distribute Electricity*. Access to infrastructure (e.g., roads, ports) and capital markets is required to distribute metals and materials and finance production and mining activities, creating a critical dependence on *Provide and Maintain Infrastructure*, *Provide Capital Markets and Investment Activities*, *Maintain Supply Chains*, *Transport Cargo and Passengers by Air*, *Transport Cargo and Passengers by Rail*, *Transport Cargo and Passengers by Road*, and *Transport Cargo and Passengers by Vessel*.

Strength of Evidence

The strength of evidence for the *Provide Metals and Materials* NCF was assessed as medium. Although there is relatively little research on most climate drivers' effect on this NCF, some studies discuss the effects that flooding and drought could have on raw material extraction and transportation. Other studies provide evidence of extreme heat's potential effects on labor productivity in mining. Reporting in the popular press suggests that flooding can result in lost profits in the sector as well. Research on sea-level rise's implications for supply chains suggests that sea-level rise could also affect the ability to transport raw materials. There is little research on wildfire (Focus FS, 2022) and extreme cold in relation to the *Provide Metals and Materials* NCF.

Table B.189. Strength of Evidence for *Provide Metals and Materials*

Strength of Evidence	
Medium	Drought; flooding; sea-level rise; tropical cyclones and hurricanes; extreme heat; severe storm systems
Low	Extreme cold; wildfire

Risk Rating Interpretations

Table B.190 provides examples of how to interpret the risk ratings for this NCF.

Table B.190. Example Interpretations of Risk Ratings for *Provide Metals and Materials*

Risk Rating	NCF Example
1: No disruption or normal operations	*Provide Metals and Materials* continues to function as it does today, with occasional delays due to weather and operational problems.
2: Minimal disruption	A few mines in flood-prone or wildfire-prone regions experience longer time periods in which they cannot operate safely, but the NCF is still able to complete all routine operations.
3: Moderate disruption	Mineral mine closures due to flooding and severe storm systems become more frequent in the upper Midwest. Sea-level rise limits port capacity at Gulf Coast ports, which are critical for mineral imports. Wildfire becomes more frequent, causing longer-lasting disruptions to mine activity in western states.
4: Major disruption	Mineral and material extraction are not viable in large portions of the country. A significant number of mines close permanently. Critical transportation nodes cease to operate because of sea-level rise. Prices for metal and minerals increase significantly in national markets.
5: Critical disruption	All mineral and material extraction is essentially halted across the country, although the products might be available as imports at very high prices.

Finance, Information Technology, and Telecommunications

Provide Capital Markets and Investment Activities

Table B.191. Scorecard for *Provide Capital Markets and Investment Activities*

	National Risk Assessment				
	Baseline	Current Emissions		High Emissions	
Provide Capital Markets and Investment Activities	Present	2050	2100	2050	2100
	1	2	1	2	1

Description: "Issue and trade securities, including debt securities (such as bonds), equities (such as stocks), and derivatives (such as options and futures); provide advisory services and related services, such as prime brokerage; maintain [and] operate organized markets and over-the-counter mechanisms for these instruments" (CISA, 2020, p. 5).

Highlights

- Markets for stocks and commodities that have specific geographic locations could be vulnerable to effects from climate change–driven events, such as flooding or tropical cyclones and hurricanes.
- Likely effects are the suspension of operations for one or more business days.
- The trend toward fully virtual exchanges mitigates this risk over time.

Synopsis

Climate change is not expected to have any appreciable effect on *Provide Capital Markets and Investment Activities*. For the most part, the geographically decentralized and technologically robust nature of the systems that provide this NCF makes disruptions from climate change highly unlikely. Nonetheless, the location of some specific assets—namely, stock and futures exchanges—in financial centers, such as New York City and Chicago, gives rise to vulnerability from flooding or hurricanes. For example, following September 11, 2001, the New York Stock Exchange closed for several days, reopening to substantial instability on September 17 (O'Brien, undated). Similarly, Hurricane Sandy led to the closure of the New York Stock Exchange for two days, October 29 and 30, 2012 (O'Brien, undated). In both cases, the disruption to the NCF was episodic, and functionality was quickly and fully restored.

The potential for this kind of disruption exists because of the geographic specificity of stock exchanges. In the wake of the COVID-19 pandemic, these institutions and their day-to-day function have become increasingly virtual (e.g., Gura, 2020). The trend toward fully virtual stock and futures exchanges not located at any specific geographic place reduces the likelihood that severe events driven by climate change will disrupt this NCF. The risk to this NCF is therefore rated higher in 2050 than 2100 because in-person activities are expected to decline over time.

Subfunctions

Because of the geographic specificity of some aspects of stock and futures exchanges, the ability to maintain and operate organized markets—specifically, the buying, selling, and issuance of shares of publicly held companies and contracts tied to commodities—faces moderately rising vulnerability to localized flooding or hurricanes. As a result, operations of such exchanges might be suspended from time to time because of disruptions from severe weather events driven by climate change. These events could cause physical damage to buildings that house operations critical to this NCF or could disrupt the ability of the responsible workforce to carry out its functions until services to an affected area are restored.

Table B.192. Subfunction Risk for *Provide Capital Markets and Investment Activities*

	Baseline	Current Emissions		High Emissions	
Subfunction	Present	2050	2100	2050	2100
Maintain and Operate Organized Markets	1	2	1	2	1

NOTE: 1 = no disruption or normal operations. 2 = minimal disruption. See Chapter 2 for further explanation of ratings.

Climate Drivers

Climate change is anticipated to drive increases in flooding and hurricanes and tropical storms across the United States, with equal or greater-than-normal effect in the Northeast, where much of the stock and futures trading presently has a physical, geographically specific presence. The effect is expected to be episodic at worst, with effects lasting only a few days within a decade.

Table B.193. Risk from Climate Drivers for *Provide Capital Markets and Investment Activities*

	Baseline	Current Emissions		High Emissions	
Climate Driver	Present	2050	2100	2050	2100
Flooding	1	2	1	2	1
Tropical cyclones and hurricanes	1	2	1	2	1

NOTE: 1 = no disruption or normal operations. 2 = minimal disruption. See Chapter 2 for further explanation of ratings.

Impact Mechanisms

The chief impact mechanism for this NCF is physical damage and disruption, with potential for concomitant workforce shortages.

Cascading Risk

Among its many cascading risks, *Provide Capital Markets and Investment Activities* depends critically on the following NCFs:

- *Distribute Electricity*

274

- *Provide Internet Routing, Access, and Connection Services*
- *Provide Wireless Access Network Services*
- *Provide Payment, Clearing, and Settlement Services.*

The immediate operation of the finance industry relies heavily on IT and digital communication, as well as the sound functioning of the banking sector, particularly the ability to facilitate payments for purchases or sales of assets. All of these functions require electricity. Communication within the industry requires functional internet services. As the workforce supplying this function increasingly telecommutes, it also relies more on mobile device communications, which require wireless access network services. These dependencies might pose a threat of disruption because market closures can lead to substantial increases in asset volatility.

Strength of Evidence

Despite the critical importance that capital markets and investment activities have for the functioning of the economy, the existing research on climate change and finance is limited and does not forecast significant disruptions to this NCF. However, historical incidents and past analogues (e.g., Hurricane Sandy) combined with the rising risks of flooding and tropical cyclones and hurricanes in the Northeast serve as examples of how an increase in the risk of episodic disruption exists as climate change unfolds.

Table B.194. Strength of Evidence for *Provide Capital Markets and Investment Activities*

Strength of Evidence	
High	Flooding; sea-level rise
Medium	Tropical cyclones and hurricanes; wildfire
Low	Severe storm systems; extreme cold; extreme heat; drought

Risk Rating Interpretations

Table B.195 provides examples of how to interpret the risk ratings for this NCF.

Table B.195. Example Interpretations of Risk Ratings for *Provide Capital Markets and Investment Activities*

Risk Rating	NCF Example
1: No disruption or normal operations	Major stock and commodity trade exchanges operate as normal.
2: Minimal disruption	Once every ten years, major stock or commodity trade exchanges experience unscheduled closures measured in days.
3: Moderate disruption	Once every five years, major stock or commodity trade exchanges experience unscheduled closures measured in days.
4: Major disruption	Once every year, major stock or commodity trade exchanges experience unscheduled closures measured in weeks.
5: Critical disruption	Major stock or commodity trade exchanges close indefinitely.

Provide Consumer and Commercial Banking Services

Table B.196. Scorecard for *Provide Consumer and Commercial Banking Services*

	National Risk Assessment				
	Baseline	Current Emissions		High Emissions	
	Present	2050	2100	2050	2100
Provide Consumer and Commercial Banking Services	1	1	1	1	1

Description: "Accept and maintain deposit accounts (e.g., checking and savings accounts) and close substitutes (e.g., short-term retail notes) from non-financial intermediaries" (CISA, 2020, p. 5).

Highlights

- Although climate change is expected to have some effect on the banking sector via other NCFs, no literature currently reports risks to *Provide Consumer and Commercial Banking Services*.
- In recent disasters that affected financial centers in the United States, this NCF was not severely disrupted.

Synopsis

Climate change is not expected to have any appreciable effect on *Provide Consumer and Commercial Banking Services*. Although research suggests that other functions of the banking sector face some risk, the geographically decentralized and technologically robust nature of the systems that underpin banking services support continuing function even as physical risks to specific locations increase through various climate drivers. Reviewing recent disasters, such as Hurricane Sandy, for example, the FDIC worked with state and federal banking agencies to support the market and encourage banks to work with borrowers affected by the storm (FDIC, 2017), leading to a strong response from the sector (American Banker, 2012). Another extreme case is, in the aftermath of September 11, 2001, when disruptions to banking resulted in missed payments between banks and liquidity shortages, the Federal Reserve responded by supplying liquidity and overall ensuring the continued stable operation of the monetary and financial system (McAndrews and Potter, 2002). In both examples, when other aspects of banking were strained, the banking sector and its regulators maintained the overall security of banking operations and ability to accept and maintain deposit accounts.

Cascading Risk

Provide Consumer and Commercial Banking Services depends critically on the following NCFs:

- *Distribute Electricity*
- *Provide Internet Routing, Access, and Connection Services*
- *Provide Payment, Clearing, and Settlement Services*
- *Provide Identity Management and Associated Trust Support Services.*

Among its many cascading risks, the immediate operation of the finance industry relies heavily on IT and digital communication, as well as the sound functioning of the banking sector, particularly the ability to facilitate payments for purchases or sales. All of these functions require electricity.

Communication within the industry requires functional internet services. Identity management is crucial to the safeguarding of access to accounts and the prevention of fraud. These dependencies might pose a threat of disruption because unexpected, widespread bank closures can induce crises of confidence and lead to bank runs.

Strength of Evidence

The evidence for climate change's expected effect on this NCF is relatively weak. In brief, although climate change's effect on the banking sector has received substantial attention, no sources note anticipated, serious effects on this NCF. Our observation is that past incidents that had effects on the banking industry in line with the climate drivers considered did not disturb this function.

Risk Rating Interpretations

Table B.197 provides examples of how to interpret the risk ratings for this NCF.

Table B.197. Example Interpretations of Risk Ratings for *Provide Consumer and Commercial Banking Services*

Risk Rating	NCF Example
1: No disruption or normal operations	*Provide Consumer and Commercial Banking Services* continues to function as it does today, with rare, localized delays due to weather and operational problems.
2: Minimal disruption	There is one unexpected bank holiday for banks, nationally, every ten years.
3: Moderate disruption	There are two to three unexpected bank holidays nationally every five years. Banks face liquidity shortages but remain solvent.
4: Major disruption	Multiple banks cease operations for a week within a single year. At least one bank fails.
5: Critical disruption	A crisis of confidence produces mass bank runs. Multiple banks fail.

Provide Funding and Liquidity Services

Table B.198. Scorecard for *Provide Funding and Liquidity Services*

	National Risk Assessment				
	Baseline	**Current Emissions**		**High Emissions**	
	Present	**2050**	**2100**	**2050**	**2100**
Provide Funding and Liquidity Services	1	2	1	2	1

Description: "Provide funding to non-financial counterparties, such as corporate or retail customers, including individual consumers" (CISA, 2020, p. 5).

Highlights

- Climate change could have some effect on this NCF through rare events that affect geographically specific sites of financial market utilities responsible for ensuring funding and liquidity services.
- In recent disasters that affected financial centers in the United States, this NCF was strained but not severely disrupted, thanks to appropriate actions by industry regulators.
- Any risk to the function is likely to be mitigated as the system becomes increasingly decentralized.

Synopsis

For the most part, the geographically decentralized and technologically robust nature of the systems that provide funding and liquidity services makes disruptions from climate change highly unlikely. Nonetheless, the location of some specific assets in financial centers, such as New York City, gives rise to vulnerability from flooding or hurricanes. During Hurricane Sandy, for example, the NCF was arguably threatened because of the inundation of various corporate bank offices in Manhattan and the inability of workers supporting the function to report to their offices. The FDIC worked with state and federal banking agencies to support the market and encourage banks to work with borrowers affected by the storm (FDIC, 2017), leading to a strong response from the sector (American Banker, 2012). In another extreme case, in the aftermath of September 11, 2001, disruptions to banking resulted in missed payments between banks and liquidity shortages. The Federal Reserve responded by supplying liquidity and overall ensuring the continued stable operation of the monetary and financial system (McAndrews and Potter, 2002). In both examples the capacity to provide funding and liquidity was strained, but the banking sector and the Federal Reserve managed to maintain the function through the crises.

Nonetheless, a 2020 report by the Commodity Futures Trading Commission (CFTC) (Climate-Related Market Risk Subcommittee, 2020) points out that this might have been a stroke of good luck. During Hurricane Sandy, a vault of the Depository Trust and Clearing Corporation flooded. This designated, systemically important clearing and settlement company took weeks to recover as it reckoned with the damage or destruction of millions of documents, including 1.7 million stock and bond certificates. Although such documents provide original records of ownership, they are increasingly anachronistic and redundant in the digital age. Thus, although a similar disruption of key funding and liquidity providers is possible, the risk to the NCF is unlikely. Risk to the NCF might rise as climate change proceeds, but any risk to the function is likely to be mitigated as the system becomes increasingly decentralized. We therefore assessed the risk of disruption as minimal, peaking at the middle of the 21st century.

Subfunctions

The subfunctions of this NCF include *Maintain Sufficient Liquidity; Regulate Liquidity and Funding Entities;* and *Establish Funding, Credit, Liquidity, and Loan Facilities.* Climate change effects pose some risk only to *Maintain Sufficient Liquidity,* which depends, in part, on the ability to protect and maintain sufficient balances via the operation of financial market utilities to support market activities.

Table B.199. Subfunction Risk for *Provide Funding and Liquidity Services*

	Baseline	Current Emissions		High Emissions	
Subfunction	Present	2050	2100	2050	2100
Maintain Sufficient Liquidity	1	2	1	2	1

NOTE: 1 = no disruption or normal operations. 2 = minimal disruption. See Chapter 2 for further explanation of ratings.

Climate Drivers

The risk to *Provide Funding and Liquidity Services* is driven mainly by the increase in likelihood of flooding and tropical cyclones and hurricanes, particularly in Chicago and New York.

Table B.200. Risk from Climate Drivers for *Provide Funding and Liquidity Services*

	Baseline	Current Emissions		High Emissions	
Climate Driver	Present	2050	2100	2050	2100
Flooding	1	2	1	2	1
Tropical cyclones and hurricanes	1	2	1	2	1

NOTE: 1 = no disruption or normal operations. 2 = minimal disruption. See Chapter 2 for further explanation of ratings.

Impact Mechanisms

The chief impact mechanism for this NCF is physical damage and disruption. Concomitant episodic workforce shortages might accompany physical damage from these events because people responsible for ensuring continued operation of the NCF might not be able to conduct their work from off-site.

Cascading Risk

Provide Funding and Liquidity Services depends critically on the following NCFs:

- *Distribute Electricity*

- *Provide Internet Routing, Access, and Connection Services*
- *Provide Payment, Clearing, and Settlement Services*
- *Provide Identity Management and Associated Trust Support Services.*

As we note elsewhere, the immediate operation of the finance industry relies heavily on IT and digital communication, as well as the sound functioning of the banking sector, particularly the ability to facilitate payments for purchases or sales. All of these functions require electricity. Communication within the industry requires functional internet services. Identity management is crucial to the safeguarding access to accounts and the prevention of fraud. These dependencies might pose a threat of disruption because unexpected, widespread bank closures can induce crises of confidence and lead to bank runs, ultimately imperiling the ability to maintain sufficient liquidity.

Strength of Evidence

The evidence for climate change's effect on this NCF is relatively weak. In brief, although climate change's effect on the banking sector has received substantial attention, no sources we found specifically mention potential effects on funding and liquidity services. Past incidents that had effects on the banking industry in line with the climate drivers considered did not appear to disturb this function. However, the CFTC (Climate-Related Market Risk Subcommittee, 2020) describes an effect of Hurricane Sandy that can be characterized as a "near miss" that illustrates the potential for disruption. Any disruption to this NCF seems likely to be episodic, and the likelihood and effect of such a disruption are increasingly diminished by the persistent trend toward distributed, virtual systems to support the function.

Table B.201. Strength of Evidence for *Provide Funding and Liquidity Services*

Strength of Evidence
Low Flooding; tropical cyclones and hurricanes

Risk Rating Interpretations

Table B.202 provides examples of how to interpret the risk ratings for this NCF.

Table B.202. Example Interpretations of Risk Ratings for *Provide Funding and Liquidity Services*

Risk Rating	NCF Example
1: No disruption or normal operations	*Provide Funding and Liquidity Services* operates normally.
2: Minimal disruption	There are occasional delays in *Provide Funding and Liquidity Services* systems, with unscheduled downtime of a few days or less per decade.
3: Moderate disruption	There are significant delays to *Provide Funding and Liquidity Services* systems annually, with disruptions of a few days or more being typical.
4: Major disruption	*Provide Funding and Liquidity Services* systems are on the verge of failure, with disruptions measured in weeks.
5: Critical disruption	*Provide Funding and Liquidity Services* systems break down entirely.

Provide Payment, Clearing, and Settlement Services

Table B.203. Scorecard for *Provide Payment, Clearing, and Settlement Services*

	National Risk Assessment				
	Baseline	Current Emissions		High Emissions	
Provide Payment, Clearing, and Settlement Services	Present	2050	2100	2050	2100
	1	2	1	2	1

Description: "Carry out processes required for the exchange of assets, including payment (transfer of funds between or among participants), clearing (transmitting, reconciling, and confirming transactions prior to settlement), and settlement (transfer of ownership and payments)" (CISA, 2020, p. 5).

Highlights

- Climate change could have some effect on this NCF through rare events that affect geographically specific sites of financial market utilities responsible for ensuring clearing and settlement services, particularly with respect to securities, commodities, and options.
- In a 2012 disaster that affected financial centers in the United States (Hurricane Sandy), one institution responsible for this NCF was damaged, but no operational effect was reported.
- Any risk to the function is likely to be mitigated by the general and continuing movement away from hard-copy, paper records toward redundant, cloud-based virtual records and decentralized operations.

Synopsis

Clearance and settlement in banking occur largely via the Clearing House Interbank Payments System and the Federal Reserve's Fedwire Funds Service networks. These systems are relatively decentralized and technologically robust (e.g., C. Cooper, Labonte, and Perkins, 2019), making disruptions from climate change highly unlikely. Nonetheless, the location of some specific, critical clearance and settlement assets in financial centers, such as New York City and Chicago, gives rise to vulnerability from flooding or tropical storm events. During Hurricane Sandy, for example, a vault of the Depository Trust and Clearing Corporation flooded: This important clearing and settlement company took weeks to recover as it reckoned with the damage or destruction of millions of documents, including 1.3 million stock and bond certificates. No operational effect was reported: The damage presented "an administrative and logistical challenge [rather] than an economic issue," and available backups in data centers were able to support recreation of vault inventory (Jonas, 2012).

According to the CFTC (Climate-Related Market Risk Subcommittee, 2020), potential for disruption of this NCF exists because of the geographic specificity of some of systemically important financial market utilities (FMUs). Under the Dodd–Frank Wall Street Reform and Consumer Protection Act (Pub. L. 111-203, 2010), the Financial Stability Oversight Council can designate an institution or system to be a systemically important FMU if

> the failure of or a disruption to the functioning of the FMU could create or increase the risk of significant liquidity or credit problems spreading among financial institutions or markets and thereby threaten the stability of the U.S. financial system. (U.S. Department of the Treasury, undated)

Although the CFTC report does not identify specific FMUs at risk, the eight currently identified FMUs (and their locations) are

- the Clearing House Payments Company, as operator of the Clearing House Interbank Payments System (Michigan, New York, North Carolina, and Texas)
- CLS Bank International (New York)
- Chicago Mercantile Exchange (Chicago)
- the Depository Trust Company, a subsidiary of the Depository Trust and Clearing Corporation (New York)
- Fixed Income Clearing Corporation, a subsidiary of the Depository Trust and Clearing Corporation (New York)
- Intercontinental Exchange Clear Credit (New York)
- National Securities Clearing Corporation, a subsidiary of the Depository Trust and Clearing Corporation (New York)
- the Options Clearing Corporation (Chicago) (U.S. Department of the Treasury, undated).

Of these FMUs, the Depository Trust Company and its subsidiaries, including the Depository Trust and Clearing Corporation, are the primary U.S. depositories responsible for clearing and settlement of securities transactions (Labonte, 2012). The locations of the FMUs are predominantly in FEMA regions 2 and 5, which are anticipated to have higher effects from flooding due to climate change than the U.S. average.[54] Physical effects on infrastructure at these locations could be simultaneously accompanied by episodic workforce shortages because people might not be able to work on-site during damage and recovery time periods. The general and continuing movement away from hard-copy, paper records toward redundant, cloud-based virtual records and decentralized operations makes this scenario less likely over time. Workforce disruptions also become less likely as the trend toward telecommuting continues. The greatest risk of such a disruption is therefore judged to be in the middle of the 21st century, when climate risks have increased but these technological and social trends might not have fully mitigated potential effects.

Subfunctions

As part of the subfunction *Provide Settlement Services*, the NCF includes *Operate the Depository Trust Company*. The *Provide Clearing Services* subfunction includes *Operate the National Securities Clearing Corporation*, *Operate the Fixed Income Clearing Corporation*, *Operate the Chicago Mercantile Exchange Clearing System*, and *Operate the Options Clearing Corporation*. These two subfunctions therefore face heightened risk from climate change because of the geographically specific locations of the entities that carry them out.

---

[54] Region 2 is New Jersey, New York, Puerto Rico, and Virgin Islands. Region 5 is Illinois, Indiana, Michigan, Minnesota, Ohio, and Wisconsin.

Table B.204. Subfunction Risk for *Provide Payment, Clearing, and Settlement Services*

	Baseline	Current Emissions		High Emissions	
Subfunction	Present	2050	2100	2050	2100
Provide Settlement Services	1	2	1	2	1
Provide Clearing Services	1	2	1	2	1

NOTE: 1 = no disruption or normal operations. 2 = minimal disruption. See Chapter 2 for further explanation of ratings.

Climate Drivers

The risk to *Provide Funding and Liquidity Services* is driven mainly by the increase in likelihood of flooding and tropical cyclones and hurricanes, particularly in Chicago and New York.

Table B.205. Risk from Climate Drivers for *Provide Payment, Clearing, and Settlement Services*

	Baseline	Current Emissions		High Emissions	
Climate Driver	Present	2050	2100	2050	2100
Flooding	1	2	1	2	1
Tropical cyclones and hurricanes	1	2	1	2	1

NOTE: 1 = no disruption or normal operations. 2 = minimal disruption. See Chapter 2 for further explanation of ratings.

Impact Mechanisms

The chief impact mechanism for this NCF is physical damage and disruption. Concomitant episodic workforce shortages might accompany physical damage from these events if people responsible for ensuring continued operation of the NCF cannot conduct their work from off-site.

Cascading Risk

Provide Payment, Clearing, and Settlement Services depends critically on the following NCFs:

- *Distribute Electricity*
- *Provide Internet Routing, Access, and Connection Services.*

As we note elsewhere, the immediate operation of the finance industry relies heavily on IT and digital communication, as well as the sound functioning of the banking sector—particularly, the ability to facilitate payments for purchases or sales. All of these functions require electricity. Communication within the industry requires functional internet services. These dependencies might pose a threat of disruption because unexpected disruption of this central function of banking could induce crises of confidence and lead to bank runs.

Strength of Evidence

The evidence for climate change's effect on this NCF is relatively weak. In brief, although climate change's effect on the banking sector has received substantial attention, very little research indicates risk to *Provide Payment, Clearing, and Settlement Services*. The CFTC (Climate-Related Market Risk Subcommittee, 2020) has described an effect of Hurricane Sandy that can be characterized as a "near miss" that illustrates the potential for disruption. Any disruption to this NCF seems likely to be episodic, and the likelihood and effect of such a disruption are increasingly diminished by the shift toward telecommuting and persistent trend toward distributed, virtual systems to support the function.

Table B.206. Strength of Evidence for *Provide Payment, Clearing, and Settlement Services*

Strength of Evidence
Low Flooding; tropical cyclones and hurricanes

Risk Rating Interpretations

Table B.207 provides examples of how to interpret the risk ratings for this NCF.

Table B.207. Example Interpretations of Risk Ratings for *Provide Payment, Clearing, and Settlement Services*

Risk Rating	NCF Example
1: No disruption or normal operations	*Provide Payment, Clearing, and Settlement Services* operates normally.
2: Minimal disruption	There are occasional delays to payment, clearing, and settlement service systems, with unscheduled downtime of a few days or less per decade.
3: Moderate disruption	There are significant delays to payment, clearing, and settlement service systems annually, with disruptions of a few days or more being typical.
4: Major disruption	Payment, clearing, and settlement service systems are on the verge of failure, with disruptions measured in weeks.
5: Critical disruption	Payment, clearing, and settlement service systems break down entirely.

286

Provide Wholesale Funding

Table B.208. Scorecard for *Provide Wholesale Funding*

	National Risk Assessment				
	Baseline	**Current Emissions**		**High Emissions**	
	Present	**2050**	**2100**	**2050**	**2100**
Provide Wholesale Funding	1	1	1	1	1

Description: "Maintain processes for lending and borrowing among financial services sector parties" (CISA, 2020, p. 6).

Highlights

- As climate change unfolds, increasing numbers of properties could be threatened by increased flooding, sea-level rise, hurricane intensity, and wildfire.
- Affected entities could experience devaluations of their properties, implying a devaluation of collateral for mortgage lenders. If affected entities cannot meet debt obligations, lenders might experience losses because the underlying property cannot be sold at a sufficiently high price for repayment.
- Although this scenario is expected to have some effects on the price of lending in coastal regions and the mortgage-backed securities market, no research contemplates overall threat to the *Replenish Loanable Funds in the Mortgage Market* subfunction or other elements of the NCF.

Synopsis

Banks use wholesale funding to finance their operations and manage risk, drawing on a wide array of sources, including banking deposits, public funds, federal home loan banks, and the Federal Reserve. For the most part, climate change is not expected to have any appreciable effect on *Provide Wholesale Funding*. The geographically decentralized and technologically robust nature of the systems that provide this function (e.g., *Provide Interbank Lending Network*) supports continuing operations even as physical risks to specific locations increase through various climate drivers. Research related to this function has identified some risk connected to mortgages and mortgage-backed securities, but the existing analyses have not contemplated a level of impact that would threaten the NCF. The main idea is that changes in natural disaster risk and the occurrence of natural hazards will lead to property devaluations, reducing the collateral value of property for mortgage lenders. If borrowers then default (whether because of climate effects or for other reasons), lenders could face losses. In the wake of Hurricane Sandy, for example, 80 percent of homes affected did not have flood insurance because the event occurred in an area not traditionally at risk of serious flooding. Mortgage delinquencies increased by 200 percent as a result (Olick, 2021).

Several recent analyses have examined the potential implications of climate change for the mortgage market and mortgage-backed securities. Becketti (2021) and Patel (2021) provide clear overviews of the mechanisms in the market. Ouazad and Kahn (2021) illustrates that, in the wake of natural disasters, lenders are likelier to approve loans that conform to securitization rules, allowing them to transfer climate-driven risk. This finding and related observations are driving the market to develop better recognition of climate risk for mortgage-backed securities as a part of environmental, social, and governance ratings (e.g., McNeil, 2020), and researchers have called for the development of appropriate policy (e.g., Keys, undated). The Federal Housing Finance Agency, which regulates the government-sponsored enterprises that develop securitization rules (i.e., the Federal National Mortgage Association [Fannie Mae] and the Federal Home Loan Mortgage Corporation [Freddie Mac]) is aware of the issue and has established an internal Climate Change and Environmental, Social, and Governance Steering Committee to develop policy (Federal Housing Finance Agency, 2023).

Given the diversity of sources of wholesale funding, this NCF was assessed as being at no risk of disruption across all scenarios and time frames.

Cascading Risk

Provide Wholesale Funding depends critically on the following NCFs:

- *Distribute Electricity*
- *Provide Internet Routing, Access, and Connection Services*
- *Provide Payment, Clearing, and Settlement Services.*

As we note elsewhere, the immediate operation of the finance industry relies heavily on IT and digital communication, as well as the sound functioning of the banking sector—particularly, the ability to facilitate payments for purchases or sales. All of these functions require electricity. Communication within the industry requires functional internet services. These dependencies might pose a threat of disruption because unexpected disruption of this central function of banking could induce crises of confidence and lead to bank runs.

Strength of Evidence

The strength of evidence for climate change's effect on this NCF is low to moderate. Climate change's effect on the banking sector has received substantial attention, and some of that research has addressed functions specifically related to this NCF. Although researchers have noted the need for adjustments and potential for market volatility and losses to lenders, risk of any level of disruption to this NCF is not contemplated.

Risk Rating Interpretations

Table B.209 provides examples of how to interpret the risk ratings for this NCF.

Table B.209. Example Interpretations of Risk Ratings for *Provide Wholesale Funding*

Risk Rating	NCF Example
1: No disruption or normal operations	*Provide Wholesale Funding* continues to function as it does today, with few if any disruptions.
2: Minimal disruption	*Provide Wholesale Funding* mechanisms experience a short time period of high volatility.
3: Moderate disruption	*Provide Wholesale Funding* is stressed to the point at which interbank lending systems are unavailable or mortgage-backed securities trading is halted for one to three days.
4: Major disruption	*Provide Wholesale Funding* is stressed to the point at which interbank lending systems are unavailable or mortgage-backed securities trading is halted for a week.
5: Critical disruption	*Provide Wholesale Funding* collapses as a financial crisis similar to the one in 2007 and 2008 unfolds.

Provide Insurance Services

Table B.210. Scorecard for *Provide Insurance Services*

	National Risk Assessment				
	Baseline	Current Emissions		High Emissions	
	Present	2050	2100	2050	2100
Provide Insurance Services	2	2	3	3	3

Description: "Operate systems and markets to transfer financial risks among parties through contractual relationships, including products for individuals, corporations, and public-sector entities" (CISA, 2020, p. 5).

Highlights

- The NCF was assessed as facing a risk of minimal disruption presently due to wildfire's effects on some property insurance markets in California. As climate change unfolds, more property will be threatened by increased flooding and sea-level rise, wildfire, and severity of tropical cyclones and hurricanes. These effects are likely to raise the cost of insurance.
- Climate change generally affects this NCF through physical damage and disruption of insured assets and demand changes.
- Eventually, in some areas, affected entities might be unable to afford insurance. As prices rise, consumers might buy less insurance, and insurers might eventually exit some markets. Unanticipated, large, correlated losses could lead to the failure of some insurers.
- The most vulnerable subfunction of this NCF is *Pool Multiple Exposures to Transfer Insurable Risk* because climate change will expose property insurers to greater losses, leading to substantial increases in property insurance premiums in some markets and potential insolvencies in the wake of natural catastrophes.
- Coastal regions, regions at increased risk of exposure to tropical cyclones and hurricanes, and regions at increased risk of wildfire are likeliest to be affected.

Synopsis

Climate change is likeliest to have effects on property insurance of all the types of insurance. As flooding, sea-level rise, and wildfire increase with climate change, property insurers will be exposed to greater losses (Bevere and Weigel, 2021). These losses could result in insolvencies, absent interventions (Kunreuther, Michel-Kerjan, and Ranger, 2013). At the same time, residents in these areas will demand more insurance because they face greater risks (Katie Baker, 2022). Yet, because insurance pricing depends on historical data generated by stable underlying processes, experts expect climate change to challenge underwriters and increase uncertainty and variability in risk estimates (Kunreuther and Michel-Kerjan, 2007). Because climate drivers tend to affect whole regions, opportunities for risk transfer within adversely affected regions are also likely to decline (Grimaldi et al., 2020).

These pressures on the property insurance market will have several likely effects on this NCF. Insurers will likely raise premiums and narrow or deny coverage in areas facing higher or more-variable risk due to climate change (Moody's Investors Service, 2018). As a result, absent market interventions in affected areas, homeowners are likelier to be underinsured or uninsured entirely (Ramnath and Jeziorski, 2021). This scenario is already beginning to play out in rural California counties facing increased wildfire risk. In 2019, the California Department of Insurance intervened to impose a one-year moratorium on nonrenewals for property insurance in wildfire-prone areas (California Department of Insurance, undated). Dixon, Tsang, and Fitts (2020) shows that premiums increased faster in high-risk wildfire areas than elsewhere in the state, leading homeowners to choose higher deductibles and lower coverage limits relative to value and increasing the share of homeowners seeking coverage through the state's high-risk pool, the California Fair Access to Insurance Requirements Plan.

Although insurance markets are generally functioning, this NCF was assessed as being at risk of minimal disruption at the baseline because of climate change's effects on wildfire risk and property insurance in California. As sea-level rise, flooding, and tropical cyclone and hurricanes increase, disruptions to this market are expected to rise but remain assessed as being at risk of minimal disruption by 2050 under the current emissions scenario. In a recent analysis, for example, Swiss Re (Holzheu et al., 2021) has forecast growth, not major disruptions, to property insurance markets and premiums. Given the elevated risk of flooding in 2050 under the high emissions scenario, we surmised that regional effects on this NCF are likelier than national effects. We therefore assessed it as at risk of moderate disruption in this scenario by the middle of the 21st century. In both scenarios, climate drivers have the potential to produce problematic correlated losses regionally through flooding, sea-level rise, tropical cyclones and hurricanes, and wildfire by the end of the century. We therefore assessed it as at risk of moderate disruption by 2100.

Subfunctions

Climate change affects this NCF through the subfunction *Pool Multiple Exposures to Transfer Insurable Risk*. Generally, insurance markets function by spreading well-understood, uncorrelated risks across large numbers of policyholders. Climate change has two effects on property insurance markets: It increases uncertainty by reducing the predictive value of historical data and creates new correlated risks where they previously did not exist. These two effects are likely to increase prices for insurance, which, in turn, could shrink the pool of policyholders to the riskier parties that have higher demand for insurance (Wagner, 2022).

As a separate and distinct effect of climate change, casualty insurance markets also face risk of mass liability for climate change (Kunreuther and Michel-Kerjan, 2007). The principal issue is that carbon-intensive firms, primarily in the fossil fuel and energy sectors, could be held liable for the damage attributable to climate change and that their insurers might have to respond to that liability. Under these circumstances, insurers could find themselves highly undercapitalized and become insolvent, dramatically affecting the market's ability to provide surety services.

Given the remote nature of this scenario, it does not affect our assessment of risk at this time but should be monitored for development. Setzer and Higham (2022) reports that the total number of climate change–related cases has more than doubled since 2015, reaching more than 2,000 cases worldwide. Although liability for climate change under the legal system as it presently exists seems unlikely, if not impossible (based on duty of care and proof of causation), climate change itself could lead to a reckoning and changes in theories of tort (Kysar, 2011). In the event that firms are held liable, costs could total into the trillions of dollars (Mitchell, Robinson, and Tahmasebi, 2021), although valuations are highly uncertain and would depend on the precise legal theory applied (Schwarze, 2007).

Climate change is likely to have appreciable business effects on other lines of insurance (e.g., business interruption, auto insurance, workers' compensation): None of the studies identified and reviewed for this analysis cites a risk of their disruption (Golnaraghi and Geneva Association Task Force on Climate Change Risk Assessment for the Insurance Industry, 2021; Holzheu et al., 2021; Luckman, 2022). Likewise, other subfunctions of *Provide Insurance Services*, including *Provide Over-the-Counter Derivatives* or *Provide Surety Services*, were assessed as not at risk from the physical effects of climate change.

Table B.211. Subfunction Risk for *Provide Insurance Services*

	Baseline	Current Emissions		High Emissions	
Subfunction	Present	2050	2100	2050	2100
Pool Multiple Exposures to Transfer Insurable Risk	2	2	3	3	3

NOTE: 2 = minimal disruption. 3 = moderate disruption. See Chapter 2 for further explanation of ratings.

Climate Drivers

Although each climate driver can cause damage to insured goods or real property, it is the increase in correlation of losses that could prove problematic for this function. This analysis focuses primarily on the potential effects on this NCF from flooding, tropical cyclones and hurricanes, sea-level rise, and wildfire. Note also that extreme heat can damage infrastructure directly. Drought can cause crop failures and other insured losses that are correlated over a very wide area (e.g., Diffenbaugh, Davenport, and Burke, 2021; Douglas, 2022).

Table B.212. Risk from Climate Drivers for *Provide Insurance Services*

	Baseline	Current Emissions		High Emissions	
Climate Driver	Present	2050	2100	2050	2100
Drought	1	2	2	2	2
Extreme cold	1	1	1	1	1
Extreme heat	1	2	2	2	2
Flooding	2	2	3	3	3
Sea-level rise	2	2	3	2	3
Severe storm systems	1	1	1	1	1
Tropical cyclones and hurricanes	2	2	3	3	3
Wildfire	2	2	3	3	3

NOTE: 1 = no disruption or normal operations. 2 = minimal disruption. 3 = moderate disruption. See Chapter 2 for further explanation of ratings.

Impact Mechanisms

Climate drivers are expected to cause increasing physical damage to insured property. Demand changes are expected to follow, with more parties seeking insurance against this damage. At the same time, increased uncertainty over natural hazard risks and exposure to greater correlation of losses in some markets are expected to raise costs for insurers. The expected net effect of these shifts in both supply and demand would be an increase in the price of insurance, presenting households with the choice to either increase expenditures on property insurance or reduce their coverage. The market would be ultimately disrupted by a lack of resources for insurers as policyholder pools shrink, risk transfer would become less effective, and insurers might exit.

Cascading Risk

Provide Insurance Services is critically dependent on the following NCFs:

- *Provide Capital Markets and Investment Activities*
- *Provide Funding and Liquidity Services*
- *Distribute Electricity*
- *Provide Internet Routing, Access, and Connection Services*
- *Provide Wireless Access Network Services.*

Insurance and reinsurance providers rely on the banking system and financial markets for the management of their capital, liquidity, and the transmission of premiums and claim payments to and from policyholders. The immediate operation of the insurance industry relies heavily on IT and digital communication. All of this requires electricity. Communication within the industry requires functional internet services, and communication with insured parties and field adjusters often depends heavily on mobile device communications, which require wireless access network services. These dependencies might not pose a major threat of disruption because the insurance industry works on a longer time scale than the duration of most electricity and communication outages. In many cases, insurers can continue to operate by shifting their computing operations to areas that are not affected by electrical outage, and delays in claim processing created by communication outages can be resolved when communication is restored, resulting in full delivery of insurance services—even if they are delayed by some days.

Strength of Evidence

Climate change's effect on insurance markets occurs primarily through physical damage to insured real property. Although the effects that physical damage has for insurance markets have been well explored (Holzheu et al., 2021), exactly how climate change effects will unfold in insurance markets (e.g., issues of liability and premiums) remains uncertain. The strength of evidence on effects to *Provide Insurance Services* are therefore driven by the strength of evidence around the climate drivers themselves. Drawing from Wuebbles et al. (2017), we rated the strength of evidence as high for flooding and sea-level rise and medium for hurricanes and wildfire. Evidence of other climate drivers' potential effects on insurance is less well established.

Table B.213. Strength of Evidence for *Provide Insurance Services*

Strength of Evidence	
High	Flooding; sea-level rise
Medium	Tropical cyclones and hurricanes; wildfire
Low	Severe storm systems; extreme cold; extreme heat; drought

Risk Rating Interpretations

Table B.214 provides examples of how to interpret the risk ratings for this NCF, with a focus on property insurance.

Table B.214. Example Interpretations of Risk Ratings for *Provide Insurance Services*

Risk Rating	NCF Example
1: No disruption or normal operations	*Provide Insurance Services* continues to function as it does today in the property market.
2: Minimal disruption	Residents in a small number of counties or metropolitan statistical areas require new state-subsidized, high-risk pools for property insurance due to increased risk of wildfire or sea-level rise.
3: Moderate disruption	Most residents of a state require new, state-subsidized, high-risk pools for property insurance due to increased risk of wildfire or sea-level rise.
4: Major disruption	Most residents in a large number of metropolitan statistical areas across multiple states require new, state-subsidized, high-risk pools for property insurance due to increased risk of wildfire or sea-level rise.
5: Critical disruption	Unsubsidized property insurance is not available in a majority of states.

Provide Internet Based Content, Information, and Communication Services

Synopsis

This NCF is associated primarily with the platforms and mechanisms for delivering digital content, information, and communication services via the internet. Climate drivers, including tropical cyclones and hurricanes, severe storm systems, flooding, and wildfire, could cause damage to physical infrastructure supporting cloud computing, storage, and other services. Facilities providing such services tend to be geographically concentrated in campuses, and these campuses are vulnerable to damage by climate drivers (Rich Miller, 2015).

Cloud service providers harden data centers against climate drivers and provide services redundantly across geographically diverse sites (Rich Miller, 2015). This NCF is therefore not at risk of disruption from any identified impact mechanisms, even physical damage across a wide geographic footprint.

Among the ecosystem of communication services, content providers, and delivery platforms represented by this NCF, no specific individual or group of entities is so critical to the NCF that the broader function or its subfunctions would be disrupted by outage during a disaster. Alternatives would almost certainly be available and sufficient until service could be restored, even in worst-case scenarios arising from climate-related emergencies.

Cascading Risk

This NCF has significant interdependencies with other NCFs associated with telecommunication networks, especially *Operate Core Network*; *Provide Internet Routing, Access, and Connection Services*; *Provide Positioning, Navigation, and Timing Services*; *Provide Identity Management and Associated Trust Support Services*; *Provide Cable Access Network Services*; *Provide Radio Broadcast Access Network Services*; *Provide Satellite Access Network Services*; *Provide Wireless Access Network Services*; and *Provide*

Wireline Access Network Services. Internet-based services cannot be provided without functioning networks, and these networks often cannot function without monitoring, control, and other services provided by this NCF. This NCF also has significant interdependencies with *Generate Electricity, Transmit Electricity,* and *Distribute Electricity.* Internet-based services cannot be provided without electrical power, and electricity providers rely on internet-based services to function.

Strength of Evidence

The *Provide Internet Based Content, Information, and Communication Services* NCF does not currently experience any meaningful effect from climate drivers. Through an initial general search of the literature using keywords related to networks, internet-based services, and climate drivers, we identified few or no relevant documents. Strength of evidence is low for assessment because hypothetical effects from climate drivers are inferred primarily from the cited literature on campuses providing cloud services (Rich Miller, 2015), as well as our judgment and subject-matter expertise.

Table B.216. Strength of Evidence for *Provide Internet Based Content, Information, and Communication Services*

	Strength of Evidence
	Low Severe storm systems; extreme cold; extreme heat; drought; tropical cyclones and hurricanes; wildfire; flooding; sea-level rise

Risk Rating Interpretations

Table B.217 provides examples of how to interpret the risk ratings for this NCF.

Table B.217. Example Interpretations of Risk Ratings for *Provide Internet Based Content, Information, and Communication Services*

Risk Rating	NCF Example
1: No disruption or normal operations	Internet-based content, information, and communication services continue to function as needed across the country.
2: Minimal disruption	Physical damage to key data centers, cloud campuses, and service providers occasionally cause disruption to services for some individuals or small organizations, but disruptions are confined to the local context.
3: Moderate disruption	Regions are unable to access key internet-based services for an extended time period.
4: Major disruption	Nationwide degradation in internet-based services infrequently occurs. Service is eventually restored, but the unreliability of key services and functions degrades their utility over time.
5: Critical disruption	Organizations and individuals across the United States abandon internet-based services because providers are unable to reliably maintain the integrity and availability of key data and services.

Provide Identity Management and Associated Trust Support Services

Table B.218. Scorecard for *Provide Identity Management and Associated Trust Support Services*

	National Risk Assessment				
	Baseline	**Current Emissions**		**High Emissions**	
	Present	**2050**	**2100**	**2050**	**2100**
Provide Identity Management and Associated Trust Support Services	1	1	1	1	1

Description: "Produce and provide technologies, services, and infrastructure to ensure the identity of, authenticate, and authorize entities and ensure confidentiality, integrity, and availability of devices, services, data, and transactions" (CISA, 2020, p. 5).

Highlights

- This NCF was assessed as being at low risk of disruption because subfunctions that could be disrupted by physical damage to infrastructure are redundantly provided, geographically diverse, and hardened against climate drivers.
- Tropical cyclones and hurricanes, severe storm systems, flooding, and wildfire could cause physical damage that could affect this NCF, though the NCF is likely to be resilient to them.

Synopsis

This NCF is primarily concerned with the centralized maintenance, storage, integrity, and provision of data and digital artifacts associated with identity and access management. The NCF is not at risk of disruption from any identified impact mechanisms, even physical damage across a wide geographic footprint. The only plausible effect would arise from physical destruction of the equipment and facilities that act as repositories for data, but such repositories can and typically do have redundant, off-site storage of key data. Any disruptions to this NCF are likeliest to derive solely from disrupted access to networked communication systems, which is addressed in other NCF assessments. No effect on this NCF is expected from any climate driver.

Cascading Risk

This NCF has significant interdependencies with other NCFs associated with telecommunication networks, especially *Operate Core Network*; *Provide Internet Based Content, Information, and Communication Services*; *Provide Internet Routing, Access, and Connection Services*; *Provide Positioning, Navigation, and Timing Services*; *Provide Cable Access Network Services*; *Provide Radio Broadcast Access Network Services*; *Provide Satellite Access Network Services*; *Provide Wireless Access Network Services*; and *Provide Wireline Access Network Services*. Identity management and trust support services cannot be provided without functioning networks, and these networks and services often rely on near-real-time identity management and trust support in their operations. This NCF also has significant upstream dependencies on *Generate Electricity*, *Transmit Electricity*, and *Distribute Electricity*.

Strength of Evidence

This NCF does not currently experience any meaningful effect from climate drivers. Strength of evidence is low for assessment because hypothetical effects from climate drivers are inferred primarily from subject-matter expertise.

Table B.219. Strength of Evidence for *Provide Identity Management and Associated Trust Support Services*

	Strength of Evidence
Low	Severe storm systems; extreme cold; extreme heat; drought; tropical cyclones and hurricanes; wildfire; flooding; sea-level rise

Risk Rating Interpretations

Table B.220 provides examples of how to interpret the risk ratings for this NCF.

Table B.220. Example Interpretations of Risk Ratings for *Provide Identity Management and Associated Trust Support Services*

Risk Rating	NCF Example
1: No disruption or normal operations	Identity management and associated trust support services continue to function as needed across the country.
2: Minimal disruption	Physical damage to key data repositories and supporting infrastructure occasionally cause disruption to services for some individuals or small organizations, but disruptions are confined to the local context.
3: Moderate disruption	Regions are unable to access key identity management services for an extended time period.
4: Major disruption	Nationwide degradation in identity management services infrequently occurs. Service is eventually restored, but the unreliability of key services and functions degrades their utility over time.
5: Critical disruption	Providers are unable to reliably maintain the integrity and availability of key identity-related data and trust support services, and services across the United States experience prolonged disruptions.

Provide Internet Routing, Access, and Connection Services

Table B.221. Scorecard for *Provide Internet Routing, Access, and Connection Services*

	National Risk Assessment				
	Baseline	Current Emissions		High Emissions	
Provide Internet Routing, Access, and Connection Services	Present	2050	2100	2050	2100
	1	1	2	1	2

Description: "Provide and operate exchange and routing infrastructure, points of presence, peering points, local access services, and capabilities that enable end users to send and receive information via the Internet" (CISA, 2020, p. 3).

Highlights

- This NCF was assessed as being at risk of minimal disruption by 2100 because climate drivers cause more-frequent local disruptions to internet routing.
- Tropical cyclones and hurricanes and wildfire can cause physical damage to facilities that route internet traffic. If these facilities are critical nodes without redundancy for the local area, routing can be disrupted.
- Internet routing services are global and highly redundant, so they are resilient to broad disruption from physical damage caused by climate drivers.

Synopsis

Internet routing infrastructure is global in nature, and it is resilient because of the many redundant physical and logical links between networked automated systems involved in routing (Hinchman, 2022). In principle, disruption to even a large number of facilities, links, or systems across a region should not appreciably affect the overall operation of routing infrastructure nationally, thanks to route redundancy. In practice, however, regional disruptions precipitated by climate drivers can degrade performance more broadly. Climate drivers or natural hazards can disrupt critical access links that are traversed by all possible routing paths from affected systems to the rest of the internet (Wu et al., 2007). This can degrade or disrupt routing to large numbers of addresses, even if local routing paths remain intact. It can also cause significant network congestion outside the affected area due to traffic being rerouted around broken links. Global network congestion can be addressed rapidly (i.e., often within a matter of hours) using existing mitigations and network controls, and progressively more local disruptions can be addressed over time as connections are repaired or new connections are made (Erjongmanee et al., 2010; Palmieri et al., 2013).

Climate drivers, especially those with a wide geographic footprint, such as hurricanes and wildfire, can degrade this NCF by causing widespread physical damage to routing infrastructure. These climate drivers can damage or destroy the facilities that function as points of presence, key network access points, and internet exchange points that facilitate routing functions. If this damage causes key routes and addresses to become unreachable, it can cause a disruption of routing to the affected region and noticeable routing degradation elsewhere (Palmieri et al., 2013).

To assess how climate-related risk to this NCF might change in the future, we first state core assumptions about how the NCF will operate through 2100. We assumed the following:

- Networked communication systems will consist of regional subnets connected by core network infrastructure.
- Networked communication systems will still use both wired and wireless (electromagnetic spectrum) communication media.

The NCF does not currently experience meaningful disruptions from climate drivers, but disruptions will become more significant and more common over time because of climate change. Wildfire and hurricanes in particular are likely to increase in severity and frequency over time in both current and high emissions scenarios. These climate drivers could begin to cause meaningful regional disruptions to this NCF by 2100. Sea-level rise might eventually require operators to relocate coastal landings for undersea cables, but it is unlikely that this will lead to any meaningful disruption to the NCF.

Key Considerations for *Provide Internet Routing, Access, and Connection Services*

In assessing this NCF, we made two core assumptions about how the NCF will operate through 2100. We assumed the following for this time period:

1. Networked communication systems will continue to consist of regional subnets redundantly connected by core network infrastructure.
2. Networked communication systems will still use both wired and wireless (electromagnetic spectrum) communication media.

Subfunctions

Three subfunctions could be affected: *Provide Access to Internet Services*, *Provide Routing Between Internet Service Providers and Customers*, and *Operate Internet Route Zone*. These three subfunctions are associated with physical infrastructure, including point-of-presence sites, key network access points, and internet exchange points, that climate drivers could damage. Although physical damage to infrastructure from climate drivers affects these subfunctions in the current emissions scenario, this damage has a meaningful effect (i.e., prolonged, complete disruption of routing) to the NCF at only the local scale. Regional effects are typically mitigated within a very short time period using existing tools and controls (Palmieri et al., 2013). We assessed that this is likely to remain the case through 2050 but that more-frequent and -intense disasters through 2100 could begin to cause more-significant regional disruptions.

Table B.222. Subfunction Risk for *Provide Internet Routing, Access, and Connection Services*

Subfunction	Baseline Present	Current Emissions 2050	Current Emissions 2100	High Emissions 2050	High Emissions 2100
Provide Access to Internet Services	1	1	2	1	2
Provide Routing Between Internet Service Providers and Customers	1	1	2	1	2
Provide Internet Hosting Services	1	1	1	1	1
Operate Internet Route Zone	1	1	2	1	2
Provide Logistics and Operational Support to the Network	1	1	1	1	1
Provide Regulations, Standards, and Licensing	1	1	1	1	1

NOTE: 1 = no disruption or normal operations. 2 = minimal disruption. See Chapter 2 for further explanation of ratings.

Climate Drivers

Because of the redundancy in the relevant routes and systems, only climate drivers that cause physical damage across a wide geographic footprint are likely to affect this NCF. Tropical cyclones and hurricanes and wildfire are therefore the only two climate drivers that are likely to be relevant. These events can cause significant damage to key facilities and equipment across a wide area, potentially disrupting key routing facilities and all available redundancies across a large region. This could therefore disrupt routing beyond damaged hyperlocal access points and affect reachability of the entire local or regional subnetwork.

Table B.223. Risk from Climate Drivers for *Provide Internet Routing, Access, and Connection Services*

Climate Driver	Baseline Present	Current Emissions 2050	Current Emissions 2100	High Emissions 2050	High Emissions 2100
Drought	1	1	1	1	1
Extreme cold	1	1	1	1	1
Extreme heat	1	1	1	1	1
Flooding	1	1	1	1	2
Sea-level rise	1	1	1	1	1
Severe storm systems	1	1	1	1	1
Tropical cyclones and hurricanes	1	1	2	1	2
Wildfire	1	1	2	1	2

NOTE: 1 = no disruption or normal operations. 2 = minimal disruption. See Chapter 2 for further explanation of ratings.

Impact Mechanisms

Physical damage to key facilities and equipment, especially points of presence, key network access points, and internet exchange points, is the only relevant impact mechanism for this NCF. Damage to such a facility can disrupt routing, and this will lead to disruptions to some end users if the facility was a critical node in the routing network for those users.

Cascading Risk

This NCF has a significant upstream dependence on *Generate Electricity*, *Transmit Electricity*, and *Distribute Electricity*. Physical damage leading to loss of electricity is often the primary cause of disruptions to network functionality (Erjongmanee and Ji, 2011).

Strength of Evidence

Strong historical evidence exists for the disruption of internet routing by geographically widespread disasters, such as hurricanes (Erjongmanee and Ji, 2011). No literature was found specifically examining wildfire's effect on routing performance, but wildfire similarly has the potential to cause physical damage to key infrastructure across wide footprints. We found no literature for the effect of other climate drivers, but this is likely because they have no current effect on core network functionality and there is little concern that they could cause meaningful disruptions.

Table B.224. Strength of Evidence for *Provide Internet Routing, Access, and Connection Services*

Strength of Evidence	
High	Tropical cyclones and hurricanes
Low	Flooding; severe storm systems; extreme cold; extreme heat; drought; sea-level rise; wildfire

Risk Rating Interpretations

Table B.225 provides examples of how to interpret the risk ratings for this NCF.

Table B.225. Example Interpretations of Risk Ratings for *Provide Internet Routing, Access, and Connection Services*

Risk Rating	NCF Example
1: No disruption or normal operations	Networked communications continue to function as expected, and any effects from climate drivers are mitigated by built-in redundancies.
2: Minimal disruption	Climate drivers cause regions to experience prolonged regional disruptions to network reachability until physical infrastructure can be repaired, but the network disruption does not extend beyond the affected region.
3: Moderate disruption	Some regions experience prolonged unreachability. Frequent natural disasters cause enough prolonged disruption to available routes to noticeably degrade global routing performance.
4: Major disruption	The frequency of climate-driven disasters overcomes routing redundancies and outpaces the capacity to recover and repair physical infrastructure, leading to time periods in which even regions that were not directly affected by the disaster experience brief but significant network congestion or inability to reach critical portions of the internet.
5: Critical disruption	The frequency and severity of climate-driven disasters make it infeasible to maintain the physical infrastructure needed to reliably route global internet traffic to most of the country.

Provide Information Technology Products and Services

Table B.226. Scorecard for *Provide Information Technology Products and Services*

	National Risk Assessment				
	Baseline	**Current Emissions**		**High Emissions**	
Provide Information Technology Products and Services	**Present**	**2050**	**2100**	**2050**	**2100**
	3	4	5	4	5

Description: "Design, develop, and distribute hardware and software products and services (including security and support services) necessary to maintain or reconstitute networks and associated services" (CISA, 2020, p. 6).

Highlights

- This NCF was assessed as being at risk of major and critical disruption in the 2050 and 2100 time frames, respectively, for both emissions scenarios.
- We expect stronger or more-frequent tropical cyclones and hurricanes, severe storm systems, wildfire, drought, and flooding, and these climate drivers will cause physical damage that disrupts this NCF.
- Geographic concentration of hardware manufacturing makes this NCF highly vulnerable because climate drivers can cause significant damage to facilities in one region that leads to global shortages in critical components.
- This NCF's vulnerability is associated primarily with the subfunction *Provide, Support, and Repair Hardware*.
- The vulnerability of this NCF is location-agnostic because geographic concentration of manufacturing can and does occur in many different regions across the world, which leaves this NCF potentially vulnerable to many climate drivers.

Synopsis

This NCF is associated with the development and provision of a wide variety of hardware and software products. Although software development and provision are geographically distributed across global and local contexts, hardware manufacturing is often geographically concentrated and thus vulnerable to natural hazards affected by climate change (Romanosky et al., 2022; Varas et al., 2021). If a climate driver caused major disruption within a region that is the site of the majority of manufacturing facilities for a given hardware product (e.g., microchips), it could globally degrade or disrupt this function. Semiconductor manufacturing serves as a notable example. Taiwan is essentially a single point of failure for semiconductor supply because it provides 92 percent of global advanced semiconductor manufacturing capacity (Varas et al., 2021). Taiwan also regularly experiences severe typhoons, such as Typhoon Morakot, which made landfall in 2009 and caused massive flooding, leading to significant loss of life and economic damage ("Billions Allocated for Reconstruction in Wake of Typhoon Morakot," 2009; Li et al., 2014; Shieh et al., 2010). This risk is not limited to international hardware manufacturing; a severe winter storm in Texas in 2021 damaged fabrication facilities, disrupting production and causing cascading disruptions in hardware manufacturing (Dahad, 2021; Kleinhans and Hess, 2021).

More-frequent or -intense natural hazards affected by climate change, such as drought, flooding, severe storm systems, tropical cyclones and hurricanes, and sea-level rise, could cause prolonged disruption to key manufacturing regions and severe shortages of key hardware products worldwide (Fusion Worldwide, 2021; Kleinhans and Hess, 2021). Climate drivers could disrupt the NCF by causing widespread physical damage to geographically concentrated manufacturing facilities for hardware products that could suspend the operation of these facilities for extended time periods. Although the actual physical damage would be local or regional, it would cause global disruption of hardware supply because of the geographic concentration of specific component manufacturing.

The geographic concentration of hardware manufacturing could change over time in response to market dynamics and efforts to build resilience, but we found conflicting assessments in literature as to whether the underlying causes of global hardware manufacturing disruptions could be moderated enough to provide meaningful resilience to climate drivers. Although there appears to be recognition from the U.S. federal government (see, for example, White House, 2022) and the semiconductor industry (Fusion Worldwide, 2021) that greater geographic diversification of hardware manufacturing is needed to build resilience, some of the factors that drive geographic concentration might be fundamental characteristics of hardware manufacturing that are difficult or impossible to change (Kleinhans and Hess, 2021; Mondschein, Welburn, and Gonzales, 2022). Therefore, it is not clear whether efforts toward geographic diversification of manufacturing will be realized or whether they would be enough to provide resilience to increasingly frequent or intense natural hazards affected by climate change. Market dynamics and world events will lead to unpredictable changes in global manufacturing concentration over time, but, for the purposes of this assessment, we assumed that regional concentrations for manufacturing of specific hardware products will continue through 2100.

Because geographic concentrations of manufacturing capacity can occur in many different regions and climates, nearly all the climate drivers have the potential to disrupt this NCF. Therefore, future disruptions should be considered location-agnostic. Drought, tropical cyclones and hurricanes, severe storm systems, and flooding have all caused significant recent disruptions (Fusion Worldwide, 2021; Kleinhans and Hess, 2021), but sea-level rise and wildfire are also plausible causes of future disruptions. Significant disruptions from these drivers can be expected through 2050, and these will intensify through 2100.

Key Considerations for *Provide Information Technology Products and Services*

The risk ratings for this NCF assume that factors intrinsic to critical hardware manufacturing incentivize geographic concentration of manufacturing facilities and that those geographic concentrations will continue to occur in the future, notwithstanding current efforts to address this issue. The NCF also exhibits significant dependencies on *Supply Water*, *Generate Electricity*, *Transmit Electricity*, and *Distribute Electricity*. Sustained loss of electricity has historically been the dominant cause of disruption for this NCF. It also exhibits a strong interdependence with *Maintain Supply Chains*.

Subfunctions

The subfunction *Provide, Support, and Repair Hardware* is likely to be the key driver of risk. The geographic concentration of hardware manufacturing capacity creates critical nodes that can be disrupted by physical damage from climate drivers. Software supply is distributed across global, regional, and local contexts and is less dependent on physical infrastructure (except through interdependencies, such as with *Distribute Electricity*) and is therefore less prone to significant disruption from climate drivers.

Table B.227. Subfunction Risk for *Provide Information Technology Products and Services*

	Baseline	Current Emissions		High Emissions	
Subfunction	Present	2050	2100	2050	2100
Provide, Support, and Repair Hardware	3	4	5	4	5
Provide, Support, and Repair Software and Services	1	1	1	1	1

NOTE: 1 = no disruption or normal operations. 3 = moderate disruption. 4 = major disruption. 5 = critical disruption. See Chapter 2 for further explanation of ratings.

Climate Drivers

Drought, flooding, sea-level rise, severe storm systems, tropical cyclones and hurricanes, and wildfire are all potential drivers of risk. Drought, flooding, severe storm systems, and tropical cyclones and hurricanes are all current causes of disruption. Drought disrupts manufacturing both by limiting water supply to facilities and via dependencies, such as electricity supply from hydroelectric plants (Kleinhans and Hess, 2021). The other climate drivers cause physical damage to facilities that degrades or disrupts their operation.

Table B.228. Risk from Climate Drivers for *Provide Information Technology Products and Services*

	Baseline	Current Emissions		High Emissions	
Climate Driver	Present	2050	2100	2050	2100
Drought	3	3	4	3	4
Extreme cold	1	1	1	1	1
Extreme heat	1	1	1	1	1
Flooding	3	3	4	4	5
Sea-level rise	1	3	4	3	4
Severe storm systems	3	3	3	3	3
Tropical cyclones and hurricanes	3	4	5	4	5
Wildfire	1	3	3	3	3

NOTE: 1 = no disruption or normal operations. 3 = moderate disruption. 4 = major disruption. 5 = critical disruption. See Chapter 2 for further explanation of ratings.

Impact Mechanisms

The primary impact mechanism is physical damage to manufacturing facilities and supporting infrastructure. We exclude assessment of *Supply Water*, *Generate Electricity*, *Transmit Electricity*, and

Distribute Electricity and broader *Maintain Supply Chains* disruptions of critical components and materials, although these are likely to be significant overlapping issues for this NCF.

Cascading Risk

This NCF has key upstream dependencies on *Generate Electricity*, *Transmit Electricity*, *Distribute Electricity*, and *Supply Water*. Loss of electrical power is typically cited as the primary cause of disruption for the NCF in the aftermath of natural hazards affected by climate change. Severe droughts leading to disruption of the *Supply Water* NCF can also affect the NCF both by disrupting the supply of water to manufacturing facilities and by a further upstream effect on electricity generation via hydroelectric power plants.

The NCF also has a close interdependence with *Maintain Supply Chains*. Manufacturing of advanced IT hardware depends to a significant degree on the ready availability of many other critical hardware components and products. Disruption of the manufacture of a single component (e.g., advanced semiconductors) can disrupt the manufacture of many other IT products that require that component. A disruption to the manufacture of that component can therefore be considered a disruption to the supply chain for a wide variety of downstream products and components.

Strength of Evidence

The strength of evidence assessing climate drivers' effects on the *Provide Information Technology Products and Services* NCF was assessed as medium. Substantial historical documentation exists explaining how drought, tropical cyclones and hurricanes, flooding, and severe storm systems disrupt this NCF. Evidence to provide insight on the effect that other climate drivers, such as sea-level rise and wildfire, can have on this NCF is limited. We inferred the effects of sea-level rise based on expected exacerbation of the effects of tropical cyclones and hurricanes, and we inferred the effects of wildfire based on the plausibility of critical manufacturing nodes located in expanding wildfire risk zones.

Table B.229. Strength of Evidence for *Provide Information Technology Products and Services*

Strength of Evidence	
High	Drought; tropical cyclones and hurricanes; flooding; severe storm systems
Low	Extreme cold; extreme heat; wildfire; sea-level rise

Risk Rating Interpretations

Table B.230 provides examples of how to interpret the risk ratings for this NCF.

Table B.230. Example Interpretations of Risk Ratings for *Provide Information Technology Products and Services*

Risk Rating	NCF Example
1: No disruption or normal operations	Supply of hardware and software products can meet all routine needs.
2: Minimal disruption	Some areas could experience price shocks on certain hardware components or other challenges, but they are able to acquire products as needed.
3: Moderate disruption	Some regions are unable to acquire certain hardware products needed to meet their routine operational needs, and extended backlogs routinely form.
4: Major disruption	Extended backlogs on key hardware products routinely exist, nationwide, and many other domestic functions and industries could become nonviable as a result.
5: Critical disruption	The country cannot acquire adequate supply in multiple key hardware product categories, with cascading devastating effects on nearly all domestic industries and functions that are highly dependent on IT.

Provide Positioning, Navigation, and Timing Services

Table B.231. Scorecard for *Provide Positioning, Navigation, and Timing Services*

	National Risk Assessment				
	Baseline	Current Emissions		High Emissions	
Provide Positioning, Navigation, and Timing Services	Present	2050	2100	2050	2100
	1	1	1	1	1

Description: "Operate and maintain public and private capabilities which enable users to determine location, orientation and time" (CISA, 2020, p. 3).

Highlights

- This NCF was assessed as being at low risk of disruption.
- Some climate drivers, such as tropical cyclones and hurricanes, severe storm systems, flooding, and wildfire, could cause significant physical damage to facilities hosting key centralized PNT resources, but these facilities are resilient to disruption from natural disasters and supported by geographically remote, redundant facilities.

Synopsis

Climate drivers are unlikely to have a significant effect on this NCF. A few centralized U.S. facilities provide key PNT services. These facilities include National Institute of Standards and Technology sites in Colorado and Maryland that provide centralized timing resources and multiple DoD ground stations that facilitate control over the satellite-based Global Positioning System (GPS). However, these resources are resilient to climate drivers thanks to geographic dispersion of key facilities and multiple on-site redundancies and fail-safes (Mason et al., 2021; Sherman et al., 2021). GPS is also a dual-use technology for civilian and military purposes, and, even if some unforeseen set of circumstances allowed climate drivers to disrupt the function, ground infrastructure would likely be rapidly reconstituted. Finally, these centralized resources are also merely important parts of an ecosystem of PNT resources that could serve as alternatives in the event of a single failure (Mason et al., 2021).

Climate drivers could cause significant physical damage to key PNT facilities, but the overall redundancy of key systems and the PNT product ecosystem would likely remain resilient to even severe damage at multiple locations. This is likely to remain the case even in future scenarios in which relevant natural hazards affected by climate change, such as severe storm systems, flooding, wildfire, and tropical cyclones and hurricanes, become much more frequent or intense.

Cascading Risk

This NCF has significant interdependencies with other NCFs associated with telecommunication networks, especially *Operate Core Network*; *Provide Internet Based Content, Information, and Communication Services*; *Provide Internet Routing, Access, and Connection Services*; *Provide Cable Access Network Services*; *Provide Radio Broadcast Access Network Services*; *Provide Satellite Access Network Services*; *Provide Wireless Access Network Services*; and *Provide Wireline Access Network Services*. The

NCF cannot provide significant PNT services to some users without functioning telecommunication network infrastructure, and these other NCFs also rely heavily on PNT resources for their operations. This NCF also has significant upstream dependencies on *Generate Electricity, Transmit Electricity,* and *Distribute Electricity.*

Strength of Evidence

This NCF does not currently experience any meaningful effect from climate drivers. Strength of evidence is low for assessment because hypothetical effects from climate drivers are primarily inferred based on literature on PNT resilience and satellite infrastructure and on subject-matter expertise.

Table B.232. Strength of Evidence for *Provide Positioning, Navigation, and Timing Services*

Strength of Evidence	
Low	Severe storm systems; extreme cold; extreme heat; drought; sea-level rise; wildfire; flooding; tropical cyclones and hurricanes

Risk Rating Interpretations

Table B.233 provides examples of how to interpret the risk ratings for this NCF.

Table B.233. Example Interpretations of Risk Ratings for *Provide Positioning, Navigation, and Timing Services*

Risk Rating	NCF Example
1: No disruption or normal operations	Fully functional PNT services are available to all users.
2: Minimal disruption	PNT services are available to all users, although some users experience degraded or intermittent service in some locations.
3: Moderate disruption	Regions experience occasional inability to access centralized PNT services, and alternative PNT services are unavailable in these regions.
4: Major disruption	PNT services with national coverage, such as GPS, experience frequent disruptions, and they are inadequate for users requiring reliable, accurate PNT resources.
5: Critical disruption	No national PNT services are reliably available. Alternative local or regional PNT resources are unreliable or otherwise inadequate to meet users' PNT needs.

Provide Radio Broadcast Access Network Services

Table B.234. Scorecard for *Provide Radio Broadcast Access Network Services*

	National Risk Assessment				
	Baseline	Current Emissions		High Emissions	
	Present	2050	2100	2050	2100
Provide Radio Broadcast Access Network Services	2	2	2	2	2

Description: "Operate over-the-air radio and television (TV) stations (operating at medium, very high, and ultra-high frequencies) that offer analog and digital audio and video programming services and data service" (CISA, 2020, p. 3).

Highlights

- This NCF was assessed as being at risk of minimal disruption from climate drivers.
- We expect stronger or more-frequent flooding, wildfire, severe storm systems, tropical cyclones and hurricanes, and sea-level rise, and these climate drivers will disrupt the NCF by damaging physical infrastructure over a wide area.
- The primary subfunction affected is *Provide Local Broadcast Services*. This NCF is primarily mitigated by local offices acting as broadcast stations, and damage to these facilities, their wired connections to the local area, and their wireless transmission equipment disrupts this function.

Synopsis

Flooding, severe storm systems, tropical cyclones and hurricanes, sea-level rise, and wildfire can all destroy physical infrastructure needed to provide radio access network services. This physical destruction can cause regional degradations in network performance and even complete disruption of services locally.

Destruction of physical infrastructure, especially local offices, transmitters, and equipment that provide communication links with other local stations and a national network, would disrupt network functionality in the local area (Kwasinski et al., 2009). Although radio broadcast services are vulnerable to being disrupted by damage to local broadcast offices, they are less vulnerable than some other functions that rely on wired connections from offices to end users. Wireless broadcast often allows radio broadcast services to remain functional in the aftermath of disasters even where many other broadcast types remain disrupted (Hugelius, Adams, and Romo-Murphy, 2019; Moody, 2009).

Flooding, tropical cyclones and hurricanes, sea-level rise, severe storm systems, and wildfire are all capable of causing physical damage to the NCF's infrastructure. This NCF currently experiences local disruptions in the aftermath of these climate drivers, and we assessed the NCF as being at risk of minimal disruption in present conditions. All these climate drivers are expected to exhibit varying increases in intensity or frequency over time. This might cause more-frequent or -prolonged local disruptions, but we assessed that the disruptions to the NCF will continue to primarily be local, not regional. Many local offices often wirelessly provide overlapping access services across regions, and nearly every station in a region would need to experience a simultaneous, prolonged disruption for a truly regional disruption to this NCF. As a result, we assessed that the risk to the NCF will be the same in future scenarios.

Subfunctions

Only one subfunction of this NCF is likely to be at risk: *Provide Local Broadcast Services*. The risk to this subfunction is primarily due to the potential for climate-driven disasters to damage or destroy local stations, cables, and equipment required to produce, receive, and transmit content to the surrounding area.

Table B.235. Subfunction Risk for *Provide Radio Broadcast Access Network Services*

	Baseline	Current Emissions		High Emissions	
Subfunction	Present	2050	2100	2050	2100
Supply End User Devices	1	1	1	1	1
Provide Transmission to End Users	1	1	1	1	1
Provide Local Broadcast Services	2	2	2	2	2
Provide National Station Broadcast Services	1	1	1	1	1
Provide Regulations, Standards, and Licensing	1	1	1	1	1

NOTE: 1 = no disruption or normal operations. 2 = minimal disruption. See Chapter 2 for further explanation of ratings.

Climate Drivers

Climate drivers that can cause physical damage to local broadcast offices, cables, and transmission equipment are likely to have an effect on this NCF. These climate drivers include tropical cyclones and hurricanes, severe storm systems, and wildfire. The NCF is experiencing local disruptions from these climate drivers in the baseline scenario (see Kuligowski et al., 2014, and Moody, 2009), and the drivers are likely to intensify in the future because of climate change.

Table B.236. Risk from Climate Drivers for *Provide Radio Broadcast Access Network Services*

	Baseline	Current Emissions		High Emissions	
Climate Driver	Present	2050	2100	2050	2100
Drought	1	1	1	1	1
Extreme cold	1	1	1	1	1
Extreme heat	1	1	1	1	1
Flooding	2	2	2	2	2
Sea-level rise	1	1	2	1	2
Severe storm systems	2	2	2	2	2
Tropical cyclones and hurricanes	2	2	2	2	2
Wildfire	2	2	2	2	2

NOTE: 1 = no disruption or normal operations. 2 = minimal disruption. See Chapter 2 for further explanation of ratings.

Impact Mechanisms

Physical damage that renders local stations, cables, and transmission equipment nonfunctional is likely to be the primary impact mechanism affecting this NCF for the foreseeable future.

Cascading Risk

This NCF has a significant upstream dependence on *Generate Electricity*, *Transmit Electricity*, and *Distribute Electricity*. The degree to which physical damage to network infrastructure, as opposed to electrical power failures, is likely to be the root cause of network failure in disaster scenarios is unclear. In an examination of root causes of network disruptions caused by Hurricane Ike in 2008, for example, researchers found that all identifiable root causes of disruption linked to inadequate electrical power, rather than physical damage (Erjongmanee and Ji, 2011). Hurricane Katrina in 2005 caused significant disruptions to telecommunication networks that were caused both by electrical power outages and physical destruction of physical infrastructure (although electrical outages were still the main cause) (Kwasinski et al., 2009). Although we consider physical damage to be the main climate-driven impact mechanism for this NCF, it is likely overshadowed by the effects of electrical power failures, which is a key dependence for this NCF.

Strength of Evidence

Strength of evidence is high for tropical cyclones and hurricanes. Previous events, such as Hurricane Katrina, provide evidence of how these climate drivers affect this NCF (Kwasinski et al., 2009; Moody, 2009). Strength of evidence is low for the remaining climate drivers. We inferred similar effects for this NCF from wildfire, severe storm systems, and flooding based on evidence of effects on other NCFs that depend on local broadcast stations (S. Anderson, Barford, and Barford, 2020; Erjongmanee and Ji, 2011; Kuligowski et al., 2014).

Table B.237. Strength of Evidence for *Provide Radio Broadcast Access Network Services*

	Strength of Evidence
High	Tropical cyclones and hurricanes
Low	Sea-level rise; severe storm systems; extreme cold; extreme heat; drought; wildfire; flooding

Risk Rating Interpretations

Table B.238 provides examples of how to interpret the risk ratings for this NCF.

Table B.238. Example Interpretations of Risk Ratings for *Provide Radio Broadcast Access Network Services*

Risk Rating	NCF Example
1: No disruption or normal operations	Radio broadcast services continue to function as expected, and any effects from climate drivers are mitigated by built-in redundancies.
2: Minimal disruption	Climate drivers cause regions to experience prolonged local disruption (i.e., the area reached by the broadcast of a small number of stations) to radio broadcast networks until physical infrastructure can be repaired.
3: Moderate disruption	Radio broadcast is disabled over a large region by significant physical damage to multiple local broadcast stations.
4: Major disruption	Most local broadcast stations are unable to consistently operate, leaving most of the country without access to radio broadcast services.
5: Critical disruption	Nearly all local stations across the country cease to continuously operate, leaving the majority of the country without access to radio broadcast services.

Provide Satellite Access Network Services

Table B.239. Scorecard for *Provide Satellite Access Network Services*

	National Risk Assessment				
	Baseline	Current Emissions		High Emissions	
	Present	2050	2100	2050	2100
Provide Satellite Access Network Services	1	1	1	1	1

Description: "Provide access to core communications network via a combination of terrestrial antenna stations and platforms orbiting Earth to relay voice, video, or data signals" (CISA, 2020, p. 3).

Highlights

- This NCF was assessed as being at very low risk of disruption from climate drivers.
- Climate drivers could cause damage to satellite gateways or user receivers, weather-related delays in space launch, and disruptions to critical satellite supply chains, but none of these is likely to involve meaningful risk of disruption for this NCF.

Synopsis

Satellite network communication is often touted as a solution to communication networks' vulnerability to natural hazards (J. Goldberg and Dalessio, 2022; Zhou et al., 2021), and climate drivers are unlikely to have any meaningful effect on satellite access network services. Satellites in earth orbit are not at risk from climate drivers, but satellite access networks make use of terrestrial infrastructure that could be affected. We identified four potential ways in which climate drivers could disrupt this NCF: (1) physical damage to satellite gateways and their wired network connections, (2) physical damage to receivers on a user's premises, (3) weather-related disruption of satellite launch services, and (4) disruption of satellite supply chains. We concluded that none of these four potential disruption mechanisms is likely to have meaningful effect primarily attributable to climate drivers.

The greatest of these risks is physical damage to satellite gateways, ground station facilities, and other physical infrastructure associated with the terrestrial network. Natural hazards could damage these facilities and disrupt satellite network access services for users. However, satellite gateways are typically constructed with geographic redundancy in mind and in places that are less prone to natural hazards (Hill, 2019). Unlike other communication infrastructure that is constrained to locations near the population it serves, providers have more freedom to construct gateways in locations that will avoid major natural hazards, such as wildfire, hurricanes, and storm surges from sea-level rise. As a result, it is likely that this physical infrastructure will be resilient to effects of climate drivers. Gateways are often constructed in southern latitudes for operational reasons, and it is possible that they would see some greater effects from extreme heat, depending on the specific location, but it is not clear whether this will have any meaningful effect in any future scenario. We also considered disruption of space launch services and satellite supply chains. We assessed that, for both of these, significant commercial adaptations are likely to occur in response to market dynamics (including effects of climate drivers) and that adaptability introduces too much uncertainty to make more-useful predictions of the effects from climate drivers. For example, if organizations anticipate frequent hurricanes or sea-level rise creating unacceptable delays in satellite launch services, they are likely to have ample lead time to invest in alternative launch facilities that are less prone to these disruptions.

Hurricanes and sea-level rise currently cause delays in space launch capabilities (e.g., see National Aeronautics and Space Administration, 2022), although such events do not have any meaningful effect on satellite access network services. More-frequent and -intense hurricanes in future scenarios are similarly unlikely to have meaningful effects, and any potential effect would likely be avoided by organizational decisions in response to market dynamics. None of the climate drivers is expected to meaningfully affect this NCF in any future scenario.

Cascading Risk

This NCF has a significant upstream dependence on *Generate Electricity*, *Transmit Electricity*, and *Distribute Electricity*. Although we considered physical damage to be the main climate-driven impact mechanism for this NCF, electrical power failures are likely to be the root cause of any network failure in disaster scenarios.

Strength of Evidence

Because this NCF does not currently experience any meaningful effect from climate drivers, we used largely our judgment and subject-matter expertise to identify plausible climate effects and infer whether they will create meaningful effects in future scenarios. As a result, the strength of evidence is characterized as low.

Table B.240. Strength of Evidence for *Provide Satellite Access Network Services*

Strength of Evidence	
Low	Severe storm systems; extreme cold; extreme heat; drought; wildfire; flooding; tropical cyclones and hurricanes; sea-level rise

Risk Rating Interpretations

Table B.241 provides examples of how to interpret the risk ratings for this NCF.

Table B.241. An Example Interpretation of Risk Ratings for *Provide Satellite Access Network Services*

Risk Rating	NCF Example
1: No disruption or normal operations	Satellite communications continue to function as expected, and any effects from climate drivers are mitigated by built-in redundancies.
2: Minimal disruption	Some locations have intermittent or disrupted satellite network coverage due to satellite gateway outages and challenges maintaining satellite constellations.
3: Moderate disruption	Damage to satellite gateways causes disruption to satellite network access over a large region until the gateway's function is restored.
4: Major disruption	Damage to satellite gateways and persistent disruption to satellite supply cause most regions of the country to have intermittent or disrupted coverage by satellite network communications for civilian uses.
5: Critical disruption	The inability to maintain satellite gateways and to supply and control satellite constellations makes it infeasible to reliably provide satellite access networks. Use of functioning networks is restricted to high-priority government functions, and they are not reliable even for those functions.

Provide Wireline Access Network Services

Table B.242. Scorecard for *Provide Wireline Access Network Services*

	National Risk Assessment				
	Baseline	Current Emissions		High Emissions	
	Present	2050	2100	2050	2100
Provide Wireline Access Network Services	2	3	3	3	3

Description: "Operate circuit- and packet-switched networks via copper, fiber, and coaxial transport media, including private enterprise data and telephony networks and the public switched telephone network (PSTN)" (CISA, 2020, p. 3).

Highlights

- Stronger or more-frequent flooding, wildfire, severe storm systems, tropical cyclones and hurricanes, and sea-level rise pose risk of moderate disruption to this NCF because they are expected to damage physical infrastructure over a wide area.
- The NCF relies on central offices housing cabling and equipment and outside plant providing connections to users and other facilities, and this physical infrastructure is vulnerable to damage from climate drivers that renders these facilities inoperable and disrupts the function.
- The primary subfunctions affected are *Provide Operation and Maintenance Services*, *Operate Provider Networks*, *Connect Local Exchange to Regional Provider Infrastructure*, *Operate Local Exchanges*, and *Provide Access to Local Exchanges*.

Synopsis

Flooding, severe storm systems, tropical cyclones and hurricanes, sea-level rise, and wildfire can all destroy physical infrastructure needed to *Provide Wireline Access Network Services*. This physical destruction can cause regional degradations in network performance and even complete disruption of services locally. In Hurricane Katrina, for example, damage and destruction of central offices and outside plant associated with the public switched telephony network disrupted wireline access network services to millions of users in the region (Kwasinski et al., 2009). In some cases, these regional disruptions lasted for more than a month. Hurricane Sandy also caused flood damage at critical facilities that disrupted wireline networks in New York for several days (Special Initiative for Rebuilding and Resiliency, 2013).

Destruction of physical infrastructure, especially local offices, cabling, and equipment that provide communication links with other local stations to the network of users, would disrupt network functionality in the local area. Operators are identifying and implementing strategies to prepare for and mitigate disruptions from these events, including hardening facilities and developing tools that allow reconfiguration of networks before and in the aftermath of natural disasters (Tornatore et al., 2016), but extensive physical damage to infrastructure across a wide region will likely continue to cause disruptions to the NCF in the future.

Flooding, tropical cyclones and hurricanes, severe storm systems, and wildfire are all capable of causing physical damage to the NCF's infrastructure. *Provide Wireline Access Network Services* experiences disruptions from these climate drivers in the baseline scenario, and we assessed the NCF as being at risk of minimal disruption in present conditions. Using the available literature, we assessed that hurricanes and wildfire could have the most-direct effects on the NCF over time.

Key Considerations for *Provide Wireline Access Network Services*

In assessing this NCF, we made two core assumptions about how the NCF will operate through 2100. We assumed the following for this time period:
1. Content and access delivered by this function are still at least partially generated locally, and delivery is still primarily provided using local stations.
2. Networked communication systems will continue to use both wired and wireless (electromagnetic spectrum) communication media.

Subfunctions

Several subfunctions experience risk from climate drivers: *Provide Operation and Maintenance Services, Operate Provider Networks, Connect Local Exchange to Regional Provider Infrastructure, Operate Local Exchanges*, and *Provide Access to Local Exchanges*. These subfunctions are all at risk of disruption from physical damage to equipment, connections, and facilities caused by natural hazards. Disruption currently occurs primarily at a local level, but increasingly intense disasters might cause disruptions over wider areas.

Table B.243. Subfunction Risk for *Provide Wireline Access Network Services*

Subfunction	Baseline Present	Current Emissions 2050	Current Emissions 2100	High Emissions 2050	High Emissions 2100
Provide Regulations, Standards, and Licensing	1	1	1	1	1
Provide Operation and Maintenance Services	2	3	3	3	3
Operate Provider Networks	2	3	3	3	3
Connect Local Exchange to Regional Provider Infrastructure	2	3	3	3	3
Operate Local Exchanges	2	3	3	3	3
Provide Access to Local Exchanges	2	3	3	3	3
Supply End-User Equipment	1	1	1	1	1

NOTE: 1 = no disruption or normal operations. 2 = minimal disruption. 3 = moderate disruption. See Chapter 2 for further explanation of ratings.

Climate Drivers

Climate drivers that can cause physical damage to local broadcast offices, cables, and transmission equipment are likely to have an effect on this NCF. These climate drivers include tropical cyclones and hurricanes, severe storm systems, and wildfire. The NCF is experiencing frequent local disruptions from these climate drivers in the current emissions scenario, and the drivers are likely to intensify in the future due to climate change.

319

Table B.244. Risk from Climate Drivers for *Provide Wireline Access Network Services*

Climate Driver	Baseline Present	Current Emissions 2050	Current Emissions 2100	High Emissions 2050	High Emissions 2100
Drought	1	1	1	1	1
Extreme cold	1	1	1	1	1
Extreme heat	1	1	2	1	2
Flooding	2	2	3	3	3
Sea-level rise	1	2	2	2	2
Severe storm systems	2	3	3	3	3
Tropical cyclones and hurricanes	2	3	3	3	3
Wildfire	2	3	3	3	3

NOTE: 1 = no disruption or normal operations. 2 = minimal disruption. 3 = moderate disruption. See Chapter 2 for further explanation of ratings.

Impact Mechanisms

Physical damage that renders local stations, cables, and transmission equipment nonfunctional is likely to be the primary impact mechanism affecting this NCF for the foreseeable future. Extreme heat could also make it challenging to maintain outdoor facilities in the future, although the degree to which this might affect NCF operations is unclear.

Cascading Risk

This NCF has a significant upstream dependence on *Generate Electricity*, *Transmit Electricity*, and *Distribute Electricity*. It is not clear to what degree physical damage to network infrastructure, as opposed to electrical power failures, is likely to be the root cause of network failure in disaster scenarios. In an examination of root causes of network disruptions caused by Hurricane Ike in 2008, for example, researchers found that all identifiable root causes of disruption linked to inadequate electrical power rather than to physical damage (Erjongmanee and Ji, 2011). Hurricane Katrina in 2005 caused significant disruptions to telecommunication networks that were caused both by electrical power outages and physical destruction of physical infrastructure (although electrical outages were still the main cause) (Kwasinski et al., 2009). Although we considered physical damage to be the main climate-driven impact mechanism for this NCF, it is likely overshadowed by the effects of electrical power failures, which is a key dependence for this NCF.

Strength of Evidence

There is ample evidence from recent events to describe climate drivers' effect on this NCF, particularly for tropical cyclones and hurricanes, severe storm systems, and wildfire (Erjongmanee and Ji, 2011; Kuligowski et al., 2014; Kwasinski et al., 2009; Mena, 2020).

Table B.245. Strength of Evidence for *Provide Wireline Access Network Services*

	Strength of Evidence
High	Tropical cyclones and hurricanes; severe storm systems; wildfire
Low	Extreme cold; extreme heat; drought; sea-level rise

Risk Rating Interpretations

Table B.246 provides examples of how to interpret the risk ratings for this NCF.

Table B.246. Example Interpretations of Risk Ratings for *Provide Wireline Access Network Services*

Risk Rating	NCF Example
1: No disruption or normal operations	Connection to wireline networks continues to function as expected, and any effects from climate drivers are addressed with mitigation strategies and built-in redundancies.
2: Minimal disruption	Climate drivers cause regions to experience prolonged disruptions to wireline networks until physical infrastructure can be repaired, but the network disruption does not extend beyond the affected region.
3: Moderate disruption	Intense events with wide geographic footprints cause disruptions to wireline networks over large regions. Damaged infrastructure is able to be repaired within a time period of weeks following the event, restoring service to the affected region.
4: Major disruption	The frequency and intensity of climate-driven disasters cause significant regional disruptions in multiple regions in quick succession.
5: Critical disruption	The frequency and severity of climate-driven hazards make it infeasible to maintain the physical infrastructure needed to provide wireline network services nationally. Multiple regions cease to provide wireline network services because repair and maintenance cannot keep pace with disaster frequency.

Provide Wireless Access Network Services

Table B.247. Scorecard for *Provide Wireless Access Network Services*

	National Risk Assessment				
	Baseline	Current Emissions		High Emissions	
	Present	2050	2100	2050	2100
Provide Wireless Access Network Services	2	3	3	3	3

Description: "Provide access to core communications network via electromagnetic wave-based technologies, including cellular phones, wireless hot spots (Wi-Fi), personal communications services, high-frequency radio, unlicensed wireless, and other commercial and private radio services" (CISA, 2020, p. 3).

Highlights

- *Provide Wireless Access Network Services* was assessed to be at risk of moderate disruption in the 2050 and 2100 time periods due to physical damage to infrastructure.
- Tropical cyclones and hurricanes and wildfire are the primary climate drivers affecting this NCF, and severe storm systems contribute to a lesser extent. These climate drivers, especially hurricanes and wildfire, already inflict physical damage on cellular sites, causing prolonged disruptions to this NCF at local levels and sometimes causing brief disruptions over large geographic regions.
- In either emissions scenario, disruptions from physical damage to assets are likely to worsen over time if wireless networks continue to rely on geographically dispersed, remote physical infrastructure in wildfire and hurricane zones.
- Three subfunctions are particularly at risk: *Provide Logistics and Operational Support to the Network*, *Operate Edge Network*, and *Provide Access to Edge Network*.

Synopsis

Climate drivers, particularly tropical cyclones and hurricanes and wildfire, can cause physical damage to cellular sites, leading to degraded or fully disrupted communications for users in the affected area. These hazards often also create conditions (e.g., debris blocking transportation routes) that inhibit access to sites needed to perform repairs, thus prolonging disruptions (Goldstein, 2017).

A cellular site typically contains a cellular tower, transceiver, and other critical equipment; it connects to a central office via a wired backhaul connection that is buried underground (S. Anderson, Barford, and Barford, 2020). Remote cell sites are part of an edge network (the networked devices and facilities physically located away from core data centers and close to points of service) that maintains connection with a core carrier network. Because many areas are served by more than one cell site, the destruction of a single cell site might not disrupt network access, but damage to cell sites and backhaul connections across a wide area can lead to local and regional disruptions to network access (S. Anderson, Barford, and Barford, 2020; FCC, 2016; Goldstein, 2017; Kwasinski et al., 2009).

Climate driver–related disruption to this NCF is the result primarily of physical damage to cell sites or backhaul connections.[55] Both types of damage have historically resulted from many of the

---

[55] Power outages are the most common cause of disruption; however, because these outages result from the NCF's dependence on other NCFs and are not a direct effect of climate drivers on *Provide Wireless Access Network Services* (S. Anderson, Barford, and Barford, 2020), these effects were not considered in the rating.

climate drivers, including wildfire, hurricanes, and severe storm systems (S. Anderson, Barford, and Barford, 2020; Goldstein, 2017). Extreme heat could also contribute to disruptions in future scenarios by making it more difficult for workers to access and maintain remote cell sites that are outdoors.

All these relevant climate drivers are expected to intensify through 2100. We assessed that this NCF will begin to experience risk of moderate disruption by 2050 (i.e., more-prolonged disruptions over wider geographic regions that damage more edge sites, and backhaul connections that could be harder to access and restore), and this will continue through 2100 as conditions worsen and more assets are damaged.

Key Considerations for *Provide Wireless Access Network Services*

In assessing this NCF, we made two core assumptions about how the NCF will operate through 2100. We assumed the following for this time period:
1. Wireless access to the internet continues to be provided for users via electromagnetic signals sent and received by distributed, remote, physical infrastructure.
2. Distributed physical infrastructure continues to utilize aboveground components that remain vulnerable to damage from climate drivers.

Subfunctions

Three subfunctions of *Provide Wireless Access Network Services* are key drivers of risk: *Provide Logistics and Operational Support to the Network*, *Operate Edge Network*, and *Provide Access to Edge Network*. These subfunctions are associated with persistent operation and maintenance of remote cell sites. The climate-driven risk to this NCF is associated primarily with physical damage to this infrastructure. These subfunctions are already facing meaningful local disruptions from climate drivers, especially hurricanes and wildfire, under current circumstances, and these disruptions will affect more regions and last longer in future scenarios.

Table B.248. Subfunction Risk for *Provide Wireless Access Network Services*

Subfunction	Baseline Present	Current Emissions 2050	Current Emissions 2100	High Emissions 2050	High Emissions 2100
Provide Regulations, Standards, and Licensing	1	1	1	1	1
Provide Logistics and Operational Support to the Network	2	3	3	3	3
Operate Carrier Data Centers	1	1	1	1	1
Operate Core Carrier Network	1	1	1	1	1
Operate Edge Network	2	3	3	3	3
Provide Access to Edge Network	2	3	3	3	3
Supply End-User Equipment	1	1	1	1	1

NOTE: 1 = no disruption or normal operations. 2 = minimal disruption. 3 = moderate disruption. See Chapter 2 for further explanation of ratings.

Climate Drivers

Tropical cyclones and hurricanes and wildfire are the primary climate drivers affecting this NCF, and severe storm systems (e.g., the derecho affecting the Midwest and mid-Atlantic states in 2012) contribute to a lesser extent (S. Anderson, Barford, and Barford, 2020; Goldstein, 2017). These events can cause physical damage to cell sites and associated equipment, disrupting their ability to provide wireless network access to users.

Table B.249. Risk from Climate Drivers for *Provide Wireless Access Network Services*

Climate Driver	Baseline Present	Current Emissions 2050	Current Emissions 2100	High Emissions 2050	High Emissions 2100
Drought	1	1	1	1	1
Extreme cold	1	1	1	1	1
Extreme heat	1	1	2	2	2
Flooding	1	1	1	1	2
Sea-level rise	1	1	2	1	2
Severe storm systems	2	2	2	2	2
Tropical cyclones and hurricanes	2	3	3	3	3
Wildfire	2	3	3	3	3

NOTE: 1 = no disruption or normal operations. 2 = minimal disruption. 3 = moderate disruption. See Chapter 2 for further explanation of ratings.

Impact Mechanisms

Physical damage to infrastructure is the primary impact mechanism affecting this NCF for the foreseeable future. Climate drivers can destroy critical equipment at cell sites or the backhaul connections to the core network, and this will disrupt the function for the surrounding area.

Cascading Risk

Provide Wireless Access Network Services has key upstream dependencies on NCFs for

- *Generate Electricity*
- *Transmit Electricity*
- *Distribute Electricity.*

Physical damage leading to loss of electricity is often the primary cause of disruptions to network functionality (S. Anderson, Barford, and Barford, 2020). Remote cell sites cannot function without electrical power, and a loss of power results in a complete disruption to wireless access network services because users cannot maintain connectivity through these sites.

Strength of Evidence

There is little evidence to describe climate drivers' effect on the *Provide Wireless Access Network Services* NCF (S. Anderson, Barford, and Barford, 2020; FCC, 2016; Goldstein, 2017). Historical examples illustrate how climate drivers that are significant drivers of risk to this NCF, such as wildfire, are likely to increase in intensity and frequency in the future. However, there is insufficient evidence that specifically indicates wildfire's effect on the operability of wireless network systems. For other climate drivers, such as tropical cyclones and hurricanes, effects on the *Provide Wireless Access Network Services* NCF are less well established.

Table B.250. Strength of Evidence for *Provide Wireless Access Network Services*

Strength of Evidence	
High	Wildfire
Low	Tropical cyclones and hurricanes; extreme cold; extreme heat; severe storm systems; drought; sea-level rise; flooding

Risk Rating Interpretations

Table B.251 provides examples of how to interpret the risk ratings for this NCF.

Table B.251. Example Interpretations of Risk Ratings for *Provide Wireless Access Network Services*

Risk Rating	NCF Example
1: No disruption or normal operations	Wireless networks and communications continue to function as expected, and natural disasters do not lead to meaningful degradation in wireless network access.
2: Minimal disruption	Climate drivers cause prolonged disruptions to wireless network access until physical infrastructure can be repaired, but the disruption is local, and any regional effects are brief.
3: Moderate disruption	Climate drivers cause disruptions to wireless network access across a wide geographic footprint that extends for several days or more.
4: Major disruption	Climate drivers caused prolonged, regional disruptions to wireless network access with such frequency that multiple regions often experience simultaneous disruptions to network access.
5: Critical disruption	The frequency and severity of climate drivers make it infeasible for some regions to maintain the physical infrastructure needed to provide wireless network access, and wireless network access cannot be ensured across most of the country.

Provide Cable Access Network Services

Table B.252. Scorecard for *Provide Cable Access Network Services*

	National Risk Assessment				
	Baseline	Current Emissions		High Emissions	
	Present	2050	2100	2050	2100
Provide Cable Access Network Services	2	3	3	3	3

Description: "Provide access to communications backbone infrastructure through fiber and coaxial network, supplying analog and digital video programming services, digital telephone service, and high-speed broadband services" (CISA, 2020, p. 2).

Highlights

- This NCF was assessed as being at risk of moderate disruption from several climate drivers.
- We expect stronger and more-frequent flooding, wildfire, severe storm systems, tropical cyclones and hurricanes, and sea-level rise, and these climate drivers will disrupt the NCF by damaging physical infrastructure over a wide area.
- The primary subfunctions affected are *Provide Transmission to End Users* and *Provide Local Broadcast Services*. This NCF is provided primarily by local offices acting as broadcast stations, and damage to these facilities, their wired connections to the local area, and their wireless transmission equipment disrupts this function.

Synopsis

Although broader, core network infrastructure across the United States is likely to be resilient to disruptions from climate drivers thanks to built-in redundancies, local or regional access networks can experience more-significant disruptions (Hinchman, 2022). Flooding, severe storm systems, tropical cyclones and hurricanes, sea-level rise, and wildfire can all destroy physical infrastructure needed to provide cable access network services. This physical destruction can cause regional degradations in network performance and even complete disruption of services locally.

Destruction of physical infrastructure, especially local offices, transmitters, and equipment that provide communication links with a national network, other local stations, and end users, would disrupt network functionality in the local area. Operators are identifying and implementing strategies to prepare for and mitigate disruptions from these events, including tools that allow reconfiguration of networks before and in the aftermath of natural disasters (Tornatore et al., 2016), but extensive physical damage to infrastructure across a wide region will likely continue to cause disruptions to the NCF in the future.

Flooding, tropical cyclones and hurricanes, sea-level rise, severe storm systems, and wildfire are all capable of causing physical damage to the NCF's infrastructure. All these events are expected to exhibit varying climate-driven increases in intensity and frequency over time. We assessed, based on the available literature, that hurricanes, sea-level rise, and wildfire could have the most-direct effects on the NCF over time. We assess that, as climate change intensifies over time, there will be proportionally greater local and regional disruptions to this NCF during these time periods.

Key Considerations for *Provide Cable Access Network Services*

In assessing this NCF, we made two core assumptions about how the NCF will operate through 2100. We assumed the following for this time period:

1. Content and access delivered by this function are still at least partially generated locally, and delivery is still provided primarily using local stations.
2. Networked communication systems will continue to use both wired and wireless (electromagnetic spectrum) communication media.

Subfunctions

Only two subfunctions of this NCF are likely to be at risk: *Provide Transmission to End Users* and *Provide Local Broadcast Services*. The risk to these subfunctions is due primarily to the potential for climate-driven natural hazards to damage or destroy local stations, cables, and equipment required to produce, receive, and transmit content to the surrounding area. These subfunctions already experience local disruptions during hurricanes (Tornatore et al., 2016), severe storms (Kuligowski et al., 2014), and wildfires (Mena, 2020).

Table B.253. Subfunction Risk for *Provide Cable Access Network Services*

	Baseline	Current Emissions		High Emissions	
Subfunction	Present	2050	2100	2050	2100
Supply End User Devices	1	1	1	1	1
Provide Transmission to End Users	2	3	3	3	3
Provide Local Broadcast Services	2	3	3	3	3
Provide National Station Broadcast Services	1	1	1	1	1
Provide Regulations, Standards, and Licensing	1	1	1	1	1

NOTE: 1 = no disruption or normal operations. 2 = minimal disruption. 3 = moderate disruption. See Chapter 2 for further explanation of ratings.

Climate Drivers

Climate drivers that can cause physical damage to local broadcast offices, cables, and transmission equipment are likely to have an effect on this NCF. These climate drivers include tropical cyclones and hurricanes, sea-level rise, severe storm systems, and wildfire. In particular, the combination of sea-level rise with tropical cyclones and hurricanes is likely to be a significant cause of disruption of this NCF in coastal communities. The NCF is experiencing frequent local disruptions from these climate drivers in the current emissions scenario, and the drivers are likely to intensify in the future because of climate change.

Table B.254. Risk from Climate Drivers for *Provide Cable Access Network Services*

Climate Driver	Baseline	Current Emissions		High Emissions	
	Present	2050	2100	2050	2100
Drought	1	1	1	1	1
Extreme cold	1	1	1	1	1
Extreme heat	1	1	2	1	2
Flooding	1	1	1	1	2
Sea-level rise	1	2	3	2	3
Severe storm systems	2	3	3	3	3
Tropical cyclones and hurricanes	2	3	3	3	3
Wildfire	2	3	3	3	3

NOTE: 1 = no disruption or normal operations. 2 = minimal disruption. 3 = moderate disruption. See Chapter 2 for further explanation of ratings.

Impact Mechanisms

Physical damage that renders local stations, cables, and transmission equipment nonfunctional is likely to be the primary impact mechanism affecting this NCF for the foreseeable future. Extreme heat could also make it challenging to maintain outdoor facilities in the future, although the extent to which this might affect NCF operations is unclear.

Cascading Risk

This NCF has a significant upstream dependence on *Generate Electricity*, *Transmit Electricity*, and *Distribute Electricity*. It is not clear to what degree physical damage to network infrastructure, as opposed to electrical power failures, is likely to be the root cause of network failure for each hazard, time frame, and scenario. In an examination of root causes of network disruptions caused by Hurricane Ike in 2008, for example, researchers found that all identifiable root causes of disruption linked to inadequate electrical power, rather than physical damage (Erjongmanee and Ji, 2011). Hurricane Katrina in 2005 caused significant disruptions to telecommunication networks that were caused both by electrical power outages and physical destruction of infrastructure (although electrical outages were still the main cause) (Kwasinski et al., 2009). Although we consider physical damage to be the main climate-driven impact mechanism for this NCF, it is likely overshadowed by the effects of electrical power failures, which is a key dependence for this NCF.

Strength of Evidence

The strength of evidence of climate drivers' effects on the *Provide Cable Access Network Services* NCF was assessed as high for wildfire, tropical cyclones and hurricanes, severe storm systems, and sea-level rise. Historical examples provide ample evidence of climate drivers' effect on this NCF (Erjongmanee and Ji, 2011; Kuligowski et al., 2014; Kwasinski et al., 2009; Mena, 2020; Tornatore et al., 2016). For example, there is considerable evidence of how wildfire and sea-level rise, both of which are significant drivers of risk for this NCF, are likely to increase in intensity and frequency in the future. Effects from other climate drivers, such as tropical cyclones and hurricanes and severe storm systems, are less well established, but both are expected to intensify with corresponding increases to risk to this NCF.

Table B.255. Strength of Evidence for *Provide Cable Access Network Services*

Strength of Evidence	
High	Wildfire; tropical cyclones and hurricanes; severe storm systems; sea-level rise
Low	Flooding; extreme cold; extreme heat; drought

Risk Rating Interpretations

Table B.256 provides examples of how to interpret the risk ratings for this NCF.

Table B.256. Example Interpretations of Risk Ratings for *Provide Cable Access Network Services*

Risk Rating	NCF Example
1: No disruption or normal operations	Connection to cable access networks continues to function as expected, and any effects from climate drivers are addressed with mitigation strategies and built-in redundancies.
2: Minimal disruption	Climate drivers cause regions to experience prolonged disruptions to cable access networks until physical infrastructure can be repaired, but the network disruption does not extend beyond the affected region.
3: Moderate disruption	Intense events with wide geographic footprints cause disruptions to cable access networks over large regions. Damaged infrastructure is repaired within a time period of weeks following the event, restoring service to the affected region.
4: Major disruption	The frequency and intensity of climate-driven natural hazard events cause significant regional disruptions in multiple regions in quick succession. The extensive damage leads to prolonged disruptions to cable access networks that even degrade national broadcast services.
5: Critical disruption	The frequency and severity of climate-driven natural hazard events make it infeasible to maintain the physical infrastructure needed to provide cable access network services nationally. Multiple regions cease to provide cable access network services because repair and maintenance cannot keep pace with disaster frequency.

Operate Core Network

Table B.257. Scorecard for *Operate Core Network*

	National Risk Assessment				
	Baseline	**Current Emissions**		**High Emissions**	
	Present	**2050**	**2100**	**2050**	**2100**
Operate Core Network	1	1	2	1	2

Description: "Maintain and operate communications backbone infrastructure for voice, video, and data transmission that connects to users through broadcasting, cable, satellite, wireless, and wireline access networks" (CISA, 2020, p. 2).

Highlights

- The core network is highly resilient to disruption from any of the climate drivers.
- In the most-extreme future scenarios, tropical cyclones and hurricanes and wildfire could minimally degrade this function by causing widespread physical damage to key infrastructure.
- The NCF's resilience is due to significant redundancy in connections and infrastructure across the United States.

Synopsis

Severe natural hazard events, such as the climate drivers included in this analysis, can cause regional disruptions to network connectivity, but the core network is resilient to these disruptions (Hinchman, 2022). Some climate drivers, such as tropical cyclones and hurricanes, wildfire, severe storm systems, and flooding, can cause damage to physical infrastructure, especially the operations centers and the high-bandwidth, long-haul cables that carry traffic between regional networks and service providers.

Severe natural hazard event disruptions to regional networks and long-haul cables are unlikely to lead to meaningful disruptions to the operation of the core network. If damage occurs to a cable or a particular region's brick-and-mortar infrastructure, core network traffic can be rerouted around the affected area using other routing pathways. Traffic that originates or terminates near the event could be disrupted, but the core network is resilient to these disruptions (Hinchman, 2022). This phenomenon was well documented in the aftermaths of Hurricanes Katrina and Ike. The region damaged by the hurricanes experienced a significant, prolonged disruption to communication networks, but disruptions could be monitored and documented because the core network continued to function (Erjongmanee and Ji, 2011; Tornatore et al., 2016).

Flooding, tropical cyclones and hurricanes, severe storm systems, and wildfire are all capable of causing physical damage to the NCF's physical infrastructure, and the frequency and severity of these events will increase over time because of climate change. Even extreme events, such as Hurricanes Ike and Katrina, however, have not led to meaningful disruptions to the operation of the core network, despite disruptions over a wide geographic area. We expect that multiple, nearly simultaneous extreme natural events in different geographic regions will cause only a minimal disruption to the routine operational needs of the core network. We assessed that, in either emissions scenario, a scenario like this is a risk only in 2100, when there are projected increases in the risks posed by wildfire, flooding, severe storm systems, tropical cyclones and hurricanes, and sea-level rise.

Key Considerations for *Operate Core Network*

In assessing this NCF, we made two core assumptions about how the NCF will operate through 2100. We assumed the following for this time period:
1. Networked communication systems will continue to consist of regional subnets redundantly connected by core network infrastructure.
2. Networked communication systems will still use both wired and wireless (electromagnetic spectrum) communication media.

Subfunctions

Three subfunctions for this NCF utilize infrastructure that could be at increasing risk of physical damage from the climate drivers: *Maintain Provider Backbone Infrastructure, Maintain Long-Haul Cables,* and *Connect Provider Networks.* The physical infrastructure associated with these subfunctions spans the country geographically, and it is therefore subject to regionally varying risks from different climate drivers, including flooding, severe storm systems, tropical cyclones and hurricanes, and wildfire. This geographic dispersion and the redundant routes for long-haul, high-bandwidth communication traffic provide intrinsic resilience to disruption caused by severe local or regional events. We could find no evidence that operation of the core network is ever meaningfully disrupted by any of the climate drivers in current conditions. We inferred that, even in the most-extreme future scenarios, none of the subfunctions is likely to experience worse than minimal disruption due to the climate drivers, individually or in combination.

Table B.258. Subfunction Risk for *Operate Core Network*

	Baseline	Current Emissions		High Emissions	
Subfunction	Present	2050	2100	2050	2100
Maintain Provider Backbone Infrastructure	1	1	2	1	2
Maintain Long-Haul Cables	1	1	2	1	2
Connect Provider Networks	1	1	2	1	2

NOTE: 1 = no disruption or normal operations. 2 = minimal disruption. See Chapter 2 for further explanation of ratings.

Climate Drivers

The climate drivers affecting this NCF are those that could damage physical infrastructure: flooding, tropical cyclones and hurricanes, and wildfire. Sea-level rise is excluded from this list, despite its potential to cause physical damage to infrastructure, because the core network and its critical physical infrastructure span primarily the interior of the United States, outside of threatened coastal areas (see, for example, Network Startup Resource Center, 2016). There are some exceptions, such as undersea cable landings and internet exchange points in Miami, Florida, and Honolulu, Hawaii, but it is not clear what effect sea-level rise will have on relevant infrastructure in these locations or whether disruption of these sites would produce any meaningful risk to the broader core network. Cable landings in coastal areas might be at risk from sea-level rise, but we considered it likely that such facilities would be relocated in the future if sea-level rise were deemed a meaningful risk. We also excluded from our analysis potential damage to undersea cables because we assumed that these are associated primarily with international communication traffic that is not strictly associated with the operation of the core U.S. network. Currently, disruptions from climate drivers with wide geographic footprints (especially hurricanes and wildfire) occur primarily in more-local or -regional subnetworks and do not create meaningful disruption to the core network (Tornatore et al., 2016).

Table B.259. Risk from Climate Drivers for *Operate Core Network*

Climate Driver	Baseline	Current Emissions		High Emissions	
	Present	2050	2100	2050	2100
Drought	1	1	1	1	1
Extreme cold	1	1	1	1	1
Extreme heat	1	1	1	1	1
Flooding	1	1	1	1	2
Sea-level rise	1	1	1	1	1
Severe storm systems	1	1	1	1	1
Tropical cyclones and hurricanes	1	1	2	1	2
Wildfire	1	1	2	1	2

NOTE: 1 = no disruption or normal operations. 2 = minimal disruption. See Chapter 2 for further explanation of ratings.

Impact Mechanisms

Significant uncertainty exists in how the NCF will operate in the future, but, given our core assumptions, physical damage is the only relevant impact mechanism for this NCF.

Cascading Risk

This NCF has key interdependencies with *Generate Electricity*, *Transmit Electricity*, and *Distribute Electricity*. Physical damage leading to loss of electricity is often the primary cause of disruptions to network functionality (Erjongmanee and Ji, 2011). The redundancy built into the core network likely makes this NCF resilient to all but the most–nationally widespread power failures, but this is nevertheless the most-important interdependence for this NCF.

Strength of Evidence

The strength of evidence of climate drivers' effects on the *Operate Core Network* NCF, based primarily on academic literature, was assessed as medium. There is ample evidence to suggest that this NCF is highly resilient to disruptions from any of the climate drivers, even those with large geographic footprints, such as hurricanes (Hinchman, 2022; Tornatore et al., 2016). Additionally, the effect that such large-scale natural disasters have on network functionality has been studied (Erjongmanee and Ji, 2011; Gomes et al., 2016; Tornatore et al., 2016). However, we found no literature on the effect of other climate drivers, such as flooding, severe storm systems, and extreme temperatures. One reason for this lack of literature could be that these climate drivers have no current effect on core network functionality and there is little concern that they could cause meaningful disruptions.

Table B.260. Strength of Evidence for *Operate Core Network*

	Strength of Evidence
Medium	Tropical cyclones and hurricanes
Low	Flooding; severe storm systems; extreme cold; extreme heat; drought; sea-level rise; wildfire

Risk Rating Interpretations

Table B.261 provides examples of how to interpret the risk ratings for this NCF.

Table B.261. Example Interpretations of Risk Ratings for *Operate Core Network*

Risk Rating	NCF Example
1: No disruption or normal operations	Networked communications continue to function as expected, and any effects from climate drivers are mitigated by built-in redundancies.
2: Minimal disruption	Climate drivers cause regions to experience prolonged disruptions to communication networks until physical infrastructure can be repaired, but the network disruption does not extend beyond the affected region.
3: Moderate disruption	Frequent occurrence of near-simultaneous disasters in multiple regions overcome system redundancies, degrading (but not fully disrupting) the performance of networked communications nationally.
4: Major disruption	The frequency of climate-driven disasters overcomes network redundancies and outpaces the capacity to recover and repair physical infrastructure, leading to short time periods in which networked communications are disrupted nationally.
5: Critical disruption	The frequency and severity of climate-driven disasters make it infeasible to maintain the physical infrastructure needed to provide networked communications, and it becomes impossible to reliably provide core network linkages connecting regional communication networks.

Abbreviations

ACC	American Chemistry Council
API	American Petroleum Institute
AR6	Sixth Assessment Report
ASCE	American Society of Civil Engineers
CBO	Congressional Budget Office
CFTC	Commodity Futures Trading Commission
CISA	Cybersecurity and Infrastructure Security Agency
COVID-19	coronavirus disease 2019
CSO	combined sewer overflow
DHS	U.S. Department of Homeland Security
DoD	U.S. Department of Defense
DOT	U.S. Department of Transportation
EHR	electronic health record
EIA	U.S. Energy Information Administration
EMS	emergency medical service
EO	executive order
EPA	U.S. Environmental Protection Agency
FAA	Federal Aviation Administration
FCC	Federal Communications Commission
FDIC	Federal Deposit Insurance Corporation
FEMA	Federal Emergency Management Agency
FMU	financial market utility
GAO	U.S. Government Accountability Office
GHG	greenhouse gas
GPS	Global Positioning System
HSOAC	Homeland Security Operational Analysis Center

IPCC	Intergovernmental Panel on Climate Change
IT	information technology
NCA4	Fourth National Climate Assessment
NCA5	Fifth National Climate Assessment
NCF	National Critical Function
NCSL	National Conference of State Legislatures
NOAA	National Oceanic and Atmospheric Administration
OMB	Office of Management and Budget
OT	operational technology
PNT	positioning, navigation, and timing
SLTT	state, local, tribal, and territorial
SPR	Strategic Petroleum Reserve
TCEQ	Texas Commission on Environmental Quality
UO_2	uranium dioxide
USDA	U.S. Department of Agriculture
USGCRP	U.S. Global Change Research Program

Bibliography

Abbott, Chris, "An Uncertain Future: Law Enforcement, National Security and Climate Change," briefing paper, Oxford Research Group, January 2008.

Abir, Mahshid, Farzad Mostashari, Parmeeth Atwal, and Nicole Lurie, "Electronic Health Records Critical in the Aftermath of Disasters," *Prehospital and Disaster Medicine*, Vol. 27, No. 6, December 2012.

Abi-Samra, Nicholas, *Power Grid Resiliency for Adverse Conditions*, Artech House, 2017.

Acar, Pinar, and Istemi Berk, "Power Infrastructure Quality and Industrial Performance: A Panel Data Analysis on OECD Manufacturing Sectors," *Energy*, Vol. 239, Part C, January 15, 2022.

ACC—*See* American Chemistry Council.

Acemoglu, Daron, Giuseppe De Feo, and Giacomo De Luca, *Weak States: Causes and Consequences of the Sicilian Mafia*, National Bureau of Economic Research, working paper 24115, December 2017.

Adapting to Rising Tides, *Adapting to Rising Tides: Contra Costa County Assessment and Adaptation Project*, March 2017.

Adelaine, Sabrina A., Mizuki Sato, Yufang Jin, and Hilary Godwin, "An Assessment of Climate Change Impacts on Los Angeles (California USA) Hospitals, Wildfires Highest Priority," *Prehospital and Disaster Medicine*, Vol. 32, No. 5, October 2017.

AECOM, *Port of Los Angeles: Sea Level Rise Adaptation Study—Final Draft*, prepared for Port of Los Angeles, September 2018.

Agnew, Robert, "Foundation for a General Strain Theory of Crime and Delinquency," *Criminology*, Vol. 30, No. 1, February 1992.

Agnew, Robert, "Dire Forecast: A Theoretical Model of the Impact of Climate Change on Crime," *Theoretical Criminology*, Vol. 16, No. 1, February 2012.

Alaska Department of Fish and Game, "Permafrost," webpage, undated. As of October 8, 2022:
https://www.adfg.alaska.gov/index.cfm?adfg=ecosystems.permafrost

Alexander-Kearns, Myriam, "Climate Change Threatens Electric Grid Reliability in the Southwest," Center for American Progress, August 21, 2015. As of November 16, 2022:
https://www.americanprogress.org/article/
climate-change-threatens-electric-grid-reliability-in-the-southwest/

Al-Farah, Rayah, Chris Redl, Yang Yang, and Jonathan Saalfield, "Jordan: Selected Issues," International Monetary Fund, Country Report 22/222, July 2022.

Allen, Thomas R., George McLeod, and Sheila Hutt, "Sea Level Rise Exposure Assessment of US East Coast Cargo Container Terminals," *Maritime Policy and Management*, Vol. 49, No. 4, 2022.

Allen-Dumas, Melissa R., Binita KC, and Colin I. Cunliff, *Extreme Weather and Climate Vulnerabilities of the Electric Grid: A Summary of Environmental Sensitivity Quantification Methods*, Oak Ridge National Laboratory, ORNL/TM-2019/1252, August 16, 2019.

American Banker, "BankThink: Sandy Response Shows How Banks Can Regain Public Trust," webpage, November 20, 2012. As of October 6, 2022:
https://www.americanbanker.com/opinion/sandy-response-shows-how-banks-can-regain-public-trust

American Chemistry Council, "Lingering Impacts from Hurricane Ida Weighed on U.S. Chemical Production in October," *PRNewswire*, December 1, 2021.

American Farm Bureau Federation, "Hurricane Ida: Direct Agricultural Impacts and Larger Implications of Flooding," webpage, September 15, 2021. As of October 8, 2022:
https://www.fb.org/market-intel/
hurricane-ida-direct-agricultural-impacts-and-larger-implications-of-flooding

American Petroleum Institute, American Gas Association, and Interstate Natural Gas Association of America, *Underground Natural Gas Storage: Integrity and Safe Operations*, July 6, 2016.

American Society of Civil Engineers, *2021 Report Card for America's Infrastructure*, circa 2021.

"Amtrak, Freight Railroads Continue Service Suspension Due to Flooding," *Progressive Railroading*, June 3, 2019.

Andersen, C. Brannon, Gregory P. Lewis, and Kenneth A. Sargent, "Influence of Wastewater-Treatment Effluent on Concentrations and Fluxes of Solutes in the Bush River, South Carolina, During Extreme Drought Conditions," *Environmental Geosciences*, Vol. 11, No. 1, March 2004.

Anderson, Henry, Claudia Brown, Lorraine L. Cameron, Megan Christenson, Kathryn C. Conlon, Samuel Dorevitch, Justin Dumas, Millicent Eidson, Aaron Ferguson, Elena Grossman, Angelina Hanson, Jeremy J. Hess, Brenda Hoppe, Jane Horton, Meredith Jagger, Stephanie Krueger, Thomas W. Largo, Giovanna M. Losurdo, Stephanie R. Mack, Colleen Moran, Cassidy Mutnansky, Kristin Raab, Shubhayu Saha, Paul J. Schramm, Asante Shipp-Hilts, Sara J. Smith, Margaret Thelen, Lauren Thie, and Robert Walker, *Climate and Health Intervention Assessment: Evidence on Public Health Interventions to Prevent the Negative Health Effects of Climate Change*, Climate and Health Technical Report Series, Climate and Health Program, National Center for Environmental Health, Division of Environmental Hazards and Health Effects, Centers for Disease Control and Prevention, 2017.

Anderson, Scott, Carol Barford, and Paul Barford, "Five Alarms: Assessing the Vulnerability of US Cellular Communication Infrastructure to Wildfires," presented at Association for Computing Machinery Internet Measurement Conference, October 2020.

Angel, Jim, Chris Swanston, Barbara Mayes Boustead, Kathryn C. Conlon, Kimberly R. Hall, Jenna L. Jorns, Kenneth E. Kunkel, M. C. Lemos, B. Lofgren, T. A. Ontl, J. Posey, K. Stone, G. Takle, and D. Todey, "Midwest," in D. R. Reidmiller, C. W. Avery, David R. Easterling, Kenneth E. Kunkel, K. L. M. Lewis, T. K. Maycock, and B. C. Stewart, eds., *Impacts, Risks, and Adaptation in the United States: Fourth National Climate Assessment*, Volume II, U.S. Global Change Research Program, 2018.

Angelo, Paul J., *Climate Change and Regional Instability in Central America: Prospects for Internal Disorder, Human Mobility, and Interstate Tensions*, Council on Foreign Relations, Center for Preventive Action, September 2022.

Animal Legal Defense Fund v. United States, complaint for declaratory and injunctive relief, U.S. District Court for the District of Oregon, Eugene Division, Case 6:18-cv-01860-MC, filed October 22, 2018.

Anshari Munir, Nur Al, Yogasmana Al Mustafa, and Fransileo Siagian, "Analysis of the Effect of Ambient Temperature and Loading on Power Transformers Ageing (Study Case of 3rd Power Transformer in Cikupa Substation)," paper presented at the 2nd International Conference on High Voltage Engineering and Power Systems, October 1–4, 2019.

Antle, John M., and Susan M. Capalbo, "Adaptation of Agricultural and Food Systems to Climate Change: An Economic and Policy Perspective," *Applied Economic Perspectives and Policy*, Vol. 32, No. 3, September 2010.

API—*See* American Petroleum Institute.

Applebaum, Katie M., Jay Graham, George M. Gray, Peter LaPuma, Sabrina A. McCormick, Amanda Northcross, and Melissa J. Perry, "An Overview of Occupational Risks from Climate Change," *Current Environmental Health Reports*, Vol. 3, No. 1, March 2016.

Armed Forces Health Surveillance Branch, "Update: Heat Illness, Active Component, U.S. Armed Forces, 2020," *Health.mil*, April 1, 2021.

ASCE—*See* American Society of Civil Engineers.

Avery, Christopher W., David R. Reidmiller, Therese (Tess) S. Carter, Kristen L. M. Lewis, and Katie Reeves, "Report Development Process," in David R. Reidmiller, Christopher W. Avery, David R. Easterling, Kenneth E. Kunkel, Kristen L. M. Lewis, Thomas K. Maycock, and Brooke C. Stewart, eds., *Impacts, Risks, and Adaptation in the United States: Fourth National Climate Assessment*, Vol. II, U.S. Global Change Research Program, 2018.

Baker, Katie, "Climate Change Will Offer Long-Term Tailwinds: Goldman Sachs," *Reinsurance News*, February 4, 2022.

Baker, Kermit, and Alexander Hermann, "Rebuilding from 2017's Natural Disasters: When, for What, and How Much?" Joint Center for Housing Studies, Harvard University, November 30, 2017.

Balbus, John, Allison Crimmins, Janet L. Gamble, David R. Easterling, Kenneth E. Kunkel, Shubhayu Saha, and Marcus C. Sarofim, "Introduction: Climate Change and Human Health," in Allison Crimmins, John Balbus, Janet L. Gamble, Charles B. Beard, Jesse E. Bell, Daniel Dodgen, Rebecca J. Eisen, Neal Fann, Michelle D. Hawkins, Stephanie C. Herring, Lesley Jantarasami, David M. Mills, Shubhayu Saha, Marcus C. Sarofim, Juli Trtanj, and Lewis Ziska, eds., *The Impacts of Climate Change on Human Health in the United States: A Scientific Assessment*, U.S. Global Change Research Program, April 2016.

Ballotpedia, "How and When Are Election Results Finalized? (2022)," webpage, undated. As of October 31, 2022:
https://ballotpedia.org/How_and_when_are_election_results_finalized%3F_(2022)?_wcsid
=90FC5BCB0FB2EE9EB74D1A48B3AE251DFA2CC6E89B460635

Barnes, Mark, "Transit Systems and Ridership Under Extreme Weather and Climate Change Stress: An Urban Transportation Agenda for Hazards Geography," *Geography Compass*, Vol. 9, No. 11, November 2015.

Barrelas, J., Q. Ren, and C. Pereira, "Implications of Climate Change in the Implementation of Maintenance Planning and Use of Building Inspection Systems," *Journal of Building Engineering*, Vol. 40, August 2021.

Basilan, Marvie, "Over 100 Flights in Puerto Rico Canceled Amid Hurricane Fiona's Onslaught," *International Business Times*, September 18, 2022.

Beach, Coral, "Infant Formula Plant Remains Closed for Flood Cleanup; No Word on How Long It Will Take," *Food Safety News*, June 29, 2022.

Becker, Austin H., Michele Acciaro, Regina Asariotis, Edgard Cabrera, Laurent Cretegny, Philippe Crist, Miguel Esteban, Andrew Mather, Steve Messner, Susumu Naruse, Adolf K. Y. Ng, Stefan Rahmstorf, Michael Savonis, Dong-Wook Song, Vladimir Stenek, and Adonis F. Velegrakis, "A Note on Climate Change Adaptation for Seaports: A Challenge for Global Ports, a Challenge for Global Society," *Climatic Change*, Vol. 120, October 2013.

Becker, Austin, Adolf K. Y. Ng, Darryn McEvoy, and Jane Mullett, "Implications of Climate Change for Shipping: Ports and Supply Chains," *WIREs Climate Change*, Vol. 9, No. 2, March–April 2018.

Beckers, Frank, and Uwe Stegemann, *A Risk-Management Approach to a Successful Infrastructure Project*, McKinsey and Company, November 1, 2013.

Becketti, Sean R., *The Impact of Climate Change on Housing and Housing Finance*, special report, Research Institute for Housing America, Mortgage Bankers Association, September 23, 2021.

Behrer, A. Patrick, and Jisung Park, "Will We Adapt? Temperature, Labor and Adaptation to Climate Change in the United States (1986–2012)," *Sense and Sustainability*, Working Paper 2, September 2017.

Belgibayeva, Adiya, Lucia Bevere, Irina Fan, James Finucane, Anja Grujovic, Thomas Holzheu, Roman Lechner, and Daniel Staib, "More Risk: The Changing Nature of P&C Insurance Opportunities to 2040," *Sigma*, No. 4, 2021.

Belvederesi, Chiara, Megan S. Thompson, and Petr E. Komers, "Statistical Analysis of Environmental Consequences of Hazardous Liquid Pipeline Accidents," *Heliyon*, Vol. 4, No. 11, November 2018.

Ben Ammar, Semir, and Martin Eling, "Common Risk Factors of Infrastructure Investments," *Energy Economics*, Vol. 49, May 2015.

Bender, Morris A., Thomas R. Knutson, Robert E. Tuleya, Joseph J. Sirutis, Gabriel A. Vecchi, Stephen T. Garner, and Isaac M. Held, "Modeled Impact of Anthropogenic Warming on the Frequency of Intense Atlantic Hurricanes," *Science*, Vol. 327, No. 5964, January 22, 2010.

Benner, Katie, and Charlie Savage, "Due Process for Undocumented Immigrants, Explained," *New York Times*, June 25, 2018.

Berg, Robbie, *Tropical Cyclone Report: Hurricane Ike (AL092008) 1–14 September 2008*, National Oceanic and Atmospheric Administration, National Hurricane Center, January 23, 2009, updated March 18, 2014.

Berke, Philip, Gavin Smith, and Ward Lyles, "Planning for Resiliency: Evaluation of State Mitigation Plans Under the Disaster Mitigation Act," *Natural Hazards Review*, Vol. 13, No. 2, May 2012.

Bevere, Lucia, and Andreas Weigel, "Natural Catastrophes in 2020: Secondary Perils in the Spotlight, but Don't Forget About Primary-Peril Risks," *Sigma*, 1/2021, March 30, 2021.

Bezgrebelna, Mariya, Kwame McKenzie, Samantha Wells, Arun Ravindran, Michael Kral, Julia Christensen, Vicky Stergiopoulos, Stephen Gaetz, and Sean A. Kidd, "Climate Change, Weather, Housing Precarity, and Homelessness: A Systematic Review of Reviews," *International Journal of Environmental Research and Public Health*, Vol. 18, No. 11, June 2021.

Biden, Joseph R., Jr., "Tackling the Climate Crisis at Home and Abroad," White House, Executive Order 14008, January 27, 2001.

Bierkandt, R., M. Auffhammer, and A. Levermann, "US Power Plant Sites at Risk of Future Sea-Level Rise," *Environmental Research Letters*, Vol. 10, No. 12, December 2015.

"Billions Allocated for Reconstruction in Wake of Typhoon Morakot," *AsiaNews*, August 20, 2009.

Birkland, Thomas A., and Carrie A. Schneider, "Emergency Management in the Courts: Trends After September 11 and Hurricane Katrina," *Justice System Journal*, Vol. 28, No. 1, 2007.

Bizo, Daniel, Rhonda Ascierto, Andy Lawrence, and Jacqueline Davis, *Uptime Institute Global Data Center Survey 2021*, Uptime Institute, September 1, 2021.

Bladon, Kevin D., Monica B. Emelko, Uldis Silins, and Micheal Stone, "Wildfire and the Future of Water Supply," *Environmental Science and Technology*, Vol. 48, No. 16, 2014.

Blanco, R. Ivan, G. Melodie Naja, Rosanna G. Rivero, and Rene M. Price, "Spatial and Temporal Changes in Groundwater Salinity in South Florida," *Applied Geochemistry*, Vol. 38, November 2013.

Blazier, Michael, "Wide-Ranging Damages [sic] Sustained in Louisiana's Kisatchie National Forest," Louisiana State University AgCenter, June 8, 2021.

Bloetscher, Frederick, Leonard Berry, Jarice Rodriguez-Seda, Nicole Hernandez Hammer, Thomas Romah, Dusan Jolovic, Barry Heimlich, and Maria Abadal Cahill, "Identifying FDOT's Physical Transportation Infrastructure Vulnerable to Sea Level Rise," *Journal of Infrastructure Systems*, Vol. 20, No. 2, June 2014.

Bloetscher, Frederick, Thomas Romah, Leonard Berry, Nicole Hernandez Hammer, and Maria Abadal Cahill, "Identification of Physical Transportation Infrastructure Vulnerable to Sea Level Rise," *Journal of Sustainable Development*, Vol. 5, No. 12, 2012.

Blom, Barry, with Jeffrey Schafer, "An Update to the Budget and Economic Outlook: 2021 to 2031," Congressional Budget Office, July 2021.

BNSF Railway, "Flooding Impacting Operations in Pacific Northwest," customer notification, November 17, 2021.

Boggess, J. M., G. W. Becker, and M. K. Mitchell, "Storm and Flood Hardening of Electrical Substations," *2014 IEEE PES T&D Conference and Exposition*, 2014.

Bolinger, Rebecca A., Vincent M. Brown, Christopher M. Fuhrmann, Karin L. Gleason, T. Andrew Joyner, Barry D. Keim, Amanda Lewis, John W. Nielsen-Gammon, Crystal J. Stiles, William Tollefson, Hannah E. Attard, and Alicia M. Bentley, "An Assessment of the Extremes and Impacts of the February 2021 South-Central U.S. Arctic Outbreak, and How Climate Services Can Help," *Weather and Climate Extremes*, Vol. 36, June 2022.

Boroush, Mark, and Ledia Guci, *Research and Development: U.S. Trends and International Comparisons*, National Science Foundation, National Science Board, NSB 2022-5, April 28, 2022.

Bousquin, Joe, "Wildfires Force Shut Down of Construction Sites in the West," *Construction Dive*, September 16, 2020.

Brown, M. E., J. M. Antle, P. Backlund, E. R. Carr, W. E. Easterling, M. K. Walsh, C. Ammann, W. Attavanich, C. B. Barrett, M. F. Bellemare, V. Dancheck, C. Funk, K. Grace, J. S. I. Ingram, H. Jiang, H. Maletta, T. Mata, A. Murray, M. Ngugi, D. Ojima, B. O'Neill, and C. Tebaldi, *Climate Change, Global Food Security, and the U.S. Food System*, U.S. Department of Agriculture, University Corporation for Atmospheric Research, and National Center for Atmospheric Research, December 2015.

Brunsdon, Chris, Jonathan Corcoran, Gary Higgs, and Andrew Ware, "The Influence of Weather on Local Geographical Patterns of Police Calls for Service," *Environment and Planning B: Urban Analytics and City Science*, Vol. 36, No. 5, October 2009.

Buchanan, Maya K., Scott Kulp, Lara Cushing, Rachel Morello-Frosch, Todd Nedwick, and Benjamin Strauss, "Sea Level Rise and Coastal Flooding Threaten Affordable Housing," *Environmental Research Letters*, Vol. 15, No. 12, December 2020.

Budryk, Zack, "Feds Cut Colorado River Allocation to Arizona, Nevada as Talks Fail," *The Hill*, August 16, 2022.

Bullock, Steven, "Climate Related Considerations in the Metals and Mining Sector," S&P Global Market Intelligence, blog post, July 7, 2020. As of October 18, 2022:
https://www.spglobal.com/marketintelligence/en/news-insights/blog/climate-related-considerations-in-the-metals-and-mining-sector

Burbidge, Rachel, "Adapting Aviation to a Changing Climate: Key Priorities for Action," *Journal of Air Transport Management*, Vol. 71, August 2018.

Bureau of Economic Analysis, "Interactive Data: Gross Output by Industry," dataset, last revised September 29, 2022. As of October 30, 2022:
https://apps.bea.gov/iTable/?reqid=150&step=2&isuri=1&categories=ugdpxind#eyJhcHBpZCI6MTUwLCJzdGVwcyI6WzEsMiwzXSwiZGF0YSI6W1siQ2F0ZWdvcmllcyIsIkdkcHHjJbmQiXSxbIlRhYmxlX0xpc3QiLCIyMTkiXV19

Bureau of Transportation Statistics, U.S. Department of Transportation, "U.S. Hazardous Materials Shipments by Transportation Mode, 2017," dataset, undated-a. As of November 2, 2022:
https://www.bts.gov/content/us-hazardous-materials-shipments-transportation-mode-2007

Bureau of Transportation Statistics, U.S. Department of Transportation, "U.S. Oil and Gas Pipeline Mileage," dataset, undated-b. As of October 10, 2022:
https://www.bts.gov/content/us-oil-and-gas-pipeline-mileage

Bureau of Transportation Statistics, U.S. Department of Transportation, *Transportation Statistics Annual Report 2020*, December 1, 2020.

Bureau of Transportation Statistics, U.S. Department of Transportation, "2022 Port Performance Freight Statistics Program: Supply-Chain Feature," January 2022.

Burillo, Daniel, Mikhail V. Chester, Stephanie Pincetl, and Eric Fournier, "Electricity Infrastructure Vulnerabilities Due to Long-Term Growth and Extreme Heat from Climate Change in Los Angeles County," *Energy Policy*, Vol. 128, May 2019.

Burillo, Daniel, Mikhail Chester, and Benjamin Ruddell, "Electric Grid Vulnerabilities to Rising Air Temperatures in Arizona," *Procedia Engineering*, Vol. 145, 2016.

Burke, Marshall, Solomon M. Hsiang, and Edward Miguel, "Climate and Conflict," *Annual Review of Economics*, Vol. 7, 2015a.

Burke, Marshall, Solomon M. Hsiang, and Edward Miguel, "Global Non-Linear Effect of Temperature on Economic Production," *Nature*, Vol. 527, November 12, 2015b.

Butke, Paul, and Scott C. Sheridan, "An Analysis of the Relationship Between Weather and Aggressive Crime in Cleveland, Ohio," *Weather, Climate, and Society*, Vol. 2, No. 2, April 2010.

Cachon, Gerard P., Santiago Gallino, and Marcelo Olivares, "Severe Weather and Automobile Assembly Productivity," Columbia Business School Research Paper 12/37, last revised December 22, 2012.

California Department of Insurance, "Mandatory One Year Moratorium on Non-Renewals," webpage, undated. As of July 21, 2023:
https://www.insurance.ca.gov/01-consumers/140-catastrophes/
MandatoryOneYearMoratoriumNonRenewals.cfm

California for All, California Natural Resources Agency, California Department of Water Resources, California Water Boards, California Environmental Protection Agency, and California Department of Food and Agriculture, *California's Water Supply Strategy: Adapting to a Hotter, Drier Future*, August 2022.

Callahan, Christopher W., and Justin S. Mankin, "Globally Unequal Effect of Extreme Heat on Economic Growth," *Science Advances*, Vol. 8, No. 43, October 28, 2022.

Calma, Justine, "Texas' Natural Gas Production Just Froze Under Pressure," *The Verge*, February 17, 2021.

Canning, Patrick, Sarah Rehkamp, Claudia Hitaj, and Christian J. Peters, *Resource Requirements of Food Demand in the United States*, U.S. Department of Agriculture, Economic Research Service, Economic Research Report 273, May 2020.

Capra, Gregory S., *Protecting Critical Rail Infrastructure*, Air University, U.S. Air Force Counterproliferation Center, Counterproliferation Papers, Future Warfare Series 38, December 2006.

Carlson, Christina, Gretchen Goldman, and Kristina Dahl, *Stormy Seas, Rising Risks: What Investors Should Know About Climate Change Impacts at Oil Refineries*, Center for Science and Democracy at the Union of Concerned Scientists, February 2015.

Carlton, Elizabeth J., Andrew P. Woster, Peter DeWitt, Rebecca S. Goldstein, and Karen Levy, "A Systematic Review and Meta-Analysis of Ambient Temperature and Diarrhoeal Diseases," *International Journal of Epidemiology*, Vol. 45, No. 1, February 2016.

Carter, Jacob, and Casey Kalman, *A Toxic Relationship: Extreme Coastal Flooding and Superfund Sites*, Union of Concerned Scientists, July 28, 2020.

Carter, Lynne, Adam Terando, Kirstin Dow, Kevin Hiers, Kenneth E. Kunkel, Aranzuzu Lascurain, Doug Marcy, Michael Osland, and Paul Schramm, "Southeast," in David R. Reidmiller, Christopher W. Avery, David R. Easterling, Kenneth E. Kunkel, Kristen L. M. Lewis, Thomas K. Maycock, and Brooke C. Stewart, eds., *Impacts, Risks, and Adaptation in the United States: Fourth National Climate Assessment*, Vol. II, U.S. Global Change Research Program, 2018.

CBO—*See* Congressional Budget Office.

Centers for Disease Control and Prevention, U.S. Department of Health and Human Services, "Industrial Water: Uses in Manufacturing and Industry," webpage, last reviewed October 11, 2016. As of January 25, 2022:
https://www.cdc.gov/healthywater/other/industrial/index.html

Central Office for Recovery, Reconstruction and Resiliency, *Transformation and Innovation in the Wake of Devastation: An Economic and Disaster Recovery Plan for Puerto Rico*, Government of Puerto Rico, August 8, 2018.

Chang, James I., and Cheng-Chung Lin, "A Study of Storage Tank Accidents," *Journal of Loss Prevention in the Process Industries*, Vol. 19, No. 1, January 2006.

Chang, Kuo-Liang, "A Reliable Waterway System Is Important to Agriculture," U.S. Department of Agriculture, Agriculture Marketing Service, February 2022.

Chatain, Léonie, Rahul Ghosh, Natalie Ambrosio Preudhomme, and Emilie Mazzacurati, "Critical Industries Have Substantial Exposure to Physical Climate Risks," Moody's ESG Solutions, November 2021.

Chavez, Steff, "Wildfires and 'Violent' Weather Leave Railroad Giant Facing $100m Bill," *Financial Times*, September 16, 2021.

Chinowsky, Paul, Jacob Helman, Sahil Gulati, James Neumann, and Jeremy Martinich, "Impacts of Climate Change on Operation of the US Rail Network," *Transport Policy*, Vol. 75, March 2019.

Chinowsky, Paul S., Jason C. Price, and James E. Neumann, "Assessment of Climate Change Adaptation Costs for the U.S. Road Network," *Global Environmental Change*, Vol. 23, No. 4, August 2013.

Choi-Schagrin, Winston, "Downpours from Ian Prompt Florida Treatment Plants to Release Waste," *New York Times*, September 29, 2022.

Chopra, Sunil, and ManMohan S. Sodhi, "Reducing the Risk of Supply Chain Disruptions," *MIT Sloan Management Review*, March 18, 2014.

CISA—*See* Cybersecurity and Infrastructure Security Agency.

City of Chicago and U.S. Environmental Protection Agency, "Climate Impacts on Agriculture and Food Supply," webpage, undated. As of October 10, 2022:
https://climatechange.chicago.gov/climate-impacts/climate-impacts-agriculture-and-food-supply

Claburn, Thomas, "Twitter Datacenter Melted Down in Labor Day Heat," *The Register*, September 13, 2022.

Climate Central, "Surging Seas: Risk Zone Map," webpage, undated. As of October 20, 2022:
https://ss2.climatecentral.org/#12/40.7298/
-74.0070?show=satellite&projections=0-K14_RCP85-SLR&level=5&unit=feet&pois=hide

Climate Central, "Surging Weather-Related Power Outages," webpage, September 13, 2022. As of October 31, 2022:
https://www.climatecentral.org/climate-matters/surging-weather-related-power-outages

Climate Change Adaptation Resource Center, U.S. Environmental Protection Agency, "Climate Adaptation and Saltwater Intrusion," last updated July 5, 2022a.

Climate Change Adaptation Resource Center, U.S. Environmental Protection Agency, "Climate Impacts on Water Utilities," webpage, last updated August 15, 2022b. As of October 30, 2022:
https://www.epa.gov/arc-x/climate-impacts-water-utilities

Climate-Related Market Risk Subcommittee, Market Risk Advisory Committee, U.S. Commodity Futures Trading Commission, *Managing Climate Risk in the U.S. Financial System*, circa 2020.

Code of Federal Regulations, Title 45, Public Welfare; Subtitle A, Department of Health and Human Services; Subchapter C, Administrative Data Standards and Related Requirements; Part 164, Security and Privacy; Subpart C, Security Standards for the Protection of Electronic Protected Health Information; Section 164.308, Administrative Safeguards.

Coffel, Ethan D., and Radley M. Horton, "Climate Change and the Impact of Extreme Temperatures on Aviation," *Weather, Climate, and Society*, Vol. 7, No. 1, January 2015.

Coffel, Ethan D., Terence R. Thompson, and Radley M. Horton, "The Impacts of Rising Temperatures on Aircraft Takeoff Performance," *Climatic Change*, Vol. 144, September 2017.

Cohen, Jeffrey P., and Catherine J. Morrison Paul, "Public Infrastructure Investment, Interstate Spatial Spillovers, and Manufacturing Costs," *Review of Economics and Statistics*, Vol. 86, No. 2, May 2004.

Colbert, Angela, "A Force of Nature: Hurricanes in a Changing Climate," National Aeronautics and Space Administration, Global Climate Change, June 1, 2022. As of November 22, 2022: https://climate.nasa.gov/news/3184/a-force-of-nature-hurricanes-in-a-changing-climate/

Cole v. Collier, memorandum and opinion setting out findings of fact and conclusions of law, U.S. District Court for the Southern District of Texas, Houston Division, entered July 19, 2017.

Collier, Stephen J., Rebecca Elliott, and Turo-Kimmo Lehtonen, "Climate Change and Insurance," *Economy and Society*, Vol. 50, No. 2, 2021.

Comes, Tina, and Bartel Van de Walle, "Measuring Disaster Resilience: The Impact of Hurricane Sandy on Critical Infrastructure Systems," *Proceedings of the 11th International ISCRAM Conference*, May 2014.

Committee on Adaptation to a Changing Climate, American Society of Civil Engineers, *Adapting Infrastructure and Civil Engineering Practice to a Changing Climate*, 2015.

Committee on Building Adaptable and Resilient Supply Chains After Hurricanes Harvey, Irma, and Maria; Office of Special Projects; Policy and Global Affairs; National Academies of Sciences, Engineering, and Medicine, *Strengthening Post-Hurricane Supply Chain Resilience: Observations from Hurricanes Harvey, Irma, and Maria*, National Academies Press, 2020.

Committee on Causes and Consequences of High Rates of Incarceration, Committee on Law and Justice, Division of Behavioral and Social Sciences and Education, National Research Council, *The Growth of Incarceration in the United States: Exploring Causes and Consequences*, National Academies Press, 2014.

Committee on Climate Change and U.S. Transportation, Division on Earth and Life Studies, Transportation Research Board, and National Research Council, *Potential Impacts of Climate Change on U.S. Transportation*, National Academies Press, Special Report 290, 2008.

Committee on Strengthening the Disaster Resilience of Academic Research Communities; Institute for Laboratory Animal Research; Board on Earth Sciences and Resources; Board on Health Sciences Policy; Division on Earth and Life Studies; Health and Medicine Division; National Academies of Sciences, Engineering, and Medicine, *Strengthening the Disaster Resilience of the Academic Biomedical Research Community: Protecting the Nation's Investment*, National Academies Press, 2017.

Commonwealth of Kentucky, "Knott County Court Operations Suspended Indefinitely Due to Historic Flooding," webpage, undated. As of September 29, 2022: https://kentucky.gov/Pages/Activity-stream.aspx?n=KentuckyCourtofJustice&prId=294

Congressional Budget Office, *Potential Increases in Hurricane Damage in the United States: Implications for the Federal Budget*, June 2016.

Cooper, Cheryl R., Marc Labonte, and David W. Perkins, *U.S. Payment System Policy Issues: Faster Payments and Innovation*, Congressional Research Service, R45927, September 23, 2019.

Cooper, Jennifer A., George W. Loomis, and Jose A. Amador, "Hell and High Water: Diminished Septic System Performance in Coastal Regions Due to Climate Change," *PLOS ONE*, Vol. 11, No. 9, 2016.

Coria, Jessica Monge, "Understanding Climate Risk: What We Learned About the Impact of Climate Risk on Affordable Housing Development," blog post, *SF Fed Blog*, March 4, 2022. As of June 1, 2023:
https://www.frbsf.org/our-district/about/sf-fed-blog/
understanding-climate-risk-impact-on-affordable-housing-development/

Cornwall, Warren, "For Scientists, Hurricane Ian Is Posing Threats—and Opportunities," *Science*, September 30, 2022.

Coyne, Justine, and Ali Oktay, "US Steel Market Keeping Watch of Logistics Issues in Wake of Hurricane Ida," *S&P Global Commodity Insights*, August 30, 2021.

Cruz, A. M., and E. Krausmann, "Damage to Offshore Oil and Gas Facilities Following Hurricanes Katrina and Rita: An Overview," *Journal of Loss Prevention in the Process Industries*, Vol. 21, No. 6, November 2008.

Cruz, Ana Maria, and Elisabeth Krausmann, "Vulnerability of the Oil and Gas Sector to Climate Change and Extreme Weather Events," *Climatic Change*, Vol. 121, 2013.

Cuijpers, Pim, Clara Miguel, Mathias Harrer, Constantin Yves Plessen, Marketa Ciharova, David Ebert, and Eirini Karyotaki, "Cognitive Behavior Therapy vs. Control Conditions, Other Psychotherapies, Pharmacotherapies and Combined Treatment for Depression: A Comprehensive Meta-Analysis Including 409 Trials with 52,702 Patients," *World Psychiatry*, Vol. 22, No. 1, February 2023.

Currie, Chris P., director, Homeland Security and Justice, U.S. Governmental Accountability Office, *FEMA Workforce: Long-Standing and New Challenges Could Affect Mission Success*, testimony before the U.S. House of Representatives Committee on Homeland Security Subcommittee on Emergency Preparedness, Response, and Recovery and Subcommittee on Oversight, Management, and Accountability, GAO-22-105631, January 20, 2022.

Cybersecurity and Infrastructure Security Agency, U.S. Department of Homeland Security, "Learn: What Are Dependencies and Why Should I Care?" webpage, undated-a. As of November 17, 2022:
https://www.cisa.gov/topics/critical-infrastructure-security-and-resilience/
resilience-services/infrastructure-dependency-primer/learn

Cybersecurity and Infrastructure Security Agency, U.S. Department of Homeland Security, "National Critical Functions," fact sheet, undated-b.

Cybersecurity and Infrastructure Security Agency, U.S. Department of Homeland Security, "Potential Hurricane Harvey Phishing Scams," alert, last revised September 20, 2017.

Cybersecurity and Infrastructure Security Agency, U.S. Department of Homeland Security, "National Critical Functions: Status Update to the Critical Infrastructure Community," July 2020.

Cybersecurity and Infrastructure Security Agency, U.S. Department of Homeland Security, "National Critical Functions: Status Update to the Critical Infrastructure Community," December 2021.

Cybersecurity and Infrastructure Security Agency, U.S. Department of Homeland Security, "Hurricane-Related Scams," alert, last revised September 30, 2022.

Dahad, Nitin, "Infineon and NXP Resume Austin Texas Fabs After Winter Storms," *EE Times*, March 3, 2021.

Daher, Daha Hassan, Léon Gaillard, Mohamed Amara, and Christophe Ménézo, "Impact of Tropical Desert Maritime Climate on the Performance of a PV Grid-Connected Power Plant," *Renewable Energy*, Vol. 125, September 2018.

Dahl, Kristina, Rachel Cleetus, Erika Spanger-Siegfried, Shana Udvardy, Astrid Caldas, and Pamela Worth, *Underwater: Rising Seas, Chronic Floods, and the Implications for US Coastal Real Estate*, Union of Concerned Scientists, June 18, 2018.

Dale, Larry, Michael Carnall, Max Wei, Gary Fitts, and Sarah Lewis McDonald, *Assessing the Impact of Wildfires on the California Electricity Grid*, California's Fourth Climate Change Assessment, California Energy Commission, CCCA4-CEC-2018-002, August 2018.

Dance, Scott, "Mississippi River Levels Are Dropping Too Low for Barges to Float," *Washington Post*, October 12, 2022.

Daniels, Jeff, "The Farm Belt Faces an Expensive Cleanup After Already-Costly Record Flooding," CNBC, March 29, 2019.

da Silva, Elizamar Ciríaco, Rejane Jurema Mansur Custódio Nogueira, Marcelle Almeida da Silva, and Manoel Bandeira de Albuquerque, "Drought Stress and Plant Nutrition," *Plant Stress*, Vol. 5, No. 1, 2011.

Davis, Stacy C., and Robert G. Boundy, *Transportation Energy Data Book*, 40th ed., Oak Ridge National Laboratory, ORNL/TM-2022/2376, February 2022, updated June 2022.

de Almeida, Beatriz Azevedo, and Ali Mostafavi, "Resilience of Infrastructure Systems to Sea-Level Rise in Coastal Areas: Impacts, Adaptation Measures, and Implementation Challenges," *Sustainability*, Vol. 8, No. 11, November 2016.

DeGood, Kevin, *A Reform Agenda for the U.S. Department of Transportation*, Center for American Progress, September 2020.

de Lima, Cicero Z., Jonathan R. Buzan, Frances C. Moore, Uris Lantz C. Baldos, Matthew Huber, and Thomas W. Hertel, "Heat Stress on Agricultural Workers Exacerbates Crop Impacts of Climate Change," *Environmental Research Letters*, Vol. 16, No. 4, April 2021.

Delaware County (Pennsylvania), "Poll Worker Pick Up Site at the Wharf Moved to Media Due to Coastal Flood Warning by the National Weather Service," webpage, undated. As of November 1, 2022: https://delcopa.gov/vote/news/wharfpickupsitemoved.html

Delevingne, Lindsay, Will Glazener, Liesbet Grégoir, and Kimberly Henderson, "Climate Risk and Decarbonization: What Every Mining CEO Needs to Know," McKinsey and Company, January 28, 2020.

Dennis, Brady, and Sarah Kaplan, "Jackson, Miss., Shows How Extreme Weather Can Trigger a Clean-Water Crisis," *Washington Post*, August 31, 2022.

Depsky, Nicholas, and Diego Pons, "Meteorological Droughts Are Projected to Worsen in Central America's Dry Corridor Throughout the 21st Century," *Environmental Research Letters*, Vol. 16, 2021.

Der Sarkissian, Rita, Jean-Marie Cariolet, Youssef Diab, and Marc Vuillet, "Investigating the Importance of Critical Infrastructures' Interdependencies During Recovery: Lessons from Hurricane Irma in Saint-Martin's Island," *International Journal of Disaster Risk Reduction*, Vol. 67, January 2022.

DeSalvo, Karen, Bob Hughes, Mary Bassett, Georges Benjamin, Michael Fraser, Sandro Galea, J. Nadine Gracia, and Jeffrey Howard, "Public Health COVID-19 Impact Assessment: Lessons Learned and Compelling Needs," National Academy of Medicine Perspectives, discussion paper, April 7, 2021.

DHS—*See* U.S. Department of Homeland Security.

DiChristopher, Tom, "Harvey Has 'Paralyzed' a Critical Part of US Manufacturing Supply Chain," CNBC, updated September 6, 2017.

Diffenbaugh, Noah S., Frances V. Davenport, and Marshall Burke, "Historical Warming Has Increased U.S. Crop Insurance Losses," *Environmental Research Letters*, Vol. 16, 2021.

Diffenbaugh, Noah S., Martin Scherer, and Robert J. Trapp, "Robust Increases in Severe Thunderstorm Environments in Response to Greenhouse Forcing," *Proceedings of the National Academy of Sciences of the United States of America*, Vol. 110, No. 41, September 23, 2013.

Division of Current Employment Statistics, U.S. Bureau of Labor Statistics, U.S. Department of Labor, "Current Employment Statistics—CES (National)," webpage, undated. As of October 3, 2022: https://www.bls.gov/ces/

Division of Foodborne, Waterborne, and Environmental Diseases; National Center for Emerging and Zoonotic Infectious Diseases; Centers for Disease Control and Prevention; U.S. Department of Health and Human Services, *Emergency Water Supply Planning Guide for Hospitals and Healthcare Facilities*, Centers for Disease Control and Prevention, U.S. Department of Health and Human Services, and American Water Works Association, 2012, updated 2019.

Dixon, Lloyd, Flavia Tsang, and Gary Fitts, *California Wildfires: Can Insurance Markets Handle the Risk?* RAND Corporation, RB-A635-1, 2020. As of October 11, 2022: https://www.rand.org/pubs/research_briefs/RBA635-1.html

DoD—*See* U.S. Department of Defense.

Dodgen, Daniel, Darrin Donato, Nancy Kelly, Annette La Greca, Joshua Morganstein, Joseph Reser, Josef Ruzek, Shulamit Schweitzer, Mark M. Shimamoto, Kimberly Thigpen Tart, and Robert Ursano, "Mental Health and Well-Being," in Allison Crimmins, John Balbus, Janet L. Gamble, Charles B. Beard, Jesse E. Bell, Daniel Dodgen, Rebecca J. Eisen, Neal Fann, Michelle D. Hawkins, Stephanie C. Herring, Lesley Jantarasami, David M. Mills, Shubhayu Saha, Marcus C. Sarofim, Juli Trtanj, and Lewis Ziska, eds., *The Impacts of Climate Change on Human Health in the United States: A Scientific Assessment*, U.S. Global Change Research Program, April 2016.

Dong, Jing, Micah Makaiwi, Navid Shafieirad, and Yundi Huang, *Modeling Multimodal Freight Transportation Network Performance Under Disruptions*, Iowa State University, Center for Transportation Research and Education, and University of Nebraska, Mid-America Transportation Center, TR 25-1121-0003-237, MATC TRB RiP No. 34770, February 1, 2015.

Dong, Xiuwen Sue; Gavin H. West, Alfreda Holloway-Beth, Xuanwen Wang, and Rosemary K. Sokas, "Heat-Related Deaths Among Construction Workers in the United States," *American Journal of Industrial Medicine*, Vol. 62, No. 12, December 2019.

Doré, Guy, Fujun Niu, and Heather Brooks, "Adaptation Methods for Transportation Infrastructure Built on Degrading Permafrost," *Permafrost and Periglacial Processes*, Vol. 27, No. 4, October–December 2016.

DOT—*See* U.S. Department of Transportation.

Douglas, Leah, "U.S. Crop Insurance Payouts Rise Sharply as Climate Change Worsens Droughts, Floods," Reuters, January 27, 2022.

Douglass, Scott L., and Bret M. Webb, *Highways in the Coastal Environment*, 3rd ed., Federal Highway Administration, Office of Bridges and Structures, Hydraulic Engineering Circular 25, FHWA-HIF-19-059, January 2020.

Doyle, Jim, Kim Hill, Debra Menk, and Richard Wallace, "Severe Weather and Manufacturing in America: Comparing the Cost of Droughts, Storms, and Extreme Temperatures with the Cost of New EPA Standards—2015 Update," Business Forward Foundation, circa September 2017.

Drought Response and Recovery Project for Water Utilities, U.S. Environmental Protection Agency, homepage, undated. As of November 2, 2022:
https://epa.maps.arcgis.com/apps/MapSeries/index.html?appid=22ce8bf3bcd742b68101d679828a00d7

Dudenhoeffer, Donald D., May R. Permann, and Milos Manic, "CIMS: A Framework for Infrastructure Interdependency Modeling and Analysis," *Proceedings of the 2006 Winter Simulation Conference*, 2006.

Dullo, Tigstu T., George K. Darkwah, Sudershan Gangrade, Mario Morales-Hernández, M. Bulbul Sharif, Alfred J. Kalyanapu, Shih-Chieh Kao, Sheikh Ghafoor, and Moetasim Ashfaq, "Assessing Climate-Change–Induced Flood Risk in the Conasauga River Watershed: An Application of Ensemble Hydrodynamic Inundation Modeling," *Natural Hazards and Earth System Sciences*, Vol. 21, No. 6, June 2, 2021.

Dzigbede, Komla D., Sarah Beth Gehl, and Katherine Willoughby, "Disaster Resiliency of U.S. Local Governments: Insights to Strengthen Local Response and Recovery from the COVID-19 Pandemic," *Public Administration Review*, Vol. 80, No. 4, July–August 2020.

Easterling, David R., Kenneth E. Kunkel, Jeff R. Arnold, Thomas R. Knutson, Allegra N. LeGrande, L. Ruby Leung, Russell S. Vose, Duane E. Waliser, and Michael F. Wehner, "Precipitation Change in the United States," in Donald J. Wuebbles, David W. Fahey, Kathy A. Hibbard, David J. Dokken, Brooke C. Stewart, and Thomas K. Maycock, eds., *Climate Science Special Report: Fourth National Climate Assessment*, Vol. I, U.S. Global Change Research Program, 2017.

Economic Research Service, U.S. Department of Agriculture, "Hurricane Impacts on Agriculture," webpage, last updated October 13, 2022. As of November 9, 2022:
https://www.ers.usda.gov/newsroom/trending-topics/hurricane-impacts-on-agriculture/

EIA—*See* U.S. Energy Information Administration.

Elkin, Elizabeth, Will Wade, and Michael Hirtzer, "Mississippi River Drought Imperils Trade on Key US Waterway," Bloomberg, updated October 6, 2022.

Elliott, Debbie, "Saltwater Is Moving Up the Mississippi River. Here's What's Being Done to Stop It," *All Things Considered*, October 27, 2022.

Enterprise Florida, "Manufacturing: Florida—The Future Is Here," brochure, undated.

EO 14008—*See* Biden, 2021.

EPA—*See* U.S. Environmental Protection Agency.

Erjongmanee, Supaporn, Chuanyi Ji, Jere Stokely, and Neale Hightower, "Large-Scale Inference of Network-Service Disruption upon Natural Disasters," in M. M. Gaber, R. R. Vatsavai, O. A. Omitaomu, J. Gama, N. V. Chawla, and A. R. Ganguly, eds., *Knowledge Discovery from Sensor Data (Sensor-KDD 2008)*, Springer, 2010.

Erjongmanee, Supaporn, and Chuanyi Ji, "Large-Scale Network-Service Disruption: Dependencies and External Factors," *IEEE Transactions on Network and Service Management*, Vol. 8, No. 4, December 2011.

Estill, Sarah, "Fires and Law Enforcement," *Community Policing Dispatch*, Vol. 11, No. 12, December 2018.

Evans, Melanie, "Cape Coral and Fort Myers Hospitals Evacuate Patients," *Wall Street Journal*, last updated October 2, 2022.

FAA—*See* Federal Aviation Administration.

Faki, Amro, Laxmi Sushama, and Guy Doré, "Regional-Scale Investigation of Pile Bearing Capacity for Canadian Permafrost Regions in a Warmer Climate," *Cold Regions Science and Technology*, Vol. 201, September 2022.

Fan, Jessie X., Ming Wen, and Neng Wan, "Built Environment and Active Commuting: Rural–Urban Differences in the U.S.," *SSM—Population Health*, Vol. 3, December 2017.

Fann, Neal, Terry M. Brennan, Patrick Dolwick, Janet L. Gamble, Vito Ilacqua, Laura Kolb, Christopher G. Nolte, Tanya L. Spero, and Lewis H. Ziska, "Air Quality Impacts," in Allison Crimmins, John Balbus, Janet L. Gamble, Charles B. Beard, Jesse E. Bell, Daniel Dodgen, Rebecca J. Eisen, Neal Fann, Michelle D. Hawkins, Stephanie C. Herring, Lesley Jantarasami, David M. Mills, Shubhayu Saha, Marcus C. Sarofim, Juli Trtanj, and Lewis Ziska, eds., *The Impacts of Climate Change on Human Health in the United States: A Scientific Assessment*, U.S. Global Change Research Program, April 2016.

Fann, Neal L., Christopher G. Nolte, Marcus C. Sarofim, Jeremy Martinich, and Nicholas J. Nassikas, "Associations Between Simulated Future Changes in Climate, Air Quality, and Human Health," *JAMA Network Open*, Vol. 4, No. 1, January 2021.

Fant, Charles, Brent Boehlert, Kenneth Strzepek, Peter Larsen, Alisa White, Sahil Gulati, Yue Li, and Jeremy Martinich, "Climate Change Impacts and Costs to U.S. Electricity Transmission and Distribution Infrastructure," *Energy*, Vol. 195, March 15, 2020.

Favata, E. A., G. Buckler, and M. Gochfeld, "Heat Stress in Hazardous Waste Workers: Evaluation and Prevention," *Occupational Medicine*, Vol. 5, No. 1, January–March 1990.

FCC—*See* Federal Communications Commission.

FDIC—*See* Federal Deposit Insurance Corporation.

Federal Aviation Administration, U.S. Department of Transportation, "Cost of Delay Estimates: 2019," APO-100, July 8, 2020.

Federal Aviation Administration, *The Economic Impact of U.S. Civil Aviation: 2020*, August 2022.

Federal Communications Commission, "Communications Status Report for Areas Impacted by Hurricane Matthew," October 11, 2016.

Federal Communications Commission, "After Storms, Watch Out for Scams," consumer guide, updated October 4, 2022.

Federal Deposit Insurance Corporation, "Hurricane Sandy: Information for Consumers and Bankers in the Affected Areas," webpage, last updated October 20, 2017. As of October 6, 2022:
https://www.fdic.gov/news/disaster/sandy.html

Federal Emergency Management Agency, U.S. Department of Homeland Security, "National Risk Index," map, undated. As of October 6, 2022:
https://hazards.fema.gov/nri/map

Federal Emergency Management Agency, U.S. Department of Homeland Security, *Design Guide for Improving Critical Facility Safety from Flooding and High Winds*, Risk Management Series, FEMA 543, January 2007.

Federal Emergency Management Agency, U.S. Department of Homeland Security, *Hurricane Ike in Texas and Louisiana: Building Performance Observations, Recommendations and Technical Guidance—Mitigation Assessment Team Report*, FEMA P-757, April 2009.

Federal Emergency Management Agency, U.S. Department of Homeland Security, "Mission Areas and Core Capabilities," webpage, last updated July 20, 2020a. As of October 4, 2022:
https://www.fema.gov/emergency-managers/national-preparedness/mission-core-capabilities

Federal Emergency Management Agency, U.S. Department of Homeland Security, *Building Codes Save: A Nationwide Study—Losses Avoided as a Result of Adopting Hazard-Resistant Building Codes*, November 2020b.

Federal Emergency Management Agency, U.S. Department of Homeland Security, *Protecting Communities and Saving Money: The Case for Adopting Building Codes*, November 2020c.

Federal Emergency Management Agency, U.S. Department of Homeland Security, "Regions, States and Territories," webpage, last updated August 29, 2022. As of January 23, 2024:
https://www.fema.gov/about/organization/regions

Federal Emergency Management Agency, U.S. Department of Homeland Security, and Assistant Secretary for Preparedness and Response, U.S. Department of Health and Human Services, *Healthcare Facilities and Power Outages: Guidance for State, Local, Tribal, Territorial, and Private Sector Partners*, August 2019.

Federal Housing Finance Agency, "Climate Change and Environmental, Social, and Governance (ESG)," webpage, last updated May 19, 2023. As of July 21, 2023:
https://www.fhfa.gov/PolicyProgramsResearch/Programs/Pages/Climate-Change-and-ESG.aspx

Federal Transit Administration, U.S. Department of Transportation, "Monthly Module Raw Data Release for August 2022," webpage, undated. As of October 11, 2022:
https://www.transit.dot.gov/ntd/data-product/monthly-module-raw-data-release

FEMA—*See* Federal Emergency Management Agency.

Financial Industry Regulatory Authority, "Business Continuity Plans and Emergency Contact Information," Financial and Operational Rule 4370, February 12, 2015.

First Street Foundation, *The First National Flood Risk Assessment: Defining America's Growing Risk*, 2020.

Fischbach, Jordan R., Linnea Warren May, Katie Whipkey, Shoshana R. Shelton, Christine Anne Vaughan, Devin Tierney, Kristin J. Leuschner, Lisa S. Meredith, and Hilary J. Peterson, *After Hurricane Maria: Predisaster Conditions, Hurricane Damage, and Recovery Needs in Puerto Rico*, Homeland Security Operational Analysis Center operated by the RAND Corporation, RR-2595-DHS, 2020. As of November 13, 2022:
https://www.rand.org/pubs/research_reports/RR2595.html

Flavelle, Christopher, "As Wildfires Get Worse, Insurers Pull Back from Riskiest Areas," *New York Times*, August 20, 2019.

Fleming, Elizabeth, Jeffrey Payne, William V. Sweet, Michael Craghan, John Haines, Juliette Finzi Hart, Heidi Stiller, and Ariana Sutton-Grier, "Coastal Effects," in David R. Reidmiller, Christopher W. Avery, David R. Easterling, Kenneth E. Kunkel, Kristen L. M. Lewis, Thomas K. Maycock, and Brooke C. Stewart, eds., *Impacts, Risks, and Adaptation in the United States: Fourth National Climate Assessment*, Vol. II, U.S. Global Change Research Program, 2018.

Flint, Amanda, and Laura Varriale, "German Steel Outlets Impacted by Heavy Rainfall, Cold-Roller Bilstein Declares Force Majeure," *S&P Global Commodity Insights*, July 15, 2021.

Focus FS, "Protecting Mining Operations from Severe and Extreme Weather," blog post, March 15, 2022. As of September 26, 2023:
https://focusfs.com/resources/blog/protecting-mining-operations-from-severe-and-extreme-weather/

Folkman, Steven, *Water Main Break Rates in the USA and Canada: A Comprehensive Study—An Asset Management Planning Tool for Water Utilities*, Utah State University, Buried Structures Laboratory, March 2018.

Food and Agriculture Organization of the United Nations, "Livestock and Climate Change," brochure, 2016.

Fortier, Claudine, and Alain Mailhot, "Climate Change Impact on Combined Sewer Overflows," *Journal of Water Resources Planning and Management*, Vol. 141, No. 5, May 2015.

Franz, Justin, "Update: Fires Impact Rail Ops in California, British Columbia," *Railfan and Railroad Magazine*, updated July 2, 2021.

Frederick, Rejane, Rebecca Cokley, Hannah Leibson, and Eliza Schultz, "Serving the Hardest Hit: Centering People with Disabilities in Emergency Planning and Response Efforts," brief, Center for American Progress, September 24, 2018.

Fusion Worldwide, "The Global Chip Shortage: A Timeline of Unfortunate Events," blog post, October 11, 2021. As of October 7, 2022:
https://info.fusionww.com/blog/the-global-chip-shortage-a-timeline-of-unfortunate-events

Gangrade, Sudershan, Shih-Chieh Kao, Tigstu T. Dullo, Alfred J. Kalyanapu, and Benjamin L. Preston, "Ensemble-Based Flood Vulnerability Assessment for Probable Maximum Flood in a Changing Environment," *Journal of Hydrology*, Vol. 576, September 2019.

Gao, Yang, Jian Lu, L. Ruby Leung, Qing Yang, Samson Hagos, and Yun Qian, "Dynamical and Thermodynamical Modulations on Future Changes of Landfalling Atmospheric Rivers over Western North America," *Geophysical Research Letters*, Vol. 42, No. 17, September 16, 2015.

Garcia, Katherine, "Offutt Air Force Base Rebuilds Three Years After Disastrous Flooding," KETV, updated March 21, 2022.

Gettinger, Andrew, "Reflections from a Health IT Perspective on Disaster Response," *HealthITbuzz*, November 15, 2017. As of October 25, 2022:
https://www.healthit.gov/buzz-blog/health-it/reflections-health-perspective-disaster-response

Glaser, John A., "Good Chemical Manufacturing Process Criteria," *Clean Technologies and Environmental Policy*, Vol. 16, February 2014.

Glick, Rich, and Matthew Christiansen, "FERC and Climate Change," *Energy Law Journal*, Vol. 40, No. 1, May 2019.

Godden, Lee, Raymond L. Ison, and Philip J. Wallis, "Water Governance in a Climate Change World: Appraising Systemic and Adaptive Effectiveness," *Water Resources Management*, Vol. 25, December 2011.

Godoy, Luis A., "Performance of Storage Tanks in Oil Facilities Damaged by Hurricanes Katrina and Rita," *Journal of Performance of Constructed Facilities*, Vol. 21, No. 6, December 2007.

Goin, Dana E., Kara E. Rudolph, and Jennifer Ahern, "Impact of Drought on Crime in California: A Synthetic Control Approach," *PLOS ONE*, Vol. 12, No. 10, 2017.

Goldberg, Jodi, and Emily Dalessio, "Spacing Out Resiliency: Why Satellite Technology Is Vital to Resilient Networks," *National Law Review*, Vol. 12, No. 90, March 31, 2022.

Goldberg, Pinelopi Koujianou, Amit Kumar Khandelwal, Nina Pavcnik, and Petia Topalova, "Imported Intermediate Inputs and Domestic Product Growth: Evidence from India," *Quarterly Journal of Economics*, Vol. 125, No. 4, November 2010.

Goldmann, Emily, and Sandro Galea, "Mental Health Consequences of Disasters," *Annual Review of Public Health*, Vol. 35, March 2014.

Goldstein, Mark L., *Telecommunications: FCC Should Improve Monitoring of Industry Efforts to Strengthen Wireless Network Resiliency*, report to the ranking member, U.S. House of Representatives Committee on Energy and Commerce, GAO-18-198, December 12, 2017.

Golnaraghi, Maryam, and Geneva Association Task Force on Climate Change Risk Assessment for the Insurance Industry, *Climate Change Risk Assessment for the Insurance Industry: A Holistic Decision-Making Framework and Key Considerations for Both Sides of the Balance Sheet*, Geneva Association, February 2021.

Gomes, Teresa, János Tapolcai, Christian Esposito, David Hutchison, Fernando Kuipers, Jacek Rak, Amaro de Sousa, Athanasios Iossifides, Rui Travanca, João André, Luísa Jorge, Lúcia Martins, Patricia Ortiz Ugalde, Alija Pašić, Dimitrios Pezaros, Simon Jouet, Stefano Secci, and Massimo Tornatore, "A Survey of Strategies for Communication Networks to Protect Against Large-Scale Natural Disasters," *2016 8th International Workshop on Resilient Networks Design and Modeling (RNDM)*, 2016.

Gomez, Brad T., Thomas G. Hansford, and George A. Krause, "The Republicans Should Pray for Rain: Weather, Turnout, and Voting in U.S. Presidential Elections," *Journal of Politics*, Vol. 69, No. 3, 2007.

Gómez, J. Alfredo, *Superfund: EPA Should Take Additional Actions to Manage Risks from Climate Change*, U.S. Government Accountability Office, GAO-20-73, October 18, 2019a.

Gómez, J. Alfredo, director, Natural Resources and Environment, U.S. Government Accountability Office, *Climate Change: Potential Economic Costs and Opportunities to Reduce Federal Fiscal Exposure*, testimony before the U.S. House of Representatives Committee on Oversight and Reform Subcommittee on Environment, GAO-20-338T, December 19, 2019b.

Gómez, J. Alfredo, *Water Infrastructure: Technical Assistance and Climate Resilience Planning Could Help Utilities Prepare for Potential Climate Change Impacts*, U.S. Government Accountability Office, GAO-20-24, January 16, 2020.

Gómez, J. Alfredo, *Chemical Accident Prevention: EPA Should Ensure Regulated Facilities Consider Risks from Climate Change*, U.S. Government Accountability Office, GAO-22-104494, February 28, 2022.

Goodman, Peter S., "'It's Not Sustainable': What America's Port Crisis Looks Like Up Close," *New York Times*, October 10, 2021, updated October 14, 2021.

Goodwin, Bradford S., Jr., and John C. Donaho, "Tropical Storm and Hurricane Recovery and Preparedness Strategies," *ILAR Journal*, Vol. 51, No. 2, 2010.

Goodwin, Gretta L., *Bureau of Prisons: Enhanced Data Capabilities, Analysis, Sharing, and Risk Assessments Needed for Disaster Preparedness*, U.S. Government Accountability Office, GAO-22-104289, February 2, 2022.

Gordon, Kate, *Risky Business: The Economic Risks of Climate Change in the United States—A Climate Risk Assessment for the United States*, Risky Business Project, 2014.

Gout, Elise, Jamil Modaffari, and Kevin DeGood, *The Compound Benefits of Greening School Infrastructure*, Center for American Progress, May 17, 2021.

Gowda, Prasanna, Jean L. Steiner, Carolyn Olson, Mark Boggess, Tracey Farrigan, and Michael A. Grusak, "Agriculture and Rural Communities," in David R. Reidmiller, Christopher W. Avery, David R. Easterling, Kenneth E. Kunkel, Kristen L. M. Lewis, Thomas K. Maycock, and Brooke C. Stewart, eds., *Impacts, Risks, and Adaptation in the United States: Fourth National Climate Assessment*, Vol. II, U.S. Global Change Research Program, 2018.

Graff Zivin, Joshua, and Matthew Neidell, "Temperature and the Allocation of Time: Implications for Climate Change," *Journal of Labor Economics*, Vol. 32, No. 1, January 2014.

Gratton, G. B., P. D. Williams, A. Padhra, and S. Rapsomanikis, "Reviewing the Impacts of Climate Change on Air Transport Operations," *Aeronautical Journal*, Vol. 126, No. 1295, January 2022.

Grenzeback, Lance R., and Andrew T. Lukmann, *Case Study of the Transportation Sector's Response to and Recovery from Hurricanes Katrina and Rita*, Transportation Research Board, 2008.

Griggs, Gary, and Kiki Patsch, "The Protection/Hardening of California's Coast: Time Are Changing," *Journal of Coastal Research*, Vol. 35, No. 5, September 2019.

Griggs, Troy, Andrew W. Lehren, Nadja Popovich, Anjali Singhvi, and Hiroko Tabuchi, "More Than 40 Sites Released Hazardous Pollutants Because of Hurricane Harvey," *New York Times*, September 8, 2017.

Grimaldi, Antonio, Kia Javanmardian, Dickon Pinner, Hamid Samandari, and Kurt Strovink, "Climate Change and P&C Insurance: The Threat and Opportunity," McKinsey and Company, November 19, 2020.

Gronewold, Andrew D., Vincent Fortin, Brent Lofgren, Anne Clites, Craig A. Stow, and Frank Quinn, "Coasts, Water Levels, and Climate Change: A Great Lakes Perspective," *Climatic Change*, Vol. 120, October 2013.

Gross, Liza, "In California's Farm Country, Climate Change Is Likely to Trigger More Pesticide Use, Fouling Waterways," *Inside Climate News*, May 10, 2021.

Gubernot, Diane M., G. Brooke Anderson, and Katherine L. Hunting, "The Epidemiology of Occupational Heat-Related Morbidity and Mortality in the United States: A Review of the Literature and Assessment of Research Needs in a Changing Climate," *International Journal of Biometeorology*, Vol. 58, No. 8, October 2014.

Guenther, Robin, and John Balbus, *Primary Protection: Enhancing Health Care Resiliency for a Changing Climate*, U.S. Department of Health and Human Services, December 2014.

Gura, David, "Silence on Wall Street: New York Stock Exchange Prepares for All-Electronic Trading," NBC News, March 23, 2020.

Hampton, Liz, "After Ida, US Energy Pipelines Off Line, Damage Being Assessed," Reuters, August 30, 2021.

Hargreaves, Steve, "Drought Strains U.S. Oil Production," *CNN Money*, July 31, 2012.

Harp, Ryan D., and Kristopher B. Karnauskas, "Global Warming to Increase Violent Crime," *Environmental Research Letters*, Vol. 15, 2020.

Hasemyer, David, "Trans-Alaska Pipeline Under Threat from Thawing Permafrost," *High Country News*, July 14, 2021.

Hayhoe, Katharine, Donald J. Wuebbles, David R. Easterling, David W. Fahey, Sarah Doherty, James P. Kossin, William V. Sweet, Russell S. Vose, and Michael F. Wehner, "Our Changing Climate," in David R. Reidmiller, Christopher W. Avery, David R. Easterling, Kenneth E. Kunkel, Kristen L. M. Lewis, Thomas K. Maycock, and Brooke C. Stewart, eds., *Impacts, Risks, and Adaptation in the United States: Fourth National Climate Assessment*, Vol. II, U.S. Global Change Research Program, 2018.

Heberger, Matthew, Heather Cooley, Pablo Herrera, Peter H. Gleick, and Eli Moore, "Potential Impacts of Increased Coastal Flooding in California Due to Sea-Level Rise," *Climatic Change*, Vol. 109, Suppl. 1, 2011.

Hegeman, Kimberly, "How Changing Climate Is Changing the Construction Industry," ForConstructionPros.com, March 7, 2019.

Heise, Rene, "NATO Is Responding to New Challenges Posed by Climate Change," *NATO Review*, April 1, 2021.

Helper, Susan, Timothy Krueger, and Howard Wial, *Locating American Manufacturing: Trends in the Geography of Production*, Metropolitan Policy Program, Brookings Institution, May 9, 2012.

Helper, Susan, and Evan Soltas, "Why the Pandemic Has Disrupted Supply Chains," blog post, Council of Economic Advisers, White House, June 17, 2021. As of November 6, 2022:
https://www.whitehouse.gov/cea/written-materials/2021/06/17/why-the-pandemic-has-disrupted-supply-chains/

Henry, Candise L., and Lincoln F. Pratson, "Differentiating the Effects of Climate Change–Induced Temperature and Streamflow Changes on the Vulnerability of Once-Through Thermoelectric Power Plants," *Environmental Science and Technology*, Vol. 53, No. 7, April 2, 2019.

Hermann, Alexander, "How Much Will Homeowners Spend to Rebuild and Repair After Hurricane Florence?" Joint Center for Housing Studies of Harvard University, September 24, 2018.

Hicke, Jeffrey A., Simone Lucatello, Linda D. Mortsch, Jackie Dawson, Mauricio Domínguez Aguilar, Carolyn A. F. Enquist, Elisabeth A. Gilmore, David S. Gutzler, Sherilee Harper, Kirstin Holsman, Elizabeth B. Jewett, Timothy A. Kohler, and Kathleen A. Miller, "North America," in Hans-Otto Pörtner, Debra C. Roberts, Melinda M. B. Tignor, Elvira Poloczanska, Katja Mintenbeck, Andrés Alegría, Marlies Craig, Stefanie Langsdorf, Sina Löschke, Vincent Möller, Andrew Okem, and Bardhyl Rama, eds., *Climate Change 2022: Impacts, Adaptation and Vulnerability—Working Group II Contribution to the Sixth Assessment Report of the Intergovernmental Panel on Climate Change*, 2022.

Hilbert, L. Brun, and Julian F. Hallai, "Natural Gas Production in Extreme Weather," *Pipeline and Gas Journal*, Vol. 248, No. 6, June 2021.

Hill, Garrett, "Satellite Internet Gateway Location Whitepaper," *X2nSat*, blog, October 23, 2019. As of October 6, 2022:
https://x2n.com/blog/satellite-internet-gateway-location-whitepaper/

Hinchman, David B., *Cybersecurity: Internet Architecture Is Considered Resilient, but Federal Agencies Continue to Address Risks*, U.S. Government Accountability Office, GAO-22-104560, March 3, 2022.

Hirschfeld, Daniella, and Brian Holland, "Sea Level Rise Adaptation Strategy for San Diego Bay: Executive Summary," ICLEI Local Governments for Sustainability, Public Agency Steering Committee, January 2012.

Hirtzer, Michael, Elizabeth Elkin, and Joe Deaux, "Dwindling Mississippi Grounds Barges, Threatens Shipments," Bloomberg, October 5, 2022, updated October 6, 2022.

Hjort, Jan, Dmitry Streletskiy, Guy Doré, Qingbai Wu, Kevin Bjella, and Miska Luoto, "Impacts of Permafrost Degradation on Infrastructure," *Nature Reviews Earth and Environment*, Vol. 3, January 2022.

Hodges, Tina, *Flooded Bus Barns and Buckled Rails: Public Transportation and Climate Change Adaptation*, U.S. Department of Transportation, Federal Transit Administration, Office of Research, Demonstration and Innovation, Report 0001, August 2011.

Holladay, Johnathan, Zia Abdullah, and Joshua Heyne, *Sustainable Aviation Fuel: Review of Technical Pathways*, U.S. Department of Energy, Office of Energy Efficiency and Renewable Energy, DOE/EE-2041, September 2020.

Holzheu, Thomas, Roman Lechner, Anja Vischer, Lucia Bevere, Daniel Staib, James Finucane, Adiya Belgibayeva, and Irina Fan, *More Risk: The Changing Nature of P&C Insurance Opportunities to 2040*, Swiss Re Institute, sigma 4/2021, September 6, 2021.

House, Darlene R., Irene Marete, and Eric M. Meslin, "To Research (or Not) That Is the Question: Ethical Issues in Research When Medical Care Is Disrupted by Political Action—A Case Study from Eldoret, Kenya," *Journal of Medical Ethics*, Vol. 42, No. 1, January 2016.

Hsiang, Solomon, Robert Kopp, Amir Jina, James Rising, Michael Delgado, Shashank Mohan, D. J. Rasmussen, Robert Muir-Wood, Paul Wilson, Michael Oppenheimer, Kate Larsen, and Trevor Houser, "Estimating Economic Damage from Climate Change in the United States," *Science*, Vol. 356, No. 6345, June 30, 2017.

Huang, Xingying, Daniel L. Swain, and Alex D. Hall, "Future Precipitation Increase from Very High Resolution Ensemble Downscaling of Extreme Atmospheric River Storms in California," *Science Advances*, Vol. 6, No. 29, July 15, 2020.

Huffstutter, P. J., and Mark Weinraub, "U.S. Farmers Face Supply Shortages, Higher Costs After Hurricane Ida," Reuters, September 22, 2021.

Hugelius, Karin, Mike Adams, and Eila Romo-Murphy, "The Power of Radio to Promote Health and Resilience in Natural Disasters: A Review," *International Journal of Environmental Research and Public Health*, Vol. 16, No. 14, July 2019.

Hummel, Michelle A., Matthew S. Berry, and Mark T. Stacey, "Sea Level Rise Impacts on Wastewater Treatment Systems Along the U.S. Coasts," *Earth's Future*, Vol. 6, No. 4, April 2018.

"Hurricane Ida: The Aftermath for Ag Infrastructure," *Farm Progress*, September 14, 2021.

Ido, Keisuke, Naoki Nakamura, and Masaharu Nakayama, "Miyagi Medical and Welfare Information Network: A Backup System for Patient Clinical Information After the Great East Japan Earthquake and Tsunami," *Tohoku Journal of Experimental Medicine*, Vol. 248, No. 1, May 2019.

In re Response to Impact of Hurricane Ida, "Omnibus Order Suspending Deadlines," General Order 2021-7, U.S. District Court, Middle District of Louisiana, September 2, 2021.

Inforum, "The National Impact of a Los Angeles and Long Beach Port Stoppage," prepared for National Association of Manufacturers, June 2022.

Interagency Security Committee, *Best Practices and Key Considerations for Enhancing Federal Facility Security and Resilience to Climate-Related Hazards*, 1st ed., December 2015.

Intergovernmental Panel on Climate Change, *Climate Change 2014: Synthesis Report—Contribution of Working Groups I, II and III to the Fifth Assessment Report of the Intergovernmental Panel on Climate Change*, 2014.

International Code Council, *2021 International Wildland–Urban Interface Code*®, 2nd printing, August 2021.

International Energy Agency, "Chemicals," webpage, undated. As of November 2, 2022: https://www.iea.org/fuels-and-technologies/chemicals

International Trade Administration, U.S. Department of Commerce, "Steel Imports Report: United States," *Global Steel Trade Monitor*, May 2020. As of October 18, 2022: https://legacy.trade.gov/steel/countries/pdfs/imports-us.pdf

Izaguirre, C., I. J. Losada, P. Camus, J. L. Vigh, and V. Stenek, "Climate Change Risk to Global Port Operations," *Nature Climate Change*, Vol. 11, January 2021.

Jacklitsch, Brenda, W. Jon Williams, Kristin Musolin, Aitor Coca, Jung-Hyun Kim, and Nina Turner, *Criteria for a Recommended Standard: Occupational Exposure to Heat and Hot Environments– Revised Criteria 2016*, U.S. Department of Health and Human Services, Centers for Disease Control and Prevention, National Institute for Occupational Safety and Health, Publication 2016-106, February 2016.

Jacobs, Jennifer M., Lia R. Cattaneo, William Sweet, and Theodore Mansfield, "Recent and Future Outlooks for Nuisance Flooding Impacts on Roadways on the U.S. East Coast," *Transportation Research Record*, Vol. 2762, No. 2, December 2018.

Jacobs, Jennifer M., Michael Culp, Lia Cattaneo, Paul Chinowsky, Anne Choate, Susanne DesRoches, Scott Douglass, and Rawlings Miller, "Transportation," in David R. Reidmiller, Christopher W. Avery, David R. Easterling, Kenneth E. Kunkel, Kristen L. M. Lewis, Thomas K. Maycock, and Brooke C. Stewart, eds., *Impacts, Risks, and Adaptation in the United States: Fourth National Climate Assessment*, Vol. II, U.S. Global Change Research Program, 2018.

Jing, Liang, Hassan M. El-Houjeiri, Jean-Christophe Monfort, Adam R. Brandt, Mohammad S. Masnadi, Deborah Gordon, and Joule A. Bergerson, "Carbon Intensity of Global Crude Oil Refining and Mitigation Potential," *Nature Climate Change*, Vol. 10, June 2020.

Johnson, Cardell, *Wildland Fire: Barriers to Recruitment and Retention of Federal Wildland Firefighters*, U.S. Government Accountability Office, GAO-23-105517, November 2022.

Johnson, Richard J., Laura G. Sánchez-Lozada, Lee S. Newman, Miguel A. Lanaspa, Henry F. Diaz, Jay Lemery, Bernardo Rodriguez-Iturbe, Dean R. Tolan, Jaime Butler-Dawson, Yuka Sato, Gabriela Garcia, Ana Andres Hernando, and Carlos A. Roncal-Jimenez, "Climate Change and the Kidney," *Annals of Nutrition and Metabolism*, Vol. 74, Suppl. 3, 2019.

Johnston, Lucy, Thomas M. Wilson, and Alexander MacKenzie, "Assisting Ph.D. Completion Following a Natural Disaster," *International Journal of Doctoral Studies*, Vol. 11, 2016.

Johnston, Peter Campbell, Jose Frazier Gomez, and Benoit Laplante, *Climate Risk and Adaptation in the Electric Power Sector*, Asian Development Bank, 2012.

Jonas, Ilaina, "DTCC Finds 1.3 Million Soaked Securities in Sandy-Flooded NY Vault," Reuters, November 14, 2012.

Jones, Alexi, "Cruel and Unusual Punishment: When States Don't Provide Air Conditioning in Prison," Prison Policy Initiative, briefing, June 18, 2019.

Jones, Matthew W., Adam Smith, Richard Betts, Josep G. Canadell, I. Colin Prentice, and Corinne Le Quéré, "Climate Change Increases the Risk of Wildfires," *ScienceBrief Review*, January 2020.

Jonkeren, Olaf, Piet Rietveld, Jos van Ommeren, and Aline te Linde, "Climate Change and Economic Consequences for Inland Waterway Transport in Europe," *Regional Environmental Change*, Vol. 14, June 2014.

Judson, Nicholas M. F., *Interdependence of the Electricity Generation System and the Natural Gas System and Implications for Energy Security*, Lincoln Laboratory, Massachusetts Institute of Technology, TR-1173, May 15, 2013.

Juliana v. United States, "Excerpts of Record Volume 1 (Pages 1–116)," U.S. Court of Appeals for the Ninth Circuit, February 1, 2019.

Kabore, Philippe, and Nicholas Rivers, "Manufacturing Output and Extreme Temperature: Evidence from Canada," Smart Prosperity Institute, Clean Economy Working Paper 20-11, November 2020.

Kafalenos, Robert S., Kenneth J. Leonard, Daniel M. Beagan, Virginia R. Burkett, Barry D. Keim, Alan Meyers, David T. Hunt, Robert C. Hyman, Michael K. Maynard, Barbara Fritsche, Russell H. Henk, Edward J. Seymour, Leslie E. Olson, Joanne R. Potter, and Michael J. Savonis, "What Are the Implications of Climate Change and Variability for Gulf Coast Transportation?" in Michael J. Savonis, Virginia R. Burkett, and Joanne R. Potter, eds., *Impacts of Climate Change and Variability on Transportation Systems and Infrastructure: Gulf Coast Study, Phase I—A Report by the U.S. Climate Change Science Program and the Subcommittee on Global Change Research*, U.S. Climate Change Science Program, Synthesis and Assessment Product 4.7, March 2008.

Kammeyer, Cora, Peter Gleick, Heather Cooley, Gregg Brill, Sonali Abraham, and Michael Cohen, "The 2021 Western Drought: What to Expect as Conditions Worsen," Pacific Institute, June 4, 2021.

Kane, Joseph W., Adie Tomer, and Caroline George, "Rethinking Climate Finance to Improve Infrastructure Resilience," Brookings Institution, June 22, 2021.

Kashiwagi, Yuzuka, Yasuyuki Todo, and Petr Matous, "Propagation of Economic Shocks Through Global Supply Chains: Evidence from Hurricane Sandy," *Review of International Economics*, Vol. 29, No. 5, November 2021.

Kavanagh, Jennifer, C. Ben Gibson, and Samantha Cherney, *Database of State Voting Laws: Preparing for Elections During a Pandemic*, RAND Corporation, TL-A112-1, 2020. As of November 12, 2022: https://www.rand.org/pubs/tools/TLA112-1.html

Keane, Brian, Gerhard Schwarz, and Peter Thernherr, "Electrical Equipment in Cold Weather Applications," *Industry Applications Society 60th Annual Petroleum and Chemical Industry Conference*, Industry Applications Society, Institute of Electrical and Electronics Engineers, 2013.

Keenan, Jesse M., and Jacob T. Bradt, "Underwaterwriting: From Theory to Empiricism in Regional Mortgage Markets in the U.S.," *Climatic Change*, Vol. 162, October 2020.

Kelleher, Suzanne Rowan, "Hurricane Ian: 3,500 Flights Canceled over Next Two Days," *Forbes*, September 28, 2022.

Kelley, Colin P., Shahrzad Mohtadi, Mark A. Cane, Richard Seager, and Yochanan Kushnir, "Climate Change in the Fertile Crescent and Implications of the Recent Syrian Drought," *Proceedings of the National Academy of Sciences*, Vol. 112, No. 11, March 17, 2015.

Keys, Benjamin J., "Can the Federal Mortgage Finance System Help Manage Climate Risk?" Wharton School, Risk Management and Decision Processes Center, Idea 6, undated.

Kintziger, Kristina W., Kahler W. Stone, Meredith A. Jagger, and Jennifer A. Horney, "The Impact of the COVID-19 Response on the Provision of Other Public Health Services in the U.S.: A Cross Sectional Study," *PLOS ONE*, Vol. 16, No. 10, October 14, 2021.

Kirgiz, Kivanc, Michelle Burtis, and David A. Lunin, "Petroleum-Refining Industry Business Interruption Losses Due to Hurricane Katrina," *Journal of Business Valuation and Economic Loss Analysis*, Vol. 4, No. 2, 2009.

Kjellstrom, Tord, Chris Freyberg, Bruno Lemke, Matthias Otto, and David Briggs, "Estimating Population Heat Exposure and Impacts on Working People in Conjunction with Climate Change," *International Journal of Biometeorology*, Vol. 62, 2018.

Klaver, Marieke, Eric Luiijf, and Albert Nieuwenhuijs, eds., *Good Practices Manual for CIP Policies*, Recommended Elements of Critical Infrastructure Protection for Policy Makers in Europe, 2011.

Kleinhans, Jan-Peter, and Julia Hess, *Understanding the Global Chip Shortages*, Stiftung Neue Verantwortung, November 2021.

Knutson, Thomas, Suzana J. Camargo, Johnny C. L. Chan, Kerry Emanuel, Chang-Hoi Ho, James Kossin, Mrutyunjay Mohapatra, Masaki Satoh, Masato Sugi, Kevin Walsh, and Liguang Wu, "Tropical Cyclones and Climate Change Assessment: Part II: Projected Response to Anthropogenic Warming," *Bulletin of the American Meteorological Society*, Vol. 101, No. 3, March 2020.

Knutson, Thomas R., John L. McBride, Johnny Chan, Kerry Emanuel, Greg Holland, Chris Landsea, Isaac Held, James P. Kossin, A. K. Srivastava, and Masato Sugi, "Tropical Cyclones and Climate Change," *Nature Geoscience*, Vol. 3, March 2010.

Knutson, Thomas R., Joseph J. Sirutis, Ming Zhao, Robert E. Tuleya, Morris Bender, Gabriel A. Vecchi, Gabriele Villarini, and Daniel Chavas, "Global Projections of Intense Tropical Cyclone Activity for the Late Twenty-First Century from Dynamic Downscaling of CMIP5/RCP4.5 Scenarios," *Journal of Climate*, Vol. 28, No. 18, September 2015.

Kobziar, Leda N., and George R. Thompson III, "Wildfire Smoke, a Potential Infectious Agent," *Science*, Vol. 370, No. 6523, December 18, 2020.

Konkel, Lindsey, "Taking the Heat: Potential Fetal Health Effects of Hot Temperatures," *Environmental Health Perspectives*, Vol. 127, No. 10, October 2019.

Korbatov, Annabella, Julia Price-Madison, Yihui Wang, and Yi Xu, *Lights Out: The Risks of Climate and Natural Disaster Related Disruptions to the Electric Grid*, John Hopkins University, School of Advanced International Studies, for Swiss Reinsurance Company, 2017.

Kossin, James P., T. Hall, Thomas R. Knutson, Kenneth E. Kunkel, R. J. Trapp, Duane E. Waliser, and Michael F. Wehner, "Extreme Storms," in Donald J. Wuebbles, David W. Fahey, Kathy A. Hibbard, David J. Dokken, Brooke C. Stewart, and Thomas K. Maycock, eds., *Climate Science Special Report: Fourth National Climate Assessment*, Vol. I, U.S. Global Change Research Program, 2017.

Kreisher, Otto, "Operation Unified Assistance," *Air and Space Forces Magazine*, April 1, 2005.

Krogmann, Uta, Lisa S. Boyles, William J. Bamka, Sumate Chaiprapat, and C. James Martel, "Biosolids and Sludge Management," *Water Environment Research*, Vol. 71, No. 5, August 1999.

Kuhfeld, Megan, Jim Soland, Karyn Lewis, and Emily Morton, "The Pandemic Has Had Devastating Impacts on Learning. What Will It Take to Help Students Catch Up?" Brookings Institution, March 3, 2022.

Kuligowski, Erica D., Franklin T. Lombardo, Long T. Phan, Marc L. Levitan, and David P. Jorgensen, *Final Report, National Institute of Standards and Technology (NIST): Technical Investigation of the May 22, 2011, Tornado in Joplin, Missouri*, National Institute of Standards and Technology, NIST NCSTAR 3, March 2014.

Kunreuther, Howard C., and Erwann O. Michel-Kerjan, *Climate Change, Insurability of Large-Scale Disasters and the Emerging Liability Challenge*, National Bureau of Economic Research, Working Paper 12821, January 2007.

Kunreuther, Howard, Erwann Michel-Kerjan, and Nicola Ranger, "Insuring Future Climate Catastrophes," *Climatic Change*, Vol. 118, May 2013.

Kwasinski, Alexis, Wayne W. Weaver, Patrick L. Chapman, and Philip T. Krein, "Telecommunications Power Plant Damage Assessment Caused by Hurricane Katrina: Site Survey and Follow-Up Results," *IEEE Systems Journal*, Vol. 3, No. 3, September 2009.

Kysar, Douglas A., "What Climate Change Can Do About Tort Law," *Environmental Law*, Vol. 41, No. 1, Winter 2011.

Labonte, Marc, *Supervision of U.S. Payment, Clearing, and Settlement Systems: Designation of Financial Market Utilities (FMUs)*, Congressional Research Service, R41529, September 10, 2012.

Lachman, Beth E., Susan A. Resetar, Nidhi Kalra, Agnes Gereben Schaefer, and Aimee E. Curtright, *Water Management, Partnerships, Rights, and Market Trends: An Overview for Army Installation Managers*, RAND Corporation, RR-933-A, 2016. As of November 10, 2022:
https://www.rand.org/pubs/research_reports/RR933.html

Lall, Upmanu, Thomas Johnson, Peter Colohan, Amir AghaKouchak, Casey Brown, Gregory McCabe, Roger Pulwarty, and A. Sankarasubramanian, "Water," in David R. Reidmiller, Christopher W. Avery, David R. Easterling, Kenneth E. Kunkel, Kristen L. M. Lewis, Thomas K. Maycock, and Brooke C. Stewart, eds., *Impacts, Risks, and Adaptation in the United States: Fourth National Climate Assessment*, Vol. II, U.S. Global Change Research Program, 2018.

LaRocco, Lori Ann, "New York Is Now the Nation's Busiest Port in a Historic Tipping Point for U.S.-Bound Trade," CNBC, September 24, 2022.

Larsen, Ethan, Daniel Hoffman, Carlos Rivera, Brian M. Kleiner, Christian Wernz, and Raj M. Ratwani, "Continuing Patient Care During Electronic Health Record Downtime," *Applied Clinical Informatics*, Vol. 10, No. 3, May 2019.

Larsen, Peter H., Kristina Hamachi LaCommare, Joseph H. Eto, and James L. Sweeney, *Assessing Changes in the Reliability of the U.S. Electric Power System*, Lawrence Berkeley National Laboratory, August 2015.

Lauland, Andrew, Benjamin Lee Preston, Kristin J. Leuschner, Michelle E. Miro, Liam Regan, Scott R. Stephenson, Rachel Steratore, Aaron Strong, Jonathan W. Welburn, and Jeffrey B. Wenger, *A Risk Assessment of National Critical Functions During COVID-19: Challenges and Opportunities*, Homeland Security Operational Analysis Center operated by the RAND Corporation, RR-A210-1, 2022. As of November 1, 2022:
https://www.rand.org/pubs/research_reports/RRA210-1.html

Laurito, Agustina, "The Effect of Home Country Natural Disasters on the Academic Outcomes of Immigrant Students in New York City," *Education Finance and Policy*, Vol. 17, No. 2, Spring 2022.

Lauritsen, Janet L., and Nicole White, *Seasonal Patterns in Criminal Victimization Trends*, U.S. Department of Justice, Office of Justice Programs, Bureau of Justice Statistics, 245959, June 2014.

Leibovici, Fernando, and Jason Dunn, "Supply Chain Bottlenecks and Inflation: The Role of Semiconductors," *Economic Synopses*, No. 28, December 16, 2021.

Letta, Marco, and Richard S. J. Tol, "Weather, Climate and Total Factor Productivity," *Environmental and Resource Economics*, Vol. 73, No. 1, May 2019.

Levenson, Laurie L., "Climate Change and the Criminal Justice System," *Environmental Law*, Vol. 51, No. 2, 2021.

Levin, Todd, Audun Botterud, W. Neal Mann, Jonghwan Kwon, and Zhi Zhou, "Extreme Weather and Electricity Markets: Key Lessons from the February 2021 Texas Crisis," *Joule*, Vol. 6, No. 1, January 19, 2022.

Levy, Barry S., Victor W. Sidel, and Jonathan A. Patz, "Climate Change and Collective Violence," *Annual Review of Public Health*, Vol. 38, 2017.

Levy, Karen, Andrew P. Woster, Rebecca S. Goldstein, and Elizabeth J. Carlton, "Untangling the Impacts of Climate Change on Waterborne Diseases: A Systematic Review of Relationships Between Diarrheal Diseases and Temperature, Rainfall, Flooding, and Drought," *Environmental Science and Technology*, Vol. 50, No. 10, May 17, 2016.

Li, Hsin-Chi, Lung-Sheng Hsieh, Liang-Chun Chen, Lee-Yaw Lin, and Wei-Sen Lin, "Disaster Investigation and Analysis of Typhoon Morakot," *Journal of the Chinese Institute of Engineers*, Vol. 37, No. 5, 2014.

Licker, Rachel, Kristina Dahl, and John T. Abatzoglou, "Quantifying the Impact of Future Extreme Heat on the Outdoor Work Sector in the United States," *Elementa: Science of the Anthropocene*, Vol. 10, No. 1, January 13, 2022.

Lindsey, Rebecca, "Understanding the Arctic Polar Vortex," National Oceanic and Atmospheric Administration, U.S. Department of Commerce, March 5, 2021.

Lindsey, Rebecca, "Climate Change: Global Sea Level," National Oceanic and Atmospheric Administration, U.S. Department of Commerce, April 19, 2022.

Linning, Shannon J., and Ian A. Silver, "Crime Fluctuations in Response to Hurricane Evacuations: Understanding the Time-Course of Crime Opportunities During Hurricane Harvey," *Natural Hazards Review*, Vol. 22, No. 3, August 2021.

Liu, Jia C., Gavin Pereira, Sarah A. Uhl, Mercedes A. Bravo, and Michelle L. Bell, "A Systematic Review of the Physical Health Impacts from the Non-Occupational Exposure to Wildfire Smoke," *Environmental Research*, Vol. 136, January 2015.

"Local Prosecution in the Era of Climate Change," *Harvard Law Review*, Vol. 135, No. 6, April 2022.

Loew, Aviva, Paulina Jaramillo, Haibo Zhai, Rahim Ali, Bart Nijssen, Yifan Cheng, and Kelly Klima, "Fossil Fuel–Fired Power Plant Operations Under a Changing Climate," *Climatic Change*, Vol. 163, No. 1, November 2020.

Lonas, Lexi, "Hurricane Ida Caused More Than $500M in Damage to Louisiana's Agriculture," *The Hill*, September 23, 2021.

Lopez-Cantu, Tania, and Constantine Samaras, "Temporal and Spatial Evaluation of Stormwater Engineering Standards Reveal Risks and Priorities Across the United States," *Environmental Research Letters*, Vol. 13, No. 7, July 2018.

Losey, Stephen, "After Massive Flood, Offutt Looks to Build a Better Base," *Air Force Times*, August 7, 2020.

Love, Geoff, Alice Soares, and Herbert Püempel, "Climate Change, Climate Variability and Transportation," *Procedia Environmental Sciences*, Vol. 1, 2010.

Luckman, Nina, "Is Weather Risk a Workers' Comp Problem?" *Risk and Insurance*, June 24, 2022.

Luther, Linda, *Disaster Debris Removal After Hurricane Katrina: Status and Associate Issues*, Congressional Research Service, RL33477, updated April 2, 2008.

Lykou, Georgia, Panagiotis Dedousis, George Stergiopoulos, and Dimitris Gritzalis, "Assessing Interdependencies and Congestion Delays in the Aviation Network," *IEEE Access*, Vol. 8, 2020.

Lyles, Ward, Philip Berke, and Gavin Smith, "A Comparison of Local Hazard Mitigation Plan Quality in Six States, USA," *Landscape and Urban Planning*, Vol. 122, February 2014.

Mallakpour, Iman, and Gabriele Villarini, "The Changing Nature of Flooding Across the Central United States," *Nature Climate Change*, Vol. 5, March 2015.

Mann, Rebecca, and Jenny Schuetz, "As Extreme Heat Grips the Globe, Access to Air Conditioning Is an Urgent Public Health Issue," Brookings Institution, July 25, 2022.

Manuel, John, "In Katrina's Wake," *Environmental Health Perspectives*, Vol. 114, No. 1, January 2006.

MarineTraffic, "What Is the Automatic Identification System (AIS)?" webpage, undated. As of October 3, 2022:
https://help.marinetraffic.com/hc/en-us/articles/
204581828-What-is-the-Automatic-Identification-System-AIS

Markon, Carl J., Stephen T. Gray, Matthew Berman, Laura Eerkes-Medrano, Thomas Hennessy, Henry P. Huntington, Jeremy Littell, Molly McCammon, Richard Thoman, and Sarah Trainor, "Alaska," in David R. Reidmiller, Christopher W. Avery, David R. Easterling, Kenneth E. Kunkel, Kristen L. M. Lewis, Thomas K. Maycock, and Brooke C. Stewart, eds., *Impacts, Risks, and Adaptation in the United States: Fourth National Climate Assessment*, Vol. II, U.S. Global Change Research Program, 2018.

Marsh, Joanna, "Northern California Wildfires Damage BNSF's Rail Infrastructure," *FreightWaves*, July 30, 2021.

Marshall, John, Jeffery R. Scott, Kyle C. Armour, J.-M. Campin, Maxwell Kelley, and Anastasia Romanou, "The Ocean's Role in the Transient Response of Climate to Abrupt Greenhouse Gas Forcing," *Climate Dynamics*, Vol. 44, April 2015.

Martin Associates, *2018 National Economic Impact of the U.S. Coastal Port System: Executive Summary*, American Association of Port Authorities, March 2019.

Marufuzzaman, Mohammad, and Sandra Duni Ekşioğlu, "Designing a Reliable and Dynamic Multimodal Transportation Network for Biofuel Supply Chains," *Transportation Science*, Vol. 51, No. 2, May 2017.

Masson-Delmotte, V., P. Zhai, A. Pirani, S. L. Connors, C. Péan, S. Berger, N. Caud, Y. Chen, L. Goldfarb, M. I. Gomis, M. Huang, K. Leitzell, E. Lonnoy, J. B. R. Matthews, T. K. Maycock, T. Waterfield, O. Yelekçi, R. Yu, and B. Zhou, eds., "Summary for Policymakers," in V. Masson-Delmotte, P. Zhai, A. Pirani, S. L. Connors, C. Péan, S. Berger, N. Caud, Y. Chen, L. Goldfarb, M. I. Gomis, M. Huang, K. Leitzell, E. Lonnoy, J. B. R. Matthews, T. K. Maycock, T. Waterfield, O. Yelekçi, R. Yu, and B. Zhou, eds., *Climate Change 2021: The Physical Science Basis—Contribution of Working Group I to the Sixth Assessment Report of the Intergovernmental Panel on Climate Change*, 2021.

Masson-Delmotte, Valérie, Panmao Zhai, Anna Pirani, Sarah L. Connors, Clotilde Péan, Yang Chen, Leah Goldfarb, Melissa I. Gomis, J. B. Robin Matthews, Sophie Berger, Mengtian Huang, Ozge Yelekçi, Rong Yu, Baiquan Zhou, Elisabeth Lonnoy, Thomas K. Maycock, Tim Waterfield, Katherine Leitzell, and Nada Caud, *Climate Change 2021: The Physical Science Basis—Contribution of Working Group I to the Sixth Assessment Report of the Intergovernmental Panel on Climate Change*, 2021.

Mason, Richard, James Bonomo, Tim Conley, Ryan Consaul, David R. Frelinger, David A. Galvan, Dahlia Anne Goldfeld, Scott A. Grossman, Brian A. Jackson, Michael Kennedy, Vernon R. Koym, Jason Mastbaum, Thao Liz Nguyen, Jenny Oberholtzer, Ellen M. Pint, Parousia Rockstroh, Melissa Shostak, Karlyn D. Stanley, Anne Stickells, Michael J. D. Vermeer, and Stephen M. Worman, *Analyzing a More Resilient National Positioning, Navigation, and Timing Capability*, Homeland Security Operational Analysis Center operated by the RAND Corporation, RR-2970-DHS, 2021. As of November 10, 2022:
https://www.rand.org/pubs/research_reports/RR2970.html

Masters, Jeff, "Third-Costliest Year on Record for Weather Disasters in 2021: $343 Billion in Damages [sic]," *Yale Climate Connections*, January 25, 2022.

Maxwell, Keely, Susan Julius, Anne Grambsch, Ann Kosmal, Libby Larson, and Nancy Sonti, "Built Environment, Urban Systems, and Cities," in David R. Reidmiller, Christopher W. Avery, David R. Easterling, Kenneth E. Kunkel, Kristen L. M. Lewis, Thomas K. Maycock, and Brooke C. Stewart, eds., *Impacts, Risks, and Adaptation in the United States: Fourth National Climate Assessment*, Vol. II, U.S. Global Change Research Program, 2018.

McAndrews, James J., and Simon M. Potter, "Liquidity Effects of the Events of September 11, 2001," *Federal Reserve Bank of New York Economic Policy Review*, November 2002.

McClellan, Matt, "University of Kentucky Rises Above Tornado Aftermath," *Nursery Management*, December 15, 2021.

McCollester, Maria, Michelle E. Miro, and Kristin Van Abel, *Building Resilience Together: Military and Local Government Collaboration for Climate Adaptation*, RAND Corporation, RR-3014-RC, 2020. As of November 10, 2022:
https://www.rand.org/pubs/research_reports/RR3014.html

McFarland, Christiana, Brooks Rainwater, Erica Grabowski, Joshua Pine, and Anita Yadavalli, "State of the Cities 2021," National League of Cities, Center for City Solutions, 2021.

McNeil, David, *Physical Risks of Climate Change: Assessing Geography of Exposure in US Residential Mortgage-Backed Securities—Fitch Ratings*, CFA Institute, October 26, 2020.

Melillo, Gianna, "Extreme Heat Is Forcing Students out of the Classroom," *The Hill*, August 31, 2022.

Melvin, April M., Peter Larsen, Brent Boehlert, James E. Neumann, Paul Chinowsky, Xavier Espinet, Jeremy Martinich, Matthew S. Baumann, Lisa Rennels, Alexandra Bothner, Dmitry J. Nicolsky, and Sergey S. Marchenko, "Climate Change Damages [sic] to Alaska Public Infrastructure and the Economics of Proactive Adaptation," *Proceedings of the National Academy of Sciences*, Vol. 114, No. 2, December 27, 2016.

Mena, Bryan, "California Wildfires Can Bring Internet Outages. Some Want Networks to Be Tougher," *San Francisco Chronicle*, September 30, 2020.

Mendhall, Elizabeth, Cullen Hendrix, Elizabeth Nyman, Paige M. Roberts, John Robison Hoopes, James R. Watson, Vicky W. Y. Lam, and U. Rashid Sumaila, "Climate Change Increases the Risk of Fisheries Conflict," *Marine Policy*, Vol. 117, July 2020.

Metallurgprom, "US Steel Restarts Blast Furnace Following a Major Flood at the Plant Gary Works," December 1, 2019.

Michalak, Julia L., Josh J. Lawler, John E. Gross, and Caitlin E. Littlefield, *A Strategic Analysis of Climate Vulnerability of National Park Resources and Values*, National Park Service, Natural Resource Report NPS/NRSS/CCRP/NRR-2021/2293, August 2021.

Micik, Katie, "River Woes: Drought Helps Bring River Transportation into Spotlight," *Progressive Farmer*, June 25, 2013.

Miller, Benjamin M., Debra Knopman, Liisa Ecola, Brian Phillips, Moon Kim, Nathaniel Edenfield, Daniel Schwam, and Diogo Prosdocimi, *U.S. Airport Infrastructure Funding and Financing: Issues and Policy Options Pursuant to Section 122 of the 2018 Federal Aviation Administration Reauthorization Act*, RAND Corporation, RR-3175-FAA, 2020. As of November 13, 2022:
https://www.rand.org/pubs/research_reports/RR3175.html

Miller, Rebecca K., and Iris Hui, "Impact of Short School Closures (1–5 Days) on Overall Academic Performance of Schools in California," *Scientific Reports*, Vol. 12, 2022.

Miller, Rich, "The Top 10 Cloud Campuses," *Data Center Frontier*, November 23, 2015.

Miller, Rich, "Uptime: Longer Data Center Outages Are Becoming More Common," *Data Center Frontier*, June 8, 2022.

Miller, Sue, Matt Reeves, Karen Bagne, and John Tanaka, "Where's the Beef? Predicting the Effects of Climate Change on Cattle Production in Western U.S. Rangelands," *Science You Can Use Bulletin*, No. 27, September–October 2017.

Milly, P. C. D., Julio Betancourt, Malin Falkenmark, Robert M. Hirsch, Zbigniew W. Kundzewicz, Dennis P. Lettenmaier, and Ronald J. Stouffer, "Stationarity Is Dead: Whither Water Management?" *Science*, Vol. 319, No. 5863, February 1, 2008.

"Miners' Profits Face an Unusual Foe: Extreme Weather," Reuters, July 29, 2022.

Miro, Michelle E., Andrew Lauland, Rahim Ali, Edward W. Chan, Richard H. Donohue, Liisa Ecola, Timothy R. Gulden, Liam Regan, Karen M. Sudkamp, Tobias Sytsma, Michael T. Wilson, and Chandler Sachs, *Assessing Risk to the National Critical Functions as a Result of Climate Change*, Homeland Security Operational Analysis Center operated by the RAND Corporation, RR-A1645-7, 2022. As of June 3, 2023:
https://www.rand.org/pubs/research_reports/RRA1645-7.html

Mitchell, Ian, Lee Robinson, and Atousa Tahmasebi, "Valuing Climate Liability," Center for Global Development note, January 13, 2021.

Moftakhari, Hamed R., Amir AghaKouchak, Brett F. Sanders, Maura Allaire, and Richard A. Matthew, "What Is Nuisance Flooding? Defining and Monitoring an Emerging Challenge," *Water Resources Research*, Vol. 54, No. 7, July 2018.

Mohafezatkar Sereshkeh, Abolfazl, and Reza Jamshidi Chenari, "Induced Settlement Reduction of Adjacent Masonry Building in Residential Constructions," *Civil Engineering Journal*, Vol. 3, No. 7, 2017.

Mohaddes, Kamiar, Ryan N. C. Ng, M. Hashem Pesaran, Mehdi Raissi, and Jui-Chung Yang, "Climate Change and Economic Activity: Evidence from U.S. States," Center for Economic Studies and ifo Institute Working Paper 9542, January 2022.

Mondoro, Alysson, Dan M. Frangopol, and Mohamed Soliman, "Optimal Risk-Based Management of Coastal Bridges Vulnerable to Hurricanes," *Journal of Infrastructure Systems*, Vol. 23, No. 3, September 2017.

Mondschein, Jared, Jonathan W. Welburn, and Daniel Gonzales, *Securing the Microelectronics Supply Chain: Four Policy Issues for the U.S. Department of Defense to Consider*, RAND Corporation, PE-A1394-1, 2022. As of November 10, 2022:
https://www.rand.org/pubs/perspectives/PEA1394-1.html

Montgomery, David, and Simon Romero, "'Such Dire Straits': Chaos Unfolds in Texas Hospitals," *New York Times*, February 18, 2021.

Moody, Reginald F., "Radio's Role During Hurricane Katrina: A Case Study of WWL Radio and the United Radio Broadcasters of New Orleans," *Journal of Radio and Audio Media*, Vol. 16, No. 2, 2009.

Moody's Investors Service, "Climate Change Risks Outweigh Opportunities for P&C (Re)Insurers," sector brief, March 15, 2018.

Moris, Francisco, *Definitions of Research and Development: An Annotated Compilation of Official Sources*, National Science Foundation, National Center for Science and Engineering Statistics, March 2018.

Morley, Michael T., "Election Emergencies: Voting in the Wake of Natural Disasters and Terrorist Attacks," *Emory Law Journal*, Vol. 67, No. 3, 2018.

Morris, Kevin, and Peter Miller, "Authority After the Tempest: Hurricane Michael and the 2018 Elections," *Journal of Politics*, Vol. 85, No. 2, April 2023.

Murphy, Caitlin, Trieu Mai, Yinong Sun, Paige Jadun, Paul Donohoo-Vallett, Matteo Muratori, Ryan Jones, and Brent Nelson, "High Electrification Futures: Impacts to the U.S. Bulk Power System," *Electricity Journal*, Vol. 33, No. 10, December 2020.

Murphy, Tom, and Laura Ungar, "Production at Bedeviled Baby Formula Factory Halted by Storm," Associated Press, June 16, 2022.

Murray, Brendan, "The Container Ship Backlog Outside Los Angeles Ports Is Almost Cleared," Bloomberg, August 30, 2022.

Nagamatsu, Shingo, Adam Rose, and Jonathan Eyer, "Return Migration and Decontamination After the 2011 Fukushima Nuclear Power Plant Disaster," *Risk Analysis*, Vol. 40, No. 4, April 2020.

Narayanan, Anu, Michael J. Lostumbo, Kristin Van Abel, Michael T. Wilson, Anna Jean Wirth, and Rahim Ali, *The Growing Exposure of Air Force Installations to Natural Disasters*, RAND Corporation, RB-A523-1, 2021. As of November 12, 2022:
https://www.rand.org/pubs/research_briefs/RBA523-1.html

Narayanan, Anu, Henry H. Willis, Jordan R. Fischbach, Drake Warren, Edmundo Molina-Perez, Chuck Stelzner, Kathleen Loa, Lauren Kendrick, Paul Sorensen, and Tom LaTourrette, *Characterizing National Exposures to Infrastructure from Natural Disasters: Data and Methods Documentation*, RAND Corporation, RR-1453/1-DHS, 2016. As of November 13, 2022:
https://www.rand.org/pubs/research_reports/RR1453z1.html

Nasr, Amro, I. Björnsson, D. Honfi, O. Larsson Ivanov, J. Johansson, and E. Kjellström, "A Review of the Potential Impacts of Climate Change on the Safety and Performance of Bridges," *Sustainable and Resilient Infrastructure*, Vol. 6, No. 3–4, 2021.

Nateghi, Roshanak, "Multi-Dimensional Infrastructure Resilience Modeling: An Application to Hurricane-Prone Electric Power Distribution Systems," *IEEE Access*, Vol. 6, 2018.

National Aeronautics and Space Administration, "NASA Kennedy Buttons Up Operations Before Hurricane Ian," webpage, September 26, 2022. As of October 7, 2022:
https://www.nasa.gov/feature/nasa-kennedy-buttons-up-operations-before-hurricane-ian

National Center for Health Statistics, Centers for Disease Control and Prevention, U.S. Department of Health and Human Services, "Electronic Medical Records/Electronic Health Records (EMRs/EHRs)," webpage, last reviewed January 17, 2023. As of September 23, 2022:
https://www.cdc.gov/nchs/fastats/electronic-medical-records.htm

National Center for O*NET Development, "O*NET OnLine Help: The Database," webpage, undated. As of October 20, 2022:
https://www.onetonline.org/help/onet/database

National Center for O*NET Development, "Chemical Equipment Operators and Tenders," Occupation 51-9011.00, updated 2023.

National Centers for Environmental Information, National Oceanic and Atmospheric Administration, "Billion-Dollar Weather and Climate Disasters," webpage, undated-a. As of June 14, 2023:
https://www.ncei.noaa.gov/access/billions/

National Centers for Environmental Information, National Oceanic and Atmospheric Administration, "Storm Events Database," undated-b.

National Conference of State Legislatures, "Election Emergencies," webpage, updated September 1, 2020. As of September 28, 2022:
https://www.ncsl.org/elections-and-campaigns/election-emergencies

National Conference of State Legislatures, *2019–2020 State Legislative Report on Natural Disasters*, updated January 19, 2021.

National Institute of Standards and Technology, U.S. Department of Commerce, *Community Resilience Planning Guide for Buildings and Infrastructure Systems*, Volume II, Special Publication 1190, May 2016.

National Integrated Drought Information System, National Oceanic and Atmospheric Administration, U.S. Department of Commerce, "Manufacturing," webpage, undated. As of November 11, 2022:
https://www.drought.gov/sectors/manufacturing

National Intelligence Council, Office of the Director of National Intelligence, *Global Trends 2040: A More Contested World*, March 2021.

National Oceanic and Atmospheric Administration, U.S. Department of Commerce, "'Average' Atlantic Hurricane Season to Reflect More Storms," webpage, last updated April 9, 2021. As of November 22, 2022:
https://www.noaa.gov/media-release/average-atlantic-hurricane-season-to-reflect-more-storms

National Research Council and Committee on Climate Change and U.S. Transportation, Division on Earth and Life Studies, Transportation Research Board, *Potential Impacts of Climate Change on U.S. Transportation*, National Academies Press, Special Report 290, 2008.

Naumova, Elena N., "Climate Change and Food Safety Risk," Tufts University, Gerald J. and Dorothy R. Friedman School of Nutrition Science and Policy, last updated June 29, 2022.

Nazarnia, Hadi, Mohammad Nazarnia, Hadi Sarmasti, and W. Olivia Wills, "A Systematic Review of Civil and Environmental Infrastructures for Coastal Adaptation to Sea Level Rise," *Civil Engineering Journal*, Vol. 6, No. 7, 2020.

NCA4 Volume I—*See* Wuebbles et al., 2017.

NCA4 Volume II—*See* Reidmiller et al., 2018.

NCSL—*See* National Conference of State Legislatures.

Nelson, Julia, and Ryan Schuchard, "Adapting to Climate Change: A Guide for the Mining Industry," Business for Social Responsibility, undated.

Network Startup Resource Center, University of Oregon, "Internet eXchange Points: North America Region," webpage, last updated March 13, 2016. As of October 3, 2022:
https://nsrc.org/ixp/NorthAmerica.html

Neumann, James E., Paul Chinowsky, Jacob Helman, Margaret Black, Charles Fant, Kenneth Strzepek, and Jeremy Martinich, "Climate Effects on US Infrastructure: The Economics of Adaptation for Rail, Roads, and Coastal Development," *Climatic Change*, Vol. 167, 2021.

Newburger, Emma, "'It Never Stops': US Farmers Now Face Extreme Heat Wave After Floods and Trade War," CNBC, updated July 20, 2019.

Ngo, Nicole S., "Urban Bus Ridership, Income, and Extreme Weather Events," *Transportation Research Part D: Transport and Environment*, Vol. 77, December 2019.

Nieuwenhuijs, Albert, Eric Luiijf, and Marieke Klaver, "Modeling Dependencies in Critical Infrastructures," in Mauricio Papa and Sujeet Shenoi, eds., *Critical Infrastructure Protection II*, Springer, 2008.

Niles, Meredith T., Richie Ahuja, Todd Barker, Jimena Esquivel, Sophie Gutterman, Martin C. Heller, Nelson Mango, Diana Portner, Rex Raimond, Cristina Tirado, and Sonja Vermeulen, "Climate Change Mitigation Beyond Agriculture: A Review of Food System Opportunities and Implications," *Renewable Agriculture and Food Systems*, Vol. 33, No. 3, 2018.

Njini, Felix, "Sibanye Cuts US Palladium Mines Output Forecast After Floods," Bloomberg, August 11, 2022.

NOAA—*See* National Oceanic and Atmospheric Administration.

North American Electric Reliability Corporation, *Hurricane Harvey Event Analysis Report*, March 2018.

Nunno, Richard, "Electrification of U.S. Railways: Pie in the Sky, or Realistic Goal?" Environmental and Energy Study Institute, May 30, 2018.

O'Brien, Shauna, "7 Events That Closed the NYSE," Dividend University, webpage, undated. As of October 2022:
https://www.dividend.com/dividend-education/7-events-that-closed-the-nyse/

Occupational Safety and Health Administration, U.S. Department of Labor, "Heat," webpage, undated. As of November 11, 2022:
https://www.osha.gov/heat-exposure

Office of Budget and Policy, Federal Transit Administration, U.S. Department of Transportation, *2019 National Transit Database: National Transit Summaries and Trends*, undated.

Office of Electricity Delivery and Energy Reliability, U.S Department of Energy, *Hurricane Sandy Situation Report 5*, October 30, 2012a.

Office of Electricity Delivery and Energy Reliability, U.S. Department of Energy, "Hurricane Sandy Situation Report 20," November 7, 2012b.

Office of Emerging Policy, U.S. Coast Guard, *Maritime Commerce Strategic Outlook*, October 2018.

Office of Energy Policy and Systems Analysis, U.S. Department of Energy, *Climate Change and the U.S. Energy Sector: Regional Vulnerabilities and Resilience Solutions*, October 2015.

Office of Inspector General, U.S. Department of Health and Human Services, *Hospital Emergency Preparedness and Response During Superstorm Sandy*, OEI-06-13-00260, September 2014.

Office of Management and Budget, *Analytical Perspectives: Budget of the U.S. Government—Fiscal Year 2023*, circa 2022.

Office of Nuclear Energy, U.S. Department of Energy, "5 Fast Facts About Spent Nuclear Fuel," webpage, October 3, 2022. As of October 31, 2022:
https://www.energy.gov/ne/articles/5-fast-facts-about-spent-nuclear-fuel

Office of Occupational Statistics and Employment Projections, U.S. Bureau of Labor Statistics, U.S. Department of Labor, "Employment Projections: National Employment Matrix—325000 Chemical Manufacturing: Employment by Industry, Occupation, and Percent Distribution, 2021 and Projected 2031," dataset, undated. As of November 9, 2022:
https://data.bls.gov/projections/nationalMatrix?queryParams=325000&ioType=i

Office of Occupational Statistics and Employment Projections, U.S. Bureau of Labor Statistics, U.S. Department of Labor, "Industry-Occupation Matrix Data, by Occupation," Table 1.8, dataset, last modified September 8, 2022. As of October 20, 2022.
https://www.bls.gov/emp/tables/industry-occupation-matrix-occupation.htm

Office of the Chief Economist, U.S. Department of Agriculture, "Climate Science and Effects," webpage, undated. As of November 9, 2022:
https://www.usda.gov/oce/energy-and-environment/climate/assessments

Office of the National Coordinator for Health Information Technology, Office of the Secretary, U.S. Department of Health and Human Services, "Adoption of Electronic Health Records by Hospital Service Type 2019–2021," Health Information Technology Quick Stat 60, April 2022.

Office of the National Coordinator for Health Information Technology, Office of the Secretary, U.S. Department of Health and Human Services, "2015 Edition," webpage, last reviewed May 5, 2023. As of June 14, 2023:
https://www.healthit.gov/topic/certification-ehrs/2015-edition

Office of the Under Secretary for Policy (Strategy, Plans, and Capabilities), U.S. Department of Defense, *Department of Defense Climate Risk Analysis*, report submitted to the National Security Council, October 2021.

Ohio Environmental Council, Power a Clean Future Ohio, and Scioto Analysis, *The Bill Is Coming Due: Calculating the Financial Cost of Climate Change to Ohio's Local Governments*, July 2022.

Olds, Hayley T., Steven R. Corsi, Deborah K. Dila, Katherine M. Halmo, Melinda J. Bootsma, and Sandra L. McLellan, "High Levels of Sewage Contamination Released from Urban Areas After Storm Events: A Quantitative Survey with Sewage Specific Bacterial Indicators," *PLOS Medicine*, Vol. 15, No. 7, 2018.

Olick, Diana, "Mortgage Market Is Unprepared for Climate Risk, Says Industry Report," CNBC, updated September 23, 2021.

OMB—*See* Office of Management and Budget.

Ouazad, Amine, and Matthew E. Kahn, "Mortgage Finance and Climate Change: Securitization Dynamics in the Aftermath of Natural Disasters," National Bureau of Economic Research, Working Paper 26322, September 2019, revised February 2021.

Oxford Analytica, "Puerto Rico's Power Grid Hit by Another Hurricane," *Expert Briefings*, 2022.

Özek, Umut, "Examining the Educational Spillover Effects of Severe Natural Disasters: The Case of Hurricane Maria," *Journal of Human Resources*, January 2021.

Painter, David S., "Oil and the American Century," *Journal of American History*, Vol. 99, No. 1, June 2012.

Pal, Govinda, Thaneswer Patel, and Trishita Banik, "Effect of Climate Change Associated Hazards on Agricultural Workers and Approaches for Assessing Heat Stress and Its Mitigation Strategies: Review of Some Research Significances," *International Journal of Current Microbiology and Applied Sciences*, Vol. 10, No. 2, February 2021.

Palmieri, Francesco, Ugo Fiore, Aniello Castiglione, Fang-Yie Leu, and Alfredo De Santis, "Analyzing the Internet Stability in Presence of Disasters," in Alfredo Cuzzocrea, Christian Kittl, Dimitris E. Simos, Edgar Weippl, and Lida Xu, eds., *Security Engineering and Intelligence Informatics: CD-ARES 2013— Lecture Notes in Computer Science*, Vol. 8128, 2013.

Pamidimukkala, Apurva, Sharareh Kermanshachi, and Sanjgna Karthick, "Impact of Natural Disasters on Construction Projects: Strategies to Prevent Cost and Schedule Overruns in Reconstruction Projects," *Proceedings of the Creative Construction e-Conference 2020*, July 1, 2020.

Park, R. Jisung, Joshua Goodman, Michael Hurwitz, and Jonathan Smith, "Heat and Learning," *American Economic Journal: Economic Policy*, Vol. 12, No. 2, May 2020.

Park, R. Jisung, Nora Pankratz, and A. Patrick Behrer, *Temperature, Workplace Safety, and Labor Market Inequality*, Institute of Labor Economics, Discussion Paper 14560, July 2021.

Parker, Lauren E., Andrew J. McElrone, Steven M. Ostoja, and Elisabeth J. Forrestel, "Extreme Heat Effects on Perennial Crops and Strategies for Sustaining Future Production," *Plant Science*, Vol. 295, June 2020.

Parlikad, A. K., and M. Jafari, "Challenges in Infrastructure Asset Management," *IFAC—Papers Online*, Vol. 49, No. 28, 2016.

Partlow, Joshua, "A California City's Water Supply Is Expected to Run out in Two Months," *Washington Post*, October 10, 2022.

Pastick, Neal J., M. Torre Jorgenson, Bruce K. Wylie, Shawn J. Nield, Kristofer D. Johnson, and Andrew O. Finley, "Distribution of Near-Surface Permafrost in Alaska: Estimates of Present and Future Conditions," *Remote Sensing of Environment*, Vol. 168, October 2015.

Patel, Ketan B., "Managing Climate Risk in Mortgage Markets: A Role for Derivatives," *Chicago Fed Letter*, No. 462, October 2021.

Paull, Jeffrey M., and Frank S. Rosenthal, "Heat Strain and Heat Stress for Workers Wearing Protective Suits at a Hazardous Waste Site," *American Industrial Hygiene Association Journal*, Vol. 48, No. 5, 1987.

Petit, Frédéric, Duane Verner, David Brannegan, William Buehring, David Dickinson, Karen Guziel, Rebecca Haffenden, Julia Phillips, and James Peerenboom, *Analysis of Critical Infrastructure Dependencies and Interdependencies*, Risk and Infrastructure Science Center, Global Sciences Division, Argonne National Laboratory, ANL/GSS-15/4, June 2015.

Pfefferbaum, Betty, Anne K. Jacobs, Richard L. Van Horn, and J. Brian Houston, "Effects of Displacement in Children Exposed to Disasters," *Current Psychiatry Reports*, Vol. 18, No. 8, August 2016.

Phillips, Anna M., "It's California Wildfire Season. But Firefighters Say Federal Hotshot Crews Are Understaffed," *Los Angeles Times*, May 18, 2021.

Phillips, Will, "Commodity Shipments Under Threat from Low Mississippi Water Levels," *Supply Management*, October 17, 2022.

Pierce, Freddie, "The Long Haul: Midwest Flooding's Supply Chain Impact," *Supply Chain Digital*, May 17, 2020.

Pinson, A. O., K. D. White, E. E. Ritchie, H. M. Conners, and Jeff R. Arnold, *DoD Installation Exposure to Climate Change at Home and Abroad*, U.S. Army Corps of Engineers, developed for the Office of the Deputy Assistant Secretary of Defense for Environment and Energy Resilience, April 19, 2021.

Pint, Ellen M., Beth E. Lachman, Jeremy M. Eckhause, and Steven Deane-Shinbrot, *Army Installation Rail Operations: Implications of Increased Outsourcing*, RAND Corporation, RR-2009-A, 2017. As of November 10, 2022:
https://www.rand.org/pubs/research_reports/RR2009.html

Pipeline and Hazardous Materials Safety Administration, U.S. Department of Transportation, "General Pipeline FAQs," webpage, last updated November 6, 2018. As of October 28, 2022:
https://www.phmsa.dot.gov/faqs/general-pipeline-faqs

Planning and Economic Studies Section, Department of Nuclear Energy, International Atomic Energy Agency, *Climate Change and Nuclear Power 2018*, 2018.

Polansek, Tom, "Monday Storm Impacted an Estimated 37.7 Million Acres of Midwest Farmland," Reuters, August 14, 2020.

Popescu, Roxana, "For Those Who Lose Homes to Wildfires, The Challenges of Rebuilding are Daunting," *Washington Post*, November 21, 2018.

Posey, John, "Climate Change Impacts on Transportation in the Midwest," white paper prepared for the U.S. Global Change Research Program, National Climate Assessment, Midwest Technical Input Report, 2012.

Praveen, Bushra, and Pritee Sharma, "A Review of Literature on Climate Change and Its Impacts on Agriculture Productivity," *Journal of Public Affairs*, Vol. 19, No. 4, November 2019.

Prelog, Andrew J., "Modeling the Relationship Between Natural Disasters and Crime in the United States," *Natural Hazards Review*, Vol. 17, No. 1, February 2016.

Preston, Benjamin Lee, Michelle E. Miro, Paul Brenner, Christopher K. Gilmore, John F. Raffensperger, Jaime Madrigano, Alexandra Huttinger, Michael Blackhurst, and David Catt, *Beyond Recovery: Transforming Puerto Rico's Water Sector in the Wake of Hurricanes Irma and Maria*, Homeland Security Operational Analysis Center operated by the RAND Corporation, RR-2608-DHS, 2020. As of November 12, 2022:
https://www.rand.org/pubs/research_reports/RR2608.html

Proctor, Caitlin R., Juneseok Lee, David Yu, Amisha D. Shah, and Andrew J. Whelton, "Wildfire Caused Widespread Drinking Water Distribution Network Contamination," *AWWA Water Science*, Vol. 2, No. 4, July–August 2020.

Public and Affordable Housing Research Corporation and National Low Income Housing Coalition, *Taking Stock: Natural Hazards and Federally Assisted Housing*, circa June 2021.

Public Law 93-523, Safe Drinking Water Act, December 16, 1974.

Public Law 94-580, Resource Conservation and Recovery Act of 1976, October 21, 1976.

Public Law 96-510, Comprehensive Environmental Response, Compensation, and Liability Act of 1980, December 11, 1980.

Public Law 99-499, Superfund Amendments and Reauthorization Act of 1986, October 17, 1986.

Public Law 104-191, Health Insurance Portability and Accountability Act of 1996, August 21, 1996.

Public Law 111-203, Dodd–Frank Wall Street Reform and Consumer Protection Act, July 21, 2010.

Qiang, Yi, "Flood Exposure of Critical Infrastructures in the United States," *International Journal of Disaster Risk Reduction*, Vol. 39, October 2019.

Quach, Katyanna, "Google, Oracle Cloud Servers Wilt in UK Heatwave, Take Down Websites," *The Register*, July 19, 2022.

Quinton, Sophie, "Lack of Federal Firefighters Hurts California Wildfire Response," *Stateline*, July 14, 2021.

Radeloff, V. C., R. B. Hammer, S. I. Stewart, J. S. Fried, S. S. Holcomb, and J. F. McKeefry, "The Wildland–Urban Interface in the United States," *Ecological Applications*, Vol. 15, No. 3, June 2005.

Radeloff, Volker C., David P. Helmers, H. Anu Kramer, Miranda H. Mockrin, Patricia M. Alexandre, Avi Bar-Massada, Van Butsic, Todd J. Hawbaker, Sebastián Martinuzzi, Alexandra D. Syphard, and Susan I. Stewart, "Rapid Growth of the US Wildland–Urban Interface Raises Wildfire Risk," *Proceedings of the National Academy of Sciences*, Vol. 115, No. 13, March 27, 2018.

Rahurkar, Saurabh, Joshua R. Vest, and Nir Menachemi, "Despite the Spread of Health Information Exchange, There Is Little Evidence of Its Impact on Cost, Use, and Quality of Care," *Health Affairs (Millwood)*, Vol. 34, No. 3, March 2015.

Ramchand, Latha, and Ramanan Krishnamoorti, "What Harvey Taught Us: Lessons from the Energy Industry," *Forbes*, November 22, 2017.

Ramnath, Shanthi P., and Will Jeziorski, "Homeowners Insurance and Climate Change," *Chicago Fed Letter*, Vol. 460, September 2021.

Ranson, Matthew, "Crime, Weather, and Climate Change," *Journal of Environmental Economics and Management*, Vol. 67, No. 3, May 2014.

Rao, Krishna, "Climate Change and Housing: Will a Rising Tide Sink All Homes?" Zillow, June 2, 2017.

Rao, Prakash, Darren Sholes, and Joe Cresko, "Evaluation of U.S. Manufacturing Subsectors at Risk of Physical Water Shortages," *Environmental Science and Technology*, Vol. 53, No. 5, 2019.

Raymond, Colin, Tom Matthews, and Radley M. Horton, "The Emergence of Heat and Humidity Too Severe for Human Tolerance," *Science Advances*, Vol. 6, No. 19, May 8, 2020.

Reed, Larrité, and Kenneth Nugent, "The Health Effects of Dust Storms in the Southwest United States," *Southwest Respiratory and Critical Care Chronicles*, Vol. 6, No. 22, 2018.

Reeves, Magen M., "Tyndall AFB Continues Rebuild Effort One Year After Hurricane Michael," U.S. Air Force, October 10, 2019.

Regan, Liam, Susan A. Resetar, Andrew Lauland, Joie D. Acosta, Rahim Ali, Edward W. Chan, Richard H. Donohue, Liisa Ecola, Timothy R. Gulden, Chelsea Kolb, Kristin J. Leuschner, Michelle E. Miro, Patricia A. Stapleton, Chandler Sachs, Chuck Story, Tobias Sytsma, and Michael T. Wilson, *Climate Adaptation Strategies for National Critical Functions*, Homeland Security Operational Analysis Center operated by the RAND Corporation, TL-A1645-1, 2023. As of December 26, 2023: https://www.rand.org/pubs/tools/TLA1645-1.html

Rehkamp, Sarah, Patrick Canning, and Catherine Birney, *Tracking the U.S. Domestic Food Supply Chain's Freshwater Use over Time*, U.S. Department of Agriculture, Economic Research Service, Economic Research Report 288, July 2021.

Reidmiller, David R., Christopher W. Avery, David R. Easterling, Kenneth E. Kunkel, Kristen L. M. Lewis, Thomas K. Maycock, and Brooke C. Stewart, eds., *Impacts, Risks, and Adaptation in the United States: Fourth National Climate Assessment*, Vol. II, U.S. Global Change Research Program, 2018.

Repanshek, Kurt, "Wildfires Show How Climate Change Is Transforming National Parks," *National Geographic*, October 14, 2020.

Resilinc, "Wildfires Are Up 30% Year over Year and Wreaking Havoc on Supply Chains," blog post, June 29, 2021. As of November 6, 2022:
https://www.resilinc.com/blog/
wildfires-are-up-30-year-over-year-and-wreaking-havoc-on-supply-chains/

Restrepo, Dan, Michael Werz, and Joel Martinez, *5 Big Ideas for U.S. Policy in the Americas*, Center for American Progress, December 2016.

Rey, Alexander J. M., D. Reide Corbett, and Ryan P. Mulligan, "Impacts of Hurricane Winds and Precipitation on Hydrodynamics in a Back-Barrier Estuary," *Journal of Geophysical Research: Oceans*, Vol. 125, No. 12, December 2020.

Rezaei, Seyedeh Nasim, Luc Chouinard, Sébastien Langlois, and Frédéric Légeron, "Analysis of the Effect of Climate Change on the Reliability of Overhead Transmission Lines," *Sustainable Cities and Society*, Vol. 27, November 2016.

Richards, Ryan, "After the Fire: Vulnerable Communities Respond and Rebuild," Center for American Progress, July 25, 2019.

Rinaldi, Steven M., James P. Peerenboom, and Terrence K. Kelly, "Identifying, Understanding, and Analyzing Critical Infrastructure Interdependencies," *IEEE Control Systems*, Vol. 21, No. 6, December 2001.

Robinne, François-Nicolas, Dennis W. Hallema, Kevin D. Bladon, Mike D. Flannigan, Gabrielle Boisramé, Christian M. Bréthaut, Stefan H. Doerr, Giuliano Di Baldassarre, Louise A. Gallagher, Amanda K. Hohner, Stuart J. Khan, Alicia M. Kinoshita, Rua Mordecai, João Pedro Nunes, Petter Nyman, Cristina Santín, Gary Sheridan, Cathelijne R. Stoof, Matthew P. Thompson, James M. Waddington, and Yu Wei, "Scientists' Warning on Extreme Wildfire Risks to Water Supply," *Hydrological Processes*, Vol. 35, No. 5, May 2021.

Robinson, Clive G., Zoë E. Wattis, Colin Dooley, and Sladjana Popovic, "Assessment of the Threat from Wildfires on Above Ground Natural Gas Facilities," *Proceedings of the 2018 12th International Pipeline Conference*, Vol. 2: *Pipeline Safety Management Systems; Project Management, Design, Construction, and Environmental Issues; Strain Based Design; Risk and Reliability; Northern Offshore and Production Pipelines*, September 24–28, 2018.

Robitzski, Dan, "Severe Drought, Heat Upended Research This Summer," *The Scientist*, October 13, 2022.

Rodriguez, Sarai, "Health Data Exchange Barriers Impede Widespread Interoperability," *EHRIntelligence*, July 19, 2022.

Roh, Eunice, and Chas Alamo, "Climate Change Impacts Across California: Workers and Employers, Legislative Analyst's Office, April 5, 2022.

Rojas-Downing, M. Melissa, A. Pouyan Nejadhashemi, Timothy Harrigan, and Sean A. Woznicki, "Climate Change and Livestock: Impacts, Adaptation, and Mitigation," *Climate Risk Management*, Vol. 16, 2017.

Romanosky, Sasha, John Bordeaux, Michael J. D. Vermeer, Jonathan W. Welburn, Aaron Strong, and Zev Winkelman, *Identifying Critical IT Products and Services*, Homeland Security Operational Analysis Center operated by the RAND Corporation, RR-A923-2, 2022. As of November 10, 2022: https://www.rand.org/pubs/research_reports/RRA923-2.html

Rose, Joel, "Post-Sandy Fixes to NYC Subways to Cost Billions," *All Things Considered*, December 6, 2012.

Rossetti, Michael A., "Potential Impacts of Climate Change on Railroads," in Joanne R. Potter, ed., *The Potential Impacts of Climate Change on Transportation: Workshop Summary*, 2003.

Rossetti, Michael A., "Analysis of Weather Events on U.S[.] Railroads," John A. Volpe National Transportation Systems Center, U.S. Department of Transportation, January 13, 2007.

Rotton, James, and Ellen G. Cohn, "Violence Is a Curvilinear Function of Temperature in Dallas: A Replication," *Journal of Personality and Social Psychology*, Vol. 78, No. 6, June 2000.

Rowan, Emily, Christopher Evans, Marybeth Riley-Gilbert, Rob Hyman, Rob Kafalenos, Brian Beucler, Beth Rodehorst, Anne Choate, and Peter Schultz, "Assessing the Sensitivity of Transportation Assets to Extreme Weather Events and Climate Change," *Transportation Research Record: Journal of the Transportation Research Board*, Vol. 2326, No. 1, January 2013.

Rübbelke, Dirk, and Stefan Vögele, "Impacts of Climate Change on European Critical Infrastructures: The Case of the Power Sector," *Environmental Science and Policy*, Vol. 14, No. 1, January 2011.

Runyon, Luke, "Amid a Megadrought, Federal Shortage Limits Loom for the Colorado River," *Morning Edition*, July 13, 2021.

Rusco, Frank, *Climate Change: Energy Infrastructure Risks and Adaptation Efforts*, U.S. Government Accountability Office, GAO-14-74, January 31, 2014.

Rusco, Frank, director, Natural Resources and Environment, U.S. Government Accountability Office, *Strategic Petroleum Reserve: Preliminary Observations on the Emergency Oil Stockpile*, testimony before the U.S. House of Representatives Committee on Energy and Commerce Subcommittee on Energy, GAO-18-209T, November 2, 2017.

Rusco, Frank, "Oil Spills in the Wake of Hurricane Ida Highlight Need for Better Federal Oversight of Offshore Oil and Gas Pipelines," *WatchBlog*, September 14, 2021. As of June 6, 2023: https://www.gao.gov/blog/ oil-spills-wake-hurricane-ida-highlight-need-better-federal-oversight-offshore-oil-and-gas-pipelines

Sabin Center for Climate Change Law, Columbia Law School, and Arnold and Porter, "U.S. Climate Change Litigation," webpage, undated. As of October 31, 2022: http://climatecasechart.com/us-climate-change-litigation/

Safo, Nova, "U.S. Forest Service Is Short Thousands of Firefighters amid Pay Raise Delay," *Marketplace*, May 10, 2022.

Saint, Bob, "Rural Distribution System Planning Using Smart Grid Technologies," *2009 IEEE Rural Electric Power Conference*, Institute of Electrical and Electronics Engineers, 2009.

Sandink, Dan, and Barbara Robinson, "Wastewater System Inflow/Infiltration and Residential Pluvial Flood Damage Mitigation in Canada," *Water*, Vol. 14, No. 11, 2022.

Sanfilippo, Madelyn R., Yan Shvartzshnaider, Irwin Reyes, Helen Nissenbaum, and Serge Egelman, "Disaster Privacy/Privacy Disaster," *Journal of the Association for Information Science and Technology*, Vol. 71, No. 9, September 2020.

Santella, Nicholas, Laura J. Steinberg, and Hatice Sengul, "Petroleum and Hazardous Material Releases from Industrial Facilities Associated with Hurricane Katrina," *Risk Analysis*, Vol. 30, No. 4, April 2010.

Sargent, John F., Jr., "U.S. Research and Development Funding and Performance: Fact Sheet," Congressional Research Service, R44307, September 13, 2022.

Sargent, John F., Jr., Eva Lipiec, Genevieve K. Croft, Daniel Morgan, Marcy E. Gallo, Kavya Sekar, Laurie A. Harris, and Jerry H. Yen, *Federal Research and Development (R&D) Funding: FY2022*, Congressional Research Service, R46869, updated January 19, 2022.

Sarvestani, Kazem, Omran Ahmadi, and Morteza Jalali Alenjareghi, "LPG Storage Tank Accidents: Initiating Events, Causes, Scenarios, and Consequences," *Journal of Failure Analysis and Prevention*, Vol. 21, August 2021.

Schoennagel, Tania, Jennifer K. Balch, Hannah Brenkert-Smith, Philip E. Dennison, Brian J. Harvey, Meg A. Krawchuk, Nathan Mietkiewicz, Penelope Morgan, Max A. Moritz, Ray Rasker, Monica G. Turner, and Cathy Whitlock, "Adapt to More Wildfire in Western North American Forests as Climate Changes," *Proceedings of the National Academy of Sciences*, Vol. 114, No. 18, April 17, 2017.

Scholz, Serena, Rebecca Composto, Kevin Inks, Zoe Dutton, and Leola Abraham, "Impact of Climate Change on Supply Chains," CNA, 2021.

Schuetz, Jenny, "Home Mortgage and Insurance Systems Encourage Development in Climate-Risky Places, and We All Pay the Price," Brookings Institution, March 9, 2022.

Schuldt, Steven J., Mathew R. Nicholson, Yaquarri A. Adams II, and Justin D. Delorit, "Weather-Related Construction Delays in a Changing Climate: A Systematic State-of-the-Art Review," *Sustainability*, Vol. 13, No. 5, March 1, 2021.

Schulze, Ernst Detlef, Carlos A. Sierra, Vincent Egenolf, Rene Woerdehoff, Roland Irslinger, Conrad Baldamus, Inge Stupak, and Hermann Spellmann, "The Climate Change Mitigation Effect of Bioenergy from Sustainably Managed Forests in Central Europe," *GCB-Bioenergy*, Vol. 12, No. 3, March 2020.

Schulze, Stefanie S., Erica C. Fischer, Sara Hamideh, and Hussam Mahmoud, "Wildfire Impacts on Schools and Hospitals Following the 2018 California Camp Fire," *Natural Hazards*, Vol. 104, October 2020.

Schwarze, Reimund, "Liability for Climate Change: The Benefits, the Costs, and the Transaction Costs," *University of Pennsylvania Law Review*, Vol. 155, No. 6, 2007.

Schwirtz, Michael, "Report Cites Large Release of Sewage from Hurricane Sandy," *New York Times*, April 30, 2013.

Secaira, Manola, "How Massive Wildfires Create Their Own Weather," CapRadio, August 5, 2022.

Seetharam, Ishuwar, *The Indirect Effects of Hurricanes: Evidence from Firm Internal Networks*, technical report, July 2018.

Seltenrich, Nate, "Between Extremes: Health Effects of Heat and Cold," *Environmental Health Perspectives*, Vol. 123, No. 11, November 2015.

Sen, Sushobhan, Haoran Li, and Lev Khazanovich, "Effect of Climate Change and Urban Heat Islands on the Deterioration of Concrete Roads," *Results in Engineering*, Vol. 16, December 2022.

Seppänen, Olli, William J. Fisk, and Quanhong Lei-Gomez, "Room Temperature and Productivity in Office Work," *Healthy Buildings 2006 Conference*, Vol. 1, 2006.

Setzer, Joana, and Catherine Higham, *Global Trends in Climate Change Litigation: 2022 Snapshot*, London School of Economics and Political Science and Grantham Research Institute on Climate Change and the Environment, June 30, 2022.

Sharma, Ajay, Santosh K. Ojha, Luben D. Dimov, Jason G. Vogel, and Jarek Nowak, "Long-Term Effects of Catastrophic Wind on Southern US Coastal Forests: Lessons from a Major Hurricane," *PLOS ONE*, Vol. 16, No. 1, 2021.

Shaw, Eric J., *Operation Unified Assistance: 2004 Sumatran Earthquake and Tsunami Humanitarian Relief*, U.S. Naval War College, May 2013.

Sheely, Zach, "Thousands of Guardsmen Supporting Hurricane Ian Response," National Guard, September 29, 2022.

Shen, Guoqiang, Xiaoyi Yan, Long Zhou, and Zhangye Wang, "Visualizing the USA's Maritime Freight Flows Using DM, LP, and AON in GIS," *ISPRS International Journal of Geo-Information*, Vol. 9, No 5, May 2020.

Shepard, Ann, "Disaster Recovery and the Electronic Health Record," *Nursing Administration Quarterly*, Vol. 41, No. 2, April–June 2017.

Sherman, Jeffrey A., Ladan Arissian, Roger C. Brown, Matthew J. Deutch, Elizabeth A. Donley, Vladislav Gerginov, Judah Levine, Glenn K. Nelson, Andrew N. Novick, Bijunath R. Patla, Thomas E. Parker, Benjamin K. Stuhl, Douglas D. Sutton, Jian Yao, William C. Yates, Victor Zhang, and Michael A. Lombardi, *A Resilient Architecture for the Realization and Distribution of Coordinated Universal Time to Critical Infrastructure Systems in the United States: Methodologies and Recommendations from the National Institute of Standards and Technology (NIST)*, National Institute of Standards and Technology, Technical Note 2187, November 2021.

Shieh, C. L., C. M. Wang, Yu-Shiu Chen, Yuan-Jung Tsai, and W. H. Tseng, "An Overview of Disasters Resulted from Typhoon Morakot in Taiwan," *Journal of Disaster Research*, Vol. 5, No. 3, January 2010.

Shield, Stephen A., Steven M. Quiring, Jordan V. Pino, and Ken Buckstaff, "Major Impacts of Weather Events on the Electrical Power Delivery System in the United States," *Energy*, Vol. 218, March 1, 2021.

Shirzaei, Manoochehr, Mostafa Khoshmanesh, Chandrakanta Ojha, Susanna Werth, Hannah Kerner, Grace Carlson, Sonam Futi Sherpa, Guang Zhai, and Jui-Chi Lee, "Persistent Impact of Spring Floods on Crop Loss in U.S. Midwest," *Weather and Climate Extremes*, Vol. 34, December 2021.

Shuai, Mao, Wang Chengzhi, Yu Shiwen, Gen Hao, Yu Jufang, and Hou Hui, "Review on Economic Loss Assessment of Power Outages," *Procedia Computer Science*, Vol. 130, 2018.

Siddik, Md Abu Bakar, Arman Shehabi, and Landon Marston, "The Environmental Footprint of Data Centers in the United States," *Environmental Research Letters*, Vol. 16, No. 6, June 2021.

Silva, C., J. Saldanha Matos, and M. J. Rosa, "Performance Indicators and Indices of Sludge Management in Urban Wastewater Treatment Plants," *Journal of Environmental Management*, Vol. 184, Part 2, No. 30, December 15, 2016.

Silverman, Joel, Angie De Groot, Holly Gell, Monica Giovachino, Kristin Koch, Leslie-Anne Levy, Elizabeth Myrus, and Dawn Thomas, *Why the Emergency Management Community Should Be Concerned About Climate Change: A Discussion of the Impact of Climate Change on Selected Natural Hazards*, CNA, 2010.

Skarha, Julianne, Meghan Peterson, Josiah D. Rich, and David Dosa, "An Overlooked Crisis: Extreme Temperature in Incarcerated Settings," *American Journal of Public Health*, Vol. 110, Suppl. 1, January 2020.

Skipper, Joseph B., Joe B. Hanna, and Brian J. Gibson, "Alabama Power Response to Katrina: Managing a Severe Service Supply Chain Disruption," *Journal of the International Academy for Case Studies*, Vol. 16, No. 2, 2010.

Slater, Louise J., and Gabriele Villarini, "Recent Trends in U.S. Flood Risk," *Geophysical Research Letters*, Vol. 43, No. 24, December 28, 2016.

Slay, Christy, and Kevin Dooley, *Improving Supply Chain Resilience to Managing Climate Change Risks*, Sustainability Consortium and HSBC, June 2020.

Sloan, Karen, "New Orleans Courthouse Damaged by Ida, Firms Go Remote, Law Schools Closed," Reuters, August 30, 2021.

Smith, Gavin, and Olivia Vila, "A National Evaluation of State and Territory Roles in Hazard Mitigation: Building Local Capacity to Implement FEMA Hazard Mitigation Grants," *Sustainability*, Vol. 12, No. 23, December 1, 2020.

Smythe, Tiffany C., *Assessing the Impacts of Hurricane Sandy on the Port of New York and New Jersey's Maritime Responders and Response Infrastructure*, University of Colorado Natural Hazards Center, May 31, 2013.

Somanathan, Eswaran, Rohini Somanathan, Anant Sudarshan, and Meenu Tewari, "The Impact of Temperature on Productivity and Labor Supply: Evidence from Indian Manufacturing," *Journal of Political Economy*, Vol. 129, No. 6, June 2021.

Souto, Laiz, Joshua Yip, Wen-Ying Wu, Brent Austgen, Erhan Kutanoglu, John Hasenbein, Zong-Liang Yang, Carey W. King, and Surya Santoso, "Power System Resilience to Floods: Modeling, Impact Assessment, and Mid-Term Mitigation Strategies," *International Journal of Electrical Power and Energy Systems*, Vol. 135, February 2022.

Spader, Jonathan, "How Much of the Damaged Housing Stock Was Rebuilt After Hurricanes Katrina and Rita?" *Housing Perspectives*, blog post, September 3, 2015.

Special Initiative for Rebuilding and Resiliency, City of New York, *A Stronger, More Resilient New York*, June 11, 2013.

Spector, June T., and Perry E. Sheffield, "Re-Evaluating Occupational Heat Stress in a Changing Climate," *Annals of Occupational Hygiene*, Vol. 58, No. 8, October 2014.

Stavros, E. Natasha, Donald McKenzie, and Narasimhan Larkin, "The Climate–Wildfire–Air Quality System: Interactions and Feedbacks Across Spatial and Temporal Scales," *WIREs Climate Change*, Vol. 5, No. 6, November–December 2014.

Stein, Robert M., "Election Administration During Natural Disasters and Emergencies: Hurricane Sandy and the 2012 Election," *Election Law Journal*, Vol. 14, No 1, March 2015.

Sternberg, Ernest, George C. Lee, and Danial Huard, "Counting Crises: US Hospital Evacuations, 1971–1999," *Prehospital and Disaster Medicine*, Vol. 19, No. 2, April–June 2004.

Stocker, Thomas F., Dahe Qin, Gian-Kasper Plattner, Melinda M. B. Tignor, Simon K. Allen, Judith Boschung, Alexander Nauels, Yu Xia, Vincent Bex, and Pauline M. Midgley, eds., *Climate Change 2013: The Physical Science Basis—Contribution of Working Group I to the Fifth Assessment Report of the Intergovernmental Panel on Climate Change*, Intergovernmental Panel on Climate Change, 2013.

Stockholm International Peace Research Institute, *SIPRI Yearbook 2017: Armaments, Disarmament and International Security*, 2017. As of November 1, 2022:
https://www.sipri.org/yearbook/2017

Stoltz, Amanda Daria, Manoj Shivlani, and Robert Glazer, "Fishing Industry Perspectives on Sea-Level Rise Risk and Adaptation," *Water*, Vol. 13, No. 8, April 2, 2021.

Stone, Geoffrey R., "Civil Liberties in Wartime," ShareAmerica, April 6, 2015.

Storm Surge Unit, National Hurricane Center, National Weather Service, National Oceanic and Atmospheric Administration, U.S. Department of Commerce, "Introduction to Storm Surge," undated.

Strauss, Benjamin H., Philip M. Orton, Klaus Bittermann, Maya K. Buchanan, Daniel M. Gilford, Robert E. Kopp, Scott Kulp, Chris Massey, Hans de Moel, and Sergey Vinogradov, "Economic Damages [sic] from Hurricane Sandy Attributable to Sea Level Rise Caused by Anthropogenic Climate Change," *Nature Communications*, Vol. 12, 2021.

Stults, Missy, "Integrating Climate Change into Hazard Mitigation Planning: Opportunities and Examples in Practice," *Climate Risk Management*, Vol. 17, 2017.

Sturgis, Linda A., Tiffany C. Smythe, and Andrew E. Tucci, *Port Recovery in the Aftermath of Hurricane Sandy: Improving Port Resiliency in the Era of Climate Change*, Center for a New American Security, August 2014.

Substance Abuse and Mental Health Services Administration, *Greater Impact: How Disasters Affect People of Low Socioeconomic Status*, Disaster Technical Assistance Center Supplemental Research Bulletin, July 2017.

Summers, Kevin, Andrea Lamper, and Kyle Buck, "National Hazards Vulnerability and the Remediation, Restoration and Revitalization of Contaminated Sites—2. RCRA Sites," *Sustainability*, Vol. 13, No. 2, January 2, 2021.

Sweeney, Kyle, Helen Wiley, and Carolyn Kousky, *The Challenge of Financial Recovery from Disasters: The Case of Florida Homeowners After Hurricane Michael*, Risk Management and Decision Processes Center, University of Pennsylvania, May 2022.

Sweet, W. V., B. D. Hamlington, R. E. Kopp, C. P. Weaver, P. L. Barnard, D. Bekaert, W. Brooks, M. Craghan, G. Dusek, T. Frederikse, G. Garner, A. S. Genz, J. P. Krasting, E. Larour, D. Marcy, J. J. Marra, J. Obeysekera, M. Osler, M. Pendleton, D. Roman, L. Schmied, W. Veatch, K. D. White, and C. Zuzak, *Global and Regional Sea Level Rise Scenarios for the United States: Updated Mean Projections and Extreme Water Level Probabilities Along U.S. Coastlines*, National Oceanic and Atmospheric Administration, National Ocean Service, Technical Report NOS 01, February 2022.

Sweet, W. V., R. Horton, R. E. Kopp, Allegra N. LeGrande, and A. Romanou, "Sea Level Rise," in Donald J. Wuebbles, David W. Fahey, Kathy A. Hibbard, David J. Dokken, Brooke C. Stewart, and Thomas K. Maycock, eds., *Climate Science Special Report: Fourth National Climate Assessment*, Vol. I, U.S. Global Change Research Program, 2017.

Tabuchi, Hiroko, Nadja Popovich, Blacki Migliozzi, and Andrew W. Lehren, "Floods Are Getting Worse, and 2,500 Chemical Sites Lie in the Water's Path," *New York Times*, February 6, 2018.

Task Force on Emergency Preparedness for Elections, National Association of Secretaries of State, *State Laws and Practices for the Emergency Management of Elections*, February 2014, updated April 2017.

TCEQ—*See* Texas Commission on Environmental Quality.

Tebaldi, Claudia, Benjamin H. Strauss, and Chris E. Zervas, "Modelling Sea Level Rise Impacts on Storm Surges Along US Coasts," *Environmental Research Letters*, Vol. 7, No. 1, 2012.

Texas Commission on Environmental Quality, *Wastewater: Hurricane Harvey Related Sanitary Sewer Overflows and Other WW Discharges*, Dataset 12.28.2017, circa 2017.

Texas Commission on Environmental Quality, *Hurricane Harvey Response 2017: After Action Report*, April 3, 2018.

Theisen, Ole Magnus, "Climate Change and Violence: Insights from Political Science," *Current Climate Change Reports*, Vol. 3, December 2017.

Thompson, Terence R., "Climate Change Impacts upon the Commercial Air Transport Industry: An Overview," *Carbon and Climate Law Review*, Vol. 10, No. 2, 2016.

Thomson, Madeleine C., and Lawrence R. Stanberry, "Climate Change and Vectorborne Diseases," *New England Journal of Medicine*, Vol. 387, November 24, 2022.

Tirado, M. C., R. Clarke, L. A. Jaykus, A. McQuatters-Gollop, and J. M. Frank, "Climate Change and Food Safety: A Review," *Food Research International*, Vol. 43, No. 7, August 2010.

Tobin, Isabelle, Wouter Greuell, Sonia Jerez, Fulco Ludwig, Robert Vautard, Michelle T. H. van Vliet, and Françoise-Marie Bréon, "Vulnerabilities and Resilience of European Power Generation to 1.5°C, 2°C and 3°C Warming," *Environmental Research Letters*, Vol. 13, No. 4, April 2018.

Tomer, Adie, and Joseph Kane, *The Great Port Mismatch: U.S. Goods Trade and International Transportation*, Global Cities Initiative, June 2015.

Tornatore, Massimo, Joao André, Péter Babarczi, Torsten Braun, Eirik Følstad, Poul Heegaard, Ali Hmaity, Marija Furdek, Luisa Jorge, Wojciech Kmiecik, Carmen Mas Machuca, Lucia Martins, Carmo Medeiros, Francesco Musumeci, Alija Pašić, Jacek Rak, Steven Simpson, Rui Travanca, and Artemios Voyiatzis, "A Survey on Network Resiliency Methodologies Against Weather-Based Disruptions," *2016 8th International Workshop on Resilient Networks Design and Modeling (RNDM)*, 2016.

Trager, Rebecca, "Hurricane Hammers US Chemical Industry," *Chemistry World*, September 1, 2017.

Trainer, Vera L., Stephanie K. Moore, Gustaaf Hallegraeff, Raphael M. Kudela, Alejandro Clement, Jorge I. Mardones, and William P. Cochlan, "Pelagic Harmful Algal Blooms and Climate Change: Lessons from Nature's Experiments with Extremes," *Harmful Algae*, Vol. 91, January 2020.

TRICARE, "Military Hospitals and Clinics," webpage, June 24, 2022. As of October 5, 2022: https://www.tricare.mil/Military-Hospitals-and-Clinics

Tudi, Muyesaier, Huada Daniel Ruan, Li Wang, Jia Lyu, Ross Sadler, Des Connell, Cordia Chu, and Dung Tri Phung, "Agriculture Development, Pesticide Application and Its Impact on the Environment," *International Journal of Environmental Research and Public Health*, Vol. 18, No. 3, February 1, 2021.

Turner, Lyle R., Adrian G. Barnett, Des Connell, and Shilu Tong, "Ambient Temperature and Cardiorespiratory Morbidity: A Systematic Review and Meta-Analysis," *Epidemiology*, Vol. 23, No. 4, July 2012.

Turner, Sean W. D., Mohamad Hejazi, Son H. Kim, Leon Clarke, and Jae Edmonds, "Climate Impacts on Hydropower and Consequences for Global Electricity Supply Investment Needs," *Energy*, Vol. 141, December 15, 2017.

Turner, Sean W. D., Kristian Nelson, Nathalie Voisin, Vincent Tidwell, Ariel Miara, Ana Dyreson, Stuart Cohen, Dan Mantena, Julie Jin, Pete Warnken, and Shih-Chieh Kao, "A Multi-Reservoir Model for Projecting Drought Impacts on Thermoelectric Disruption Risk Across the Texas Power Grid," *Energy*, Vol. 231, September 15, 2021.

"UK Royal Air Force Halts Flights at Brize Norton Base Due to Heatwave," Reuters, July 18, 2022.

Underwood, E., "Global Drought Clustering Could Mean Big Losses for Mining," *Eos*, March 28, 2017.

United Kingdom Research and Innovation, "Survey Findings of the Impact of COVID-19 on Researchers," August 5, 2021.

United Nations Conference on Trade and Development, *Review of Maritime Transport 2021*, United Nations Publications, 2021.

United Nations Environment Programme, *Spreading Like Wildfire: The Rising Threat of Extraordinary Landscape Fires—A UNEP Rapid Response Assessment*, 2022.

United Nations Environment Programme, World Meteorological Organization, and United Nations Convention to Combat Desertification, *Global Assessment of Sand and Dust Storms*, 2016.

U.S. Army Public Health Center, *2020 Health of the Force: Create a Healthier Force for Tomorrow*, undated.

U.S. Census Bureau, U.S. Department of Commerce, "USA Trade® Online," database, undated. As of October 18, 2022:
https://usatrade.census.gov/

U.S. Climate Resilience Toolkit, "Food Safety and Nutrition," webpage, last modified August 6, 2021. As of October 10, 2022:
https://toolkit.climate.gov/topics/food-resilience/food-safety-and-nutrition

U.S. Cluster Mapping Project, Institute for Strategy and Competitiveness, Harvard Business School, "Production Technology and Heavy Machinery," webpage, undated. As of December 14, 2021:
https://clustermapping.us/cluster/production_technology_and_heavy_machinery

U.S. Courts, "Courts Help Each Other Survive Hurricane Season," webpage, January 2018. As of September 28, 2022:
https://www.uscourts.gov/news/2018/01/18/courts-help-each-other-survive-hurricane-season

U.S. Courts, "Hurricane Ida Courthouse Closures," webpage, August 30, 2021, updated August 31, 2021. As of September 28, 2022:
https://www.uscourts.gov/news/2021/08/30/hurricane-ida-courthouse-closures

U.S. Department of Agriculture, *Action Plan for Climate Adaptation and Resilience*, August 2021.

U.S. Department of Defense, *Base Structure Report: Fiscal Year 2015 Baseline—A Summary of the Real Property Inventory Data*, undated-a.

U.S. Department of Defense, *Base Structure Report: Fiscal Year 2018 Baseline—A Summary of the Real Property Inventory Data*, undated-b.

U.S. Department of Homeland Security, Homeland Infrastructure Foundation-Level Data, "Nonferrous Metal Processing Plants," dataset, December 1, 2005, updated January 18, 2018. As of November 9, 2022:
https://hifld-geoplatform.opendata.arcgis.com/datasets/
geoplatform::nonferrous-metal-processing-plants/explore?location=25.589241%2C-100.348365%2C4.00

U.S. Department of Homeland Security, Homeland Infrastructure Foundation-Level Data, "Transmission Lines," dataset, April 4, 2022. As of November 16, 2022:
https://hifld-geoplatform.opendata.arcgis.com/datasets/electric-power-transmission-lines/
explore?location=40.173735%2C-111.104115%2C4.29

U.S. Department of the Treasury, "Designations," webpage, undated. As of October 6, 2022:
https://home.treasury.gov/policy-issues/financial-markets-financial-institutions-and-fiscal-service/fsoc/
designations

U.S. Department of Transportation, "Positioning, Navigation and Timing (PNT) and Spectrum Management," webpage, undated. As of October 3, 2022:
https://www.transportation.gov/pnt

U.S. Department of Transportation, *Supply Chain Assessment of the Transportation Industrial Base: Freight and Logistics*, February 2022.

U.S. Energy Information Administration, U.S. Department of Energy, "Winter Supply Disruptions from Well Freeze-Offs Can Rival Effects of Summer Storms," *Today in Energy*, October 6, 2011.

U.S. Energy Information Administration, U.S. Department of Energy, *Marine Fuel Choice for Ocean-Going Vessels Within Emissions Control Areas*, June 2015a.

U.S. Energy Information Administration, U.S. Department of Energy, "The Basics of Underground Natural Gas Storage," webpage, November 15, 2015b. As of October 7, 2022:
https://www.eia.gov/naturalgas/storage/basics/

U.S. Energy Information Administration, U.S. Department of Energy, "Map of All Energy Infrastructure and Resources in the U.S.," webpage, August 12, 2021a. As of October 7, 2022:
https://atlas.eia.gov/apps/5039a1a01ec34b6bbf0ab4fd57da5eb4/explore

U.S. Energy Information Administration, U.S. Department of Energy, "Hurricane Ida Caused at Least 1.2 Million Electricity Customers to Lose Power," *Today in Energy*, September 15, 2021b.

U.S. Energy Information Administration, U.S. Department of Energy, "Capacity of Operable Petroleum Refineries by State as of January 1, 2022," Table 3, *Refinery Capacity Report with Data for January 1, 2022*, June 21, 2022a. As of October 7, 2022:
https://www.eia.gov/petroleum/refinerycapacity/table3.pdf

U.S. Energy Information Administration, U.S. Department of Energy, "Oil and Petroleum Products Explained: Refining Crude Oil—Top 10 U.S. Refineries Operable Capacity," webpage, last updated June 21, 2022b. As of October 7, 2022:
https://www.eia.gov/energyexplained/oil-and-petroleum-products/
refining-crude-oil-refinery-rankings.php

U.S. Energy Information Administration, U.S. Department of Energy, "Electricity Explained: Electricity Generation, Capacity, and Sales in the United States," webpage, last updated July 15, 2022c. As of November 17, 2022:
https://www.eia.gov/energyexplained/electricity/electricity-in-the-us-generation-capacity-and-sales.php

U.S. Energy Information Administration, U.S. Department of Energy, "Oil and Petroleum Products Explained: Where Our Oil Comes From," webpage, last updated September 16, 2022d. As of October 10, 2022:
https://www.eia.gov/energyexplained/oil-and-petroleum-products/where-our-oil-comes-from.php

U.S. Energy Information Administration, U.S. Department of Energy, "Natural Gas Explained: Where Our Natural Gas Comes From," webpage, updated October 18, 2022e. As of October 10, 2022:
https://www.eia.gov/energyexplained/natural-gas/where-our-natural-gas-comes-from.php

U.S. Environmental Protection Agency, "Location of U.S. Facilities," webpage, last updated February 20, 2016. As of October 18, 2022:
https://archive.epa.gov/sectors/web/html/map-7.html

U.S. Environmental Protection Agency, "Agriculture," in *Multi-Model Framework for Quantitative Sectoral Impacts Analysis: A Technical Report for the Fourth National Climate Assessment*, EPA 430-R-17-001, 2017.

U.S. Environmental Protection Agency, "EPA Settles with City of Pittsburgh, PWSA on Stormwater Violations," news release, February 5, 2021.

U.S. Environmental Protection Agency, "Climate Change Indicators in the United States," webpage, last updated August 1, 2022a. As of November 22, 2022:
https://www.epa.gov/climate-indicators/

U.S. Environmental Protection Agency, "Climate Adaptation Profile: Continental Steel Corp.," last updated August 23, 2022b.

U.S. Food and Drug Administration, U.S. Department of Health and Human Services, "Safety of Food and Animal Food Crops Affected by Hurricanes, Flooding, and Power Outages," webpage, June 1, 2022. As of October 10, 2022:
https://www.fda.gov/food/food-safety-during-emergencies/
safety-food-and-animal-food-crops-affected-hurricanes-flooding-and-power-outages

U.S. Geological Survey, U.S. Department of the Interior, "Mineral Resources Data System (MRDS)," webpage, undated. As of October 18, 2022:
https://mrdata.usgs.gov/mrds/

U.S. Global Change Research Program, "Glossary," undated.

U.S. Nuclear Regulatory Commission, "Dry Cask Storage," webpage, last updated May 3, 2021. As of October 31, 2022:
https://www.nrc.gov/waste/spent-fuel-storage/dry-cask-storage.html

USDA—*See* U.S. Department of Agriculture.

USGCRP—*See* U.S. Global Change Research Program.

Van Houtven, George, Michael Gallaher, Jared Woollacott, and Emily Decker, *Act Now or Pay Later: The Costs of Climate Inaction for Ports and Shipping*, RTI International, March 1, 2022.

van Vliet, Michelle T. H., Justin Sheffield, David Wiberg, and Eric F. Wood, "Impacts of Recent Drought and Warm Years on Water Resources and Electricity Supply Worldwide," *Environmental Research Letters*, Vol. 11, No. 12, December 2016.

Varas, Antonio, Raj Varadarajan, Jimmy Goodrich, and Falan Yinug, *Strengthening the Global Semiconductor Supply Chain in an Uncertain Era*, Semiconductor Industry Association, April 2021.

Vartanian, Charlie, Rich Bauer, Leo Casey, Clyde Loutan, David Narang, and Vishal Patel, "Ensuring System Reliability: Distributed Energy Resources and Bulk Power System Considerations," *IEEE Power and Energy Magazine*, Vol. 16, No. 6, November–December 2018.

Verbruggen, Aviel, and Mohamed Al Marchohi, "Views on Peak Oil and Its Relation to Climate Change Policy," *Energy Policy*, Vol. 38, No. 10, October 2010.

Vermeulen, Sonja J., Bruce M. Campbell, and John S. I. Ingram, "Climate Change and Food Systems," *Annual Review of Environment and Resources*, Vol. 37, November 2012.

Verschuur, J., E. E. Koks, and J. W. Hall, "Port Disruptions Due to Natural Disasters: Insights into Port and Logistics Resilience," *Transportation Research Part D: Transport and Environment*, Vol. 85, August 2020.

Vilchis, Ernesto, Jon D. Haveman, Howard J. Shatz, Stephen S. Cohen, Peter Gordon, Matthew C. Hipp, Seth K. Jacobson, Edward E. Leamer, James E. Moore II, Qisheng Pan, Harry W. Richardson, Jay Stowsky, Christopher Thornberg, and Amy B. Zegart, *Protecting the Nation's Seaports: Balancing Security and Cost*, Public Policy Institute of California, June 2006.

Villarini, Gabriele, Radoslaw Goska, James A. Smith, and Gabriel A. Vecchi, "North Atlantic Tropical Cyclones and U.S. Flooding," *Bulletin of the American Meteorological Society*, Vol. 95, No. 9, September 2014.

Virginia Secretary of Natural and Historic Resources, "Coastal Adaptation and Resilience Master Plan," webpage, undated. As of December 14, 2023: https://www.naturalresources.virginia.gov/initiatives/resilience--coastal-adaptation/

Vose, Russell S., David R. Easterling, Kenneth E. Kunkel, Allegra N. LeGrande, and Michael F. Wehner, "Temperature Changes in the United States," in Donald J. Wuebbles, David W. Fahey, Kathy A. Hibbard, David J. Dokken, Brooke C. Stewart, and Thomas K. Maycock, eds., *Climate Science Special Report: Fourth National Climate Assessment*, Vol. I, U.S. Global Change Research Program, 2017.

Waddell, Braeden, "Report: Global Temperature Rise Will Lead to Increased Disasters Without Action," *U.S. News and World Report*, August 18, 2021.

Waddell, Samantha L., Dushyantha T. Jayaweera, Mehdi Mirsaeidi, John C. Beier, and Naresh Kumar, "Perspectives on the Health Effects of Hurricanes: A Review and Challenges," *International Journal of Environmental Research and Public Health*, Vol. 18, No. 5, March 2021.

Wagner, Katherine R. H., "Adaptation and Adverse Selection in Markets for Natural Disaster Insurance," *American Economic Journal: Economic Policy*, Vol. 14, No. 3, August 2022.

Wakiyama, Takako, and Eric Zusman, "The Impact of Electricity Market Reform and Subnational Climate Policy on Carbon Dioxide Emissions Across the United States: A Path Analysis," *Renewable and Sustainable Energy Reviews*, Vol. 149, October 2021.

Walker, David M., comptroller general of the United States, U.S. Government Accountability Office, *Hurricane Katrina: GAO's Preliminary Observations Regarding Preparedness, Response, and Recovery*, testimony before the U.S. Senate Committee on Homeland Security and Governmental Affairs, GAO-06-442T, March 8, 2006.

Wallischeck, Eric, *GPS Dependencies in Transportation: An Inventory of Global Positioning System Dependencies in the Transportation Sector, Best Practices for Improved Robustness of GPS Devices, and Potential Alternative Solutions for Positioning, Navigation and Timing*, U.S. Department of Transportation, Office of the Assistant Secretary for Research and Technology, John A. Volpe National Transportation Systems Center, DOT-VNTSC-NOAA-16-01, 2016.

Wang, Sun Ling, Eldon Ball, Richard Nehring, Ryan Williams, and Truong Chau, "Impacts of Climate Change and Extreme Weather on US Agricultural Productivity: Evidence and Projection," in Wolfram Schlenker, ed., *Agricultural Productivity and Producer Behavior*, National Bureau of Economic Research, 2018.

Wang, Tianni, Zhuohua Qu, Zaili Yang, Timothy Nichol, Geoff Clarke, and Ying-En Ge, "Climate Change Research on Transportation Systems: Climate Risks, Adaptation and Planning," *Transportation Research, Part D: Transport and Environment*, Vol. 88, November 2020.

Wang, Xiao-jun, Jian-yun Zhang, Shamsuddin Shahid, En-hong Guan, Yong-xiang Wu, Juan Gao, and Rui-min He, "Adaptation to Climate Change Impacts on Water Demand," *Mitigation and Adaptation Strategies for Global Change*, Vol. 21, January 2016.

Wasley, Emily, K. Jacobs, and J. Weiss, *Mapping Climate Exposure and Climate Information Needs to Water Utility Business Functions*, Water Research Foundation, April 1, 2020.

Waugh, William L., Jr., "EMAC, Katrina, and the Governors of Louisiana and Mississippi," *Public Administration Review*, Vol. 67, No. S1, December 2007.

Wehner, Michael F., Jeff R. Arnold, Thomas Knutson, Kenneth E. Kunkel, and Allegra N. LeGrande, "Droughts, Floods, and Wildfires," in Donald J. Wuebbles, David W. Fahey, Kathy A. Hibbard, David J. Dokken, Brooke C. Stewart, and Thomas K. Maycock, eds., *Climate Science Special Report: Fourth National Climate Assessment*, Vol. I, U.S. Global Change Research Program, 2017.

Welburn, Jonathan W., and Aaron M. Strong, "Systemic Cyber Risk and Aggregate Impacts," *Risk Analysis*, Vol. 42, No. 8, August 2022.

Wellborn, Clay H., *General Services Administration Federal Facilities Affected by Hurricane Katrina*, Congressional Research Service, RS22281, version 14, November 23, 2007.

White House, "CHIPS and Science Act Will Lower Costs, Create Jobs, Strengthen Supply Chains, and Counter China," fact sheet, August 9, 2022.

Wiener, Sarah S., Nora L. Álvarez-Berríos, and Angela B. Lindsey, "Opportunities and Challenges for Hurricane Resilience on Agricultural and Forest Land in the US, Southeast and Caribbean," *Sustainability*, Vol. 12, No. 4, February 2, 2020.

Wigglesworth, Alex, "Hellish Fires, Low Pay, Trauma: California's Forest Service Firefighters Face a Morale Crisis," *Los Angeles Times*, June 14, 2022.

Wilde, Matthew, "Storms Pound Midwest Crops, Farms," *Progressive Farmer*, August 31, 2021.

Williams, A. Park, Benjamin I. Cook, and Jason E. Smerdon, "Rapid Intensification of the Emerging Southwestern North American Megadrought in 2020–2021," *Nature Climate Change*, Vol. 12, March 2022.

Willis, Henry H., Anu Narayanan, Jordan R. Fischbach, Edmundo Molina-Perez, Chuck Stelzner, Kathleen Loa, and Lauren Kendrick, *Current and Future Exposure of Infrastructure in the United States to Natural Hazards*, RAND Corporation, RR-1453-DHS, 2016. As of November 12, 2022: https://www.rand.org/pubs/research_reports/RR1453.html

Wilson, Bradley, Eric Tate, and Christopher T. Emrich, "Flood Recovery Outcomes and Disaster Assistance Barriers for Vulnerable Populations," *Frontiers in Water*, Vol. 3, December 7, 2021.

Wingard, G. Lynn, Sarah E. Bergstresser, Bethany L. Stackhouse, Miriam C. Jones, Marci E. Marot, Kristen Hoefke, Andre Daniels, and Katherine Keller, "Impacts of Hurricane Irma on Florida Bay Islands, Everglades National Park, USA," *Estuaries and Coasts*, Vol. 43, July 2020.

Woloszyn, Molly, Dennis Todey, Doug Kluck, Ray Wolf, and Mike Timlin, "Drought Status Update for the Midwest," National Integrated Drought Information System, National Oceanic and Atmospheric Administration, June 10, 2021.

Wong, Karen K., Jianrong Shi, Hongjiang Gao, Yenlik A. Zheteyeva, Kimberly Lane, Daphne Copeland, Jennifer Hendricks, LaFrancis McMurray, Kellye Sliger, Jeanette J. Rainey, and Amra Uzicanin, "Why Is School Closed Today? Unplanned K–12 School Closures in the United States, 2011–2013," *PLOS ONE*, Vol. 9, No. 12, 2014.

World Health Organization, *Chemical Releases Caused by Natural Hazard Events and Disasters: Information for Public Health Authorities*, 2018.

World Shipping Council, "Containers Lost at Sea: 2022 Update," undated.

Wright, Kevin M., and Christopher Hogan, *The Potential Impacts of Global Sea Level Rise on Transportation Infrastructure*, Part 1: *Methodology*, U.S. Department of Transportation, Center for Climate Change and Environmental Forecasting, October 2008.

Wu, Jian, Ying Zhang, Z. Morley Mao, and Kang G. Shin, "Internet Routing Resilience to Failures: Analysis and Implications," *CoNEXT '07: Proceedings of the 2007 ACM CoNEXT Conference*, December 2007.

Wuebbles, Donald J., David W. Fahey, Kathy A. Hibbard, David J. Dokken, Brooke C. Stewart, and Thomas K. Maycock, eds., *Climate Science Special Report: Fourth National Climate Assessment*, Vol. I, U.S. Global Change Research Program, 2017.

Xia, Wenyi, and Robin Lindsey, "Port Adaptation to Climate Change and Capacity Investments Under Uncertainty," *Transportation Research Part B: Methodological*, Vol. 152, October 2021.

Yesudian, Aaron N., and Richard J. Dawson, "Global Analysis of Sea Level Rise Risk to Airports," *Climate Risk Management*, Vol. 31, 2021.

Zabbey, Nenibarini, and Gustaf Olsson, "Conflicts: Oil Exploration and Water," *Global Challenges*, Vol. 1, No. 5, August 15, 2017.

Zamuda, Craig, Bryan Mignone, Dan Bilello, K. C. Hallett, Courtney Lee, Jordan Macknick, Robin Newmark, and Daniel Steinberg, *U.S. Energy Sector Vulnerabilities to Climate Change and Extreme Weather*, U.S. Department of Energy, Office of Policy and International Affairs and National Renewable Energy Laboratory, July 2013.

Zhang, Peng, Olivier Deschenes, Kyle Meng, and Junjie Zhang, "Temperature Effects on Productivity and Factor Reallocation: Evidence from a Half Million Chinese Manufacturing Plants," *Journal of Environmental Economics and Management*, Vol. 88, March 2018.

Zhang, Yi, and Jasmine Siu Lee Lam, "Estimating Economic Losses of Industry Clusters Due to Port Disruptions," *Transportation Research Part A: Policy and Practice,* Vol. 91, September 2016.

Zhang, Yuqiang, and Drew T. Shindell, "Costs from Labor Losses Due to Extreme Heat in the USA Attributable to Climate Change," *Climatic Change,* Vol. 164, 2021.

Zhao, Jerry Zhirong, Camila Fonseca-Sarmiento, and Jie Tan, *America's Trillion-Dollar Repair Bill: Capital Budgeting and the Disclosure of State Infrastructure Needs,* Volcker Alliance, working paper, November 2019.

Zhou, Jianguo, Changjia Zhou, Yuqin Kang, and Shenghui Tu, "Integrated Satellite–Ground Post-Disaster Emergency Communication Networking Technology," *Natural Hazards Research,* Vol. 1, No. 1, March 2021.

Zimmerman, Rae, "Social Implications of Infrastructure Network Interactions," *Journal of Urban Technology,* Vol. 8, No. 3, 2001.

Zohrabian, Anineh, and Kelly T. Sanders, "Assessing the Impact of Drought on the Emissions- and Water-Intensity of California's Transitioning Power Sector," *Energy Policy,* Vol. 123, December 2018.

Zong, Huixin, Peter Brimblecombe, Li Sun, Peng Wei, Kin-Fai Ho, Qingli Zhang, Jing Cai, Haidong Kan, Mengyuan Chu, Wenwei Che, Alexis Kai-Hon Lau, and Zhi Ning, "Reducing the Influence of Environmental Factors on Performance of a Diffusion-Based Personal Exposure Kit," *Sensors,* Vol. 21, No. 14, July 2, 2021.

Zouboulis, Anastasios, and Athanasia Tolkou, "Effect of Climate Change in Wastewater Treatment Plants: Reviewing the Problems and Solutions," in Sangam Shrestha, Anil K. Anal, P. Abdul Salam, and Michael van der Valk, eds., *Managing Water Resources Under Climate Uncertainty: Examples from Asia, Europe, Latin America, and Australia,* Springer Cham, November 4, 2014.

Zubaidi, Salah L., Sandra Ortega-Martorell, Hussein Al-Bugharbee, Ivan Olier, Khalid S. Hashim, Sadik Kamel Gharghan, Patryk Kot, and Rafid Al-Khaddar, "Urban Water Demand Prediction for a City That Suffers from Climate Change and Population Growth: Gauteng Province Case Study," *Water,* Vol. 12, No. 7, July 2020.

Zuloaga, Scott, and Vijay Vittal, "Integrated Electric Power/Water Distribution System Modeling and Control Under Extreme Mega Drought Scenarios," *IEEE Transactions on Power Systems,* Vol. 36, No. 1, January 2021.